Transforming Global Health

Korydon H. Smith • Pavani Kalluri Ram
Editors

Transforming Global Health

Interdisciplinary Challenges, Perspectives, and Strategies

 Springer

Editors
Korydon H. Smith
School of Architecture and Planning
University at Buffalo
State University of New York (SUNY)
Buffalo, NY, USA

Pavani Kalluri Ram
United States Agency for International
Development
Washington, DC, USA

ISBN 978-3-030-32114-7   ISBN 978-3-030-32112-3   (eBook)
https://doi.org/10.1007/978-3-030-32112-3

This Springer imprint is published by the registered company Springer Nature Switzerland AG
The registered company address is: Gewerbestrasse 11, 6330 Cham, Switzerland

Foreword

Informing, Inspiring, and Mobilizing New Talent to Address Global Health Inequity

The editors of this book shine a bright light on disciplines, approaches, and issues that have been neglected in global health for far too long: oral health, supply chains, disability, nutrition, media, urbanization, and more.

The editors' primary purpose in writing this book was not to facilitate additional academic publications in scientific journals; rather, it is a challenge to all of us, particularly academia, to identify ways to increase our impact on the pressing global and local challenges before us. It summons an increase in our efforts to inform policymakers, create intersectoral and interdisciplinary partnerships, build capacity in low-resource institutions, and, of course, advance knowledge.

Transforming Global Health dares academia to reform and strengthen its contract with society.

The book's practical approach is rooted in the history and circumstances that befell the city in which the editors' home institution resides. Buffalo, New York, an affluent global hub in its heyday, faced increasing poverty and health disparities from the latter part of the twentieth century. Throughout this period, the University at Buffalo (UB) remained committed to supporting local communities while maintaining its eye on global challenges. In 2014, its leadership took this one step further. It chose to tackle grand global challenges through the creation of a Community of Excellence in Global Health Equity. This structure was a mechanism for the university not to do "business as usual" but to connect neglected disciplines in global health, such as architecture, urban planning, and engineering, with the traditional biomedical disciplines that dominate the field. They also made certain that the well-being of people, especially those with the greatest need, was central to the work.

The Consortium of Universities for Global Health (CUGH), the world's largest academic global health organization, has taken a similar path to that of UB. CUGH is dedicated to mobilizing academia's capabilities to improve the health of people and the planet. Impact oriented, it has used its convening power to build interdisciplinary relationships with institutions across Africa, Latin America, Asia, and Europe. Working across education, research, service, and advocacy, CUGH assists its members in strengthening their work. CUGH is mission driven to improve the well-being of people and the health of the planet. Like UB, CUGH works to reduce inequity, build capacity, neutralize power imbalances, share knowledge, work across disciplines, and bring sectors together.

This textbook, UB's innovative work, and CUGH's many activities are all efforts to reform academia and practice; to engage other sectors and utilize universities' capabilities across education, research, and service; and to have a greater impact on local and global challenges. Taking this forward-leaning approach beyond the "ivory tower" will provide new research and training opportunities for students and faculty. By creating new ways to tackle public challenges, academia will show its value to other sectors. This approach, in turn, could attract new, much-needed sources of funds from governments, the private sector, foundations, and NGOs that profoundly improve lives.

The editors share a framework that can be used by any organization to understand and overcome barriers to global health equity. This "Spiral of Progress" approach consists of seven steps to help decipher the intricacies and unlock the creative solutions to a wide spectrum of grand challenges.

Transforming Global Health is an important companion to existing texts in the field. Readers, from policymakers and implementers to students and faculty, will find it a highly relevant guide to understanding what is needed to improve global health equity and achieve the Sustainable Development Goals. The book will inform, inspire, and mobilize individuals across disciplines to collaborate and address the pressing global health challenges we all face.

Keith Martin
Consortium of Universities for Global Health
Washington, DC, USA

Preface

Introduction: The Audacity to Achieve Global Health Equity

Global health is undeniably a complex topic. And overcoming health disparities—across ages, across genders, across economic ranks, across *all* countries—may very well be impossible. Yet we, the handful of contributors in this book and the thousands of academics and practitioners around the world invested in global health work, pursue this grand ambition with intellectual vigor, expressive passion, and mindful optimism (and frankly, a fair dose of uncertainty, frustration, and skepticism) [1]. This book comes from a lofty vision (and concerted actions) to transform global health—to change how it is investigated, taught, and practiced, even conceptualized—in order to achieve equity.

For researchers; professionals, and teachers, the challenges are apparent, though the solutions are not. For you, we hope that this book opens up new ways of thinking about long-standing and emergent global health issues. For students new to the subject of global health, the range of challenges, and the intricacies of each of them, may not be as clear. For you, we provide a wide-reaching, yet interesting and accessible, look at the factors that influence human health across the world, and we draw on the perspectives of many disciplines.

Here is the backstory on how this came to be.

In 2014, the United Nations was finalizing its Sustainable Development Goals (SDGs), the aggressive agenda to end poverty, eliminate hunger, realize gender equality, and improve the economic, environmental, and social conditions of people across more than 190 member countries. Three key initiatives guided the launch of the SDGs:

1. the foundational work of the Millennium Development Goals, which guided global efforts from 2000 to 2015
2. a UN summit in Brazil in June 2012 titled "The Future We Want," which reaffirmed the commitment to human rights and sustainable development
3. a 30-member Open Working Group, who drafted the goals

Advancing equity, improving wellness, and stewarding planetary health emerged as core aims, and implementation continues to evolve through the "High-level Political Forum on Sustainable Development" [2].

Concurrently, the World Health Organization focused significant attention on the growing outbreak of Ebola in western Africa. First reported on 23 March 2014, the outbreak would become, by the end of the year, the largest Ebola epidemic in history. While the vast majority of the 11,325 deaths were constrained to Guinea, Liberia, and Sierra Leone, confirmed cases in Italy, Spain, the United Kingdom, and the United States illustrated how seemingly distant outbreaks could cross not only borders but also oceans. Health economists and public health scholars have estimated the economic and social burden of the outbreak on affected countries to be greater than $50 billion. Nevertheless, for their courageous, timely, and effective efforts to stop the spread of disease, *Time* magazine named The Ebola Fighters as their 2014 "Person of the Year" [3].

The "DNA" of This Book

While these two globally significant events played out, a much more modest initiative began in Buffalo, the second largest city in New York State. A former global powerhouse—a hub of commerce and innovation, with more millionaires per capita than any city in the United States in 1900—Buffalo experienced precipitous economic and population decline throughout the second part of the twentieth century. Poverty, racial segregation, and educational and health disparities grew; yet the University at Buffalo—The State University of New York remained steadfast in both supporting local communities and maintaining global ambitions. In 2014, the Provost and Executive Vice President of Academic Affairs, Charles Zukoski, along with the Vice President for Research, Alexander Cartwright (now Chancellor of the University of Missouri), launched a signature initiative.

Their aspiration was to leverage the university's standing as the most comprehensive public research university in the northeastern United States and to propel higher-impact research through a series of new Communities of Excellence. Zukoski solicited "audacious" proposals across all faculty ranks and disciplines to take on "grand, global challenges." Following a yearlong competitive process, the university inaugurated three new interdisciplinary communities, including the Community of Excellence in Global Health Equity. It was to be co-led by the editors of this book—Ram, then a faculty member in the Department of Epidemiology and Environmental Health, trained as a physician and former employee of the US Centers for Disease Control and Prevention; and Smith, a faculty member in the Department of Architecture with training in both architecture and higher education leadership. They worked alongside Li Lin, a faculty member in Industrial and Systems Engineering who studies efficacies (and inefficiencies) in healthcare settings, and Samina Raja, a faculty member in Urban and Regional Planning who studies food systems.

In the landscape of global health, this was a curious mix.

A number of truly outstanding universities have an exceptional history of global health research and education—the London School of Hygiene and Tropical Medicine, Johns Hopkins University, Emory University, and the University of Washington, to name only a few. Their initiatives commonly reside in robust schools of medicine or public health, among other health sciences; and their reputations are built on influential research, often in the domain of infectious diseases, such as the foundational work on Ebola vaccine development at Yale University in the 1990s. The Community of Excellence in Global Health Equity at the University at Buffalo sought to complement these long-standing legacies in two important ways [4].

Founding Principles

The Community of Excellence in Global Health Equity launched with four guiding principles. These principles sought to transform both the academy and work in global health. For the academy, the principles signaled the importance of collaborating with nonacademic stakeholders toward greater societal impacts; for the domain of global health, the principles affirmed the importance of involving a wider range of disciplines. The principles are:

1. *We are equity driven.* We work on major, yet understudied, global health disparities, for which UB possesses leadership capacity—such as providing clean water and sanitation for the 800 million people with disabilities in LMICs.
2. *We are impact oriented.* We provide policymakers, practitioners, and dissemination organizations with knowledge and tools to do their best work, while simultaneously delivering traditional scholarly outputs—such as developing solutions to the causes and health effects of air pollution.
3. *We are stakeholder responsive.* We work with international partners to define problems, carry out research, and find solutions—such as collaborating with international organizations and communities throughout the world to improve newborn survival and health.
4. *We are Health-APEX, together.* We leverage underutilized disciplines, bringing architecture, planning, engineering, and other cross-synergizing (APEX) disciplines together with the health sciences to sharpen the prying edge to otherwise intractable problems—such as physicians and social workers working with architects and engineers to improve health access for resettled refugees.

First, our focus was less on producing traditional scholarly outputs, e.g., journal articles, and more on practical applications. Since our inception, we have sought to inform public policymakers, bolster partnerships with governments and international nongovernment organizations, build knowledge and capacity among practitioners, and give voice to the areas that desperately need more funding. A clear example resides in a collaboration with the World Health Organization and UNICEF, through the Joint Monitoring Programme, which measures progress toward the attainment of the SDGs. The partnership culminated in the publishing of the first ever tool for assessing the accessibility of water, sanitation, and hygiene facilities in schools for children with disabilities, who have lower graduation rates due in part to poor school design [5].

Second, given the complexity of nearly all global health challenges, we saw a need to involve many more disciplines. Awareness has grown about the interconnectedness of income inequality, gender inequality, education inequality, and other factors with health. Commensurately, disease and disability are not solely biological problems; climate, ecology, political and social structures, technologies, and systems all play a role in health and well-being.

To cultivate food equity across the globe would require people from the fields of community health, communications, education, mathematics, political science, and regional planning. To ensure the health and well-being of refugees would require architects, family physicians, industrial engineers, linguists, and social workers. To combat antimicrobial resistance in the built and natural environments where "superbugs" emerge would require expertise from chemistry, environmental engineering, geography, and medicine.

As interdisciplinary teams formed around these and other topics, all of them working with international partners, we realized the importance of sharing their work with practitioners, researchers, and students. As such, these teams, and their corresponding span of disciplines, are represented in this book—ranging from anthropology to biochemistry, dental medicine to history, management to urban planning, among many others. This collaborative approach was embedded in the guiding principles of the Community of Excellence in Global Health Equity: Principle 3, stakeholder responsiveness, and Principle 4, linking the health sciences with the social sciences, applied sciences, and the humanities.

During the founding of this collaborative center, we also developed a conceptual framework that would both recognize the complexity of and provide a phased approach to achieving health equity. We call this the "Spiral of Progress," which spans from *understanding* barriers to equity to *overcoming* them, with seven progressive steps:

1. characterizing the baseline/needs
2. researching solutions
3. piloting options
4. translating prototypes to implementation
5. assessing impacts
6. adapting solutions for other cultural and geographic contexts
7. evaluating residual barriers (and re-characterizing the baseline/needs) (Fig. 1)

Because there is an existing history of global health research, solution finding, and implementation of solutions, not all problems require beginning at step one. In many cases, the challenge resides not in finding solutions but in scaling them up, broadening their reach and impacts. This tool provided a way of analyzing problems and focusing the work of scholarly teams, who have a tendency to linger at step one rather than pushing forward to other phases of development and implementation.

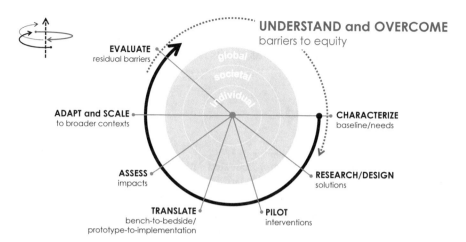

Fig. 1 Spiral of progress. Source: Community for Global Health Equity, graphic design: Korydon Smith

Additional Context

Just as the Community of Excellence in Global Health Equity does not seek to replicate the work of scholars at other institutions, this book is a companion, not competitor, to the robust contributions of other compilations, including offerings by Paul Farmer, Richard Skolnik, and others. We do not deliver a history of global health or an overview of traditional global health concepts, as others have already done this (and done it well!). Our goal is to expand the representation of topics and disciplines, including such broad-reaching issues as climate change, disability rights, gender-based violence, and end-of-life care. Many topics, and the people they negatively affect, have literally been hidden from view, such as the underground e-waste recycling system in several south Asian countries. Other issues, like pharmaceutical supply chains, have enormous impacts on health, but remain proportionally underrepresented in global health texts due to the focus on the development, rather than distribution, of medicines. Some of the topics covered, like breakthroughs in genetics, are new to global health, while others, like nutrition, have a long history in global health research and practice but not from the disciplinary perspectives represented here [6].

This book is a call to college students across majors to think broadly and creatively about how they might contribute to the drive toward global health equity.

Global health is not merely the domain of physicians, nurses, and public health workers (though they are, of course, essential), and health disparities are not only the concern of low-income nations. Ensuring health equity is critical for people living in all settings, as localized outbreaks can become global pandemics, threatening health, security, welfare, and economies worldwide. Likewise, health disparities occur in nearly all high-income countries, like the urban poor of the United States, while equity is achieved in places, like Cuba, with far fewer resources. This is why

Goal 17 of the SDGs, strengthening global partnerships across sectors, was established. However, it is among the most deceptively complex goals to achieve. Working across disciplines and across sectors is difficult.

Many Challenges Ahead

The grand challenge of global health inequity "is one of the defining issues of the twenty-first century, attracting unprecedented levels of interest" from academics, practitioners, political and business leaders, and students alike. Twenty-five percent of the world lives in informal settlements ("slums"). Improving these people's lives requires collaboration between experts in water and sanitation, public health, housing and neighborhood development, education, economics, political science, sociology, and so on. The same is true for the roughly 11 million children under the age of five who die each year, largely to preventable causes; and for the nearly 70 million refugees currently displaced by violence; and for the one billion people with disabilities worldwide. Like the educational mission of the Community of Excellence in Global Health Equity, this book strives to "motivate and capacitate the next generation of problem solvers" [7].

At one time, infectious diseases, like malaria, were the leading causes of death worldwide. More recently, noncommunicable illnesses, like heart disease and stroke, have climbed to the top. And, in the near future, it is anticipated that a highly varied set of environmentally driven conditions—such as "super bugs," air pollution, and natural- and human-caused disasters—will be the greatest threats to humans and other species. Intelligent, creative, collaborative, and, yes, audacious efforts must be pursued now. They must be pursued on behalf of the urban poor, children, refugees, and other vulnerable populations, as all of our futures grow more vulnerable unless we transform our collective efforts in global health.

Buffalo, NY, USA   Korydon H. Smith
Washington, DC, USA   Pavani Kalluri Ram

References

1. The World Health Organization (WHO) defines *health inequity* as "unjust differences in health between persons of different social groups."
2. For information on the development and launch of the SDGs, see: United Nations, "Sustainable Development Goals," https://sustainabledevelopment.un.org/sdgs; United Nations, "Millennium Development Goals," http://www.un.org/millenniumgoals/; United Nations, "Future We Want: Outcome Document," https://sustainabledevelopment.un.org/futurewewant. html; United Nations, "Open Working Group Proposal for Sustainable Development Goals," https://sustainabledevelopment.un.org/owg.html; United Nations, "High-level Political Forum on Sustainable Development," https://www.un.org/sustainabledevelopment/monitoring-and-progress-hlpf/.

3. Huber C, Finelli L, Stevens W. The economic and social burden of the 2014 Ebola outbreak in West Africa. J. Infect. Dis. 2018;218:S698–704.
4. Schnell MJ, Buonocore L, Kretzschmar E, Johnson E, Rose JK. Foreign glycoproteins expressed from recombinant vesicular stomatitis viruses are incorporated efficiently into virus particles. Proc. Natl. Acad. Sci. 1996;93:11359–65.
5. World Health Organization and UNICEF. Core questions and indicators for monitoring WASH in schools in the sustainable development goals. Geneva, Switzerland: WHO; 2016.
6. Global health texts include, but are not limited to: Birn A-E, Pillay Y, Holtz TH. Textbook of international health: global health in a dynamic world. Oxford: Oxford University Press; 2009; Farmer P, Kleinman A, Kim J, Basilico M, editors. Reimagining global health: an introduction. Berkeley: University of California Press; 2013; Glassman A, Temin M. Millions saved: new cases of proven success in global health. Washington, DC: Center for Global Development; 2016; Levine R. Case studies in global health: millions saved. Burlington, MA: Jones & Bartlett Learning; 2007; Merson M, Black R, Mills AJ. Global health: diseases, programs, systems, and policies. Burlington, MA: Jones & Bartlett Learning; 2011; Perlman D, Roy A, editors. The practice of international health: a case-based orientation. Oxford: Oxford University Press; 2008; and Skolnik, R. Global health 101. Burlington, MA: Jones & Bartlett Learning; 2011.
7. The quotation regarding global health is from Dare AJ, Grimes CE, Gillies R, Greenberg SLM, Hagander L, Meara JG. Global surgery: defining an emerging global health field. Lancet 2014;384:2245–7. Statistics are contemporaneous to the time of publishing this book and were taken from: The World Bank, the World Health Organization, and the UNHCR.

Acknowledgments

The editors of this book are grateful for the inspiration, insights, and hard work of many people on this project. First, we would like to thank all of the contributors to the book. They provided critical feedback during the conceptualization of the work, coached one another as a peer writing network, and took on the risk and challenge of expanding the global health discourse. Second, our gratitude goes to Nicole Little, graduate student in the School of Architecture and Planning at the University at Buffalo, for her tireless and detailed work in managing the project schedule and offering support to authors related to graphics, citations, and content. This book is stronger for her involvement. Third, we thank Jessica Scates and Lisa Vahapoğlu, dedicated staff of the Community of Excellence in Global Health Equity, for all their efforts in the past four years in stewarding projects, building and nurturing partnerships, assisting in the work of this book, and providing the positive energy it takes to do ambitious global health work. Fourth, we are appreciative of our student review panel—Biplab Bhattacharya, Jacob Kobler, Snigdha Motadaka, Justin Rokisky, and Rachael Steegman—who, in addition to Ms. Little and Ms. Scates, reviewed chapter drafts and delivered important feedback to the author teams. This has helped make each chapter more accessible and engaging for readers from diverse backgrounds. Fifth, for their fiscal and intellectual support of the Community of Excellence in all its work, including this book, we thank our university leadership—President Satish Tripathi, Provost Charles Zukoski, Vice President for Research Venu Govindaraju, and Associate Vice President for Research Advancement Chitra Rajan—as well as our decanal advisors—Dean Liesl Folks, Dean Robin Schulze, Dean Robert Shibley, and Dean Jean Wactawski-Wende. Equally, we thank Kasia Kordas and Samina Raja, incoming co-directors of the Community of Excellence, for continuing to shepherd the work forward. Sixth, we thank Keith Martin, Executive Director of the Consortium of Universities for Global Health, for writing the foreword and continuing to provide an important voice and venue for universities and stakeholders around the world. Seventh, we extend our appreciation to the dedicated staff of Springer for facilitating both the conceptualization and production of this work. Lastly, we are grateful for the kindness, candor, and support of our families (and their patience with our travel and work schedules!). They offer the levity and grace that it takes to persist in intensely challenging, immensely important work like that of improving global health equity.

Contents

Contributors

Diana Aga University at Buffalo, State University of New York (SUNY), Buffalo, NY, USA

Syed Ishtiaque Ahmed University of Toronto, Toronto, ON, Canada

Nirupam Aich University at Buffalo, State University of New York (SUNY), Buffalo, NY, USA

Jared Aldstadt University at Buffalo, State University of New York (SUNY), Buffalo, NY, USA

Sharmistha Bagchi-Sen University at Buffalo, State University of New York (SUNY), Buffalo, NY, USA

Sara K. Berkelhamer University at Buffalo, State University of New York (SUNY), Buffalo, NY, USA

Biplab Bhattacharya University at Buffalo, State University of New York (SUNY), Buffalo, NY, USA

Emmanuel Frimpong Boamah University at Buffalo, State University of New York (SUNY), Buffalo, NY, USA

James J. Bono University at Buffalo, State University of New York (SUNY), Buffalo, NY, USA

Rainer W. Bussmann Missouri Botanical Garden, St. Louis, MO, USA

Anne H. Burnidge University at Buffalo, State University of New York (SUNY), NY, USA

Michael Canty University at Buffalo, State University of New York (SUNY), Buffalo, NY, USA

Inés Castro Universidad Nacional de Trujillo, Trujillo, Peru

Paul Coseo Arizona State University, Tempe, AZ, USA

Erica Danfrekua Dickson 37 Military Hospital, Accra, Ghana

Hanita P. Djaya University at Buffalo, State University of New York (SUNY), Buffalo, NY, USA

Sarah E. Dumas New York City Department of Health and Mental Hygiene, New York, NY, USA

Anirban Dutta University at Buffalo, State University of New York (SUNY), Buffalo, NY, USA

Joseph E. Gambacorta University at Buffalo, State University of New York (SUNY), Buffalo, NY, USA

Mayar L. Ganoza-Yupanqui Universidad Nacional de Trujillo, Trujillo, Peru

Jean Golding University of Bristol, Bristol, UK

Kim Griswold Primary Care Research Institute, Buffalo, NY, USA

Emma Seyram Hamenoo University of Ghana, Accra, Ghana

Zoé Hamstead University at Buffalo, State University of New York (SUNY), Buffalo, NY, USA

James M. Harris University at Buffalo, State University of New York (SUNY), Buffalo, NY, USA

Azfar Hussain Grand Valley State University, Grand Rapids, MI, USA

Shamim Islam University at Buffalo, State University of New York (SUNY), Buffalo, NY, USA

Ali Kadhum BestSelf Behavioral Health, Buffalo, NY, USA

Jacob D. Kathman University at Buffalo, State University of New York (SUNY), Buffalo, NY, USA

Brendan T. Kerr University at Buffalo, State University of New York (SUNY), Buffalo, NY, USA

Katarzyna Kordas University at Buffalo, State University of New York (SUNY), Buffalo, NY, USA

Felix Lam Clinton Health Access Initiative, Boston, MA, USA

Young Seop Lee Gensler, Los Angeles, CA, USA

Melinda Lemke University at Buffalo, State University of New York (SUNY), Buffalo, NY, USA

Gonzalo Malca-Garcia Missouri Botanical Garden, St. Louis, MO, USA

Nadine S. Murshid University at Buffalo, State University of New York (SUNY), Buffalo, NY, USA

Sheela Patel SPARC (Society for the Promotion of Area Resource Centers), Mumbai, India

Sarahmona M. Przybyla University at Buffalo, State University of New York (SUNY), Buffalo, NY, USA

Pavani Kalluri Ram United States Agency for International Development, Washington, DC, USA

Michael Rembis University at Buffalo, State University of New York (SUNY), Buffalo, NY, USA

Thomas Russo University at Buffalo, State University of New York (SUNY), Buffalo, NY, USA

Tara Sabo-Attwood University of Florida, Gainesville, FL, USA

Jessica Scates University at Buffalo, State University of New York (SUNY), Buffalo, NY, USA

Torsten Schunder University at Buffalo, State University of New York (SUNY), Buffalo, NY, USA

Roseanne C. Schuster Arizona State University, Tempe, AZ, USA

Douglas Sharon Missouri Botanical Garden, St. Louis, MO, USA

University at Buffalo, State University of New York (SUNY), Buffalo, NY, USA

Dorothy Siaw-Asamoah University at Buffalo, State University of New York (SUNY), Buffalo, NY, USA

Shahana Siddiqui University of Amsterdam, Amsterdam, The Netherlands

Greg Sloditskie MBS Consulting, Walnut Creek, CA, USA

Laura E. Smith University at Buffalo, State University of New York (SUNY), Buffalo, NY, USA

Jin Young Song University at Buffalo, State University of New York (SUNY), Buffalo, NY, USA

Elizabeth Stellrecht University at Buffalo, State University of New York (SUNY), Buffalo, NY, USA

Jennifer A. Surtees Genome, Environment and Microbiome Community of Excellence, University at Buffalo, State University of New York (SUNY), Buffalo, NY, USA

Deborah Waldrop University at Buffalo, State University of New York (SUNY), Buffalo, NY, USA

David G. White University of Tennessee, Knoxville, TN, USA

Gail R. Willsky University at Buffalo, State University of New York (SUNY), Buffalo, NY, USA

Sera L. Young Northwestern University, Evanston, IL, USA

About the Editors

Korydon H. Smith, EdD, M.Arch, is professor and chair of the Department of Architecture and Co-lead of the Community Excellence in Global Health Equity at the University at Buffalo – State University of New York.

Pavani Kalluri Ram, MD, is a senior medical advisor for Child Health with the US Agency for International Development (USAID), and formerly associate professor in the School of Public Health and Health Professions and co-director of the Community of Excellence in Global Health Equity at the University at Buffalo – State University of New York.

Chapter 1
Governing to Deliver Safe and Affordable Water: Perspectives from Urban Planning and Public Policy

Emmanuel Frimpong Boamah

Introduction

> Throughout history, and especially over the past century, [water] has been ill-governed. [1]

Severe water stress is now a global crisis. The water crisis in Flint (Michigan), Toledo (Ohio), and Cape Town (South Africa) is a minutiae of the global water stress story: one in four of the world's 500 largest cities is water stressed, including cities such as Jakarta (Indonesia), Bangalore (India), Mexico City (Mexico), Istanbul (Turkey), Cairo (Egypt), Miami (Florida), São Paulo (Brazil), Beijing (China), and Tokyo (Japan) [2]. By 2030, water stress will be the reality for almost half of the world's population, which will force governments to spend around $200 billion annually to mitigate this stress [3]. Apart from its public health and economic implications, water stress presents untold geopolitical ramifications. Mark Twain rightly opined, "whisky is for drinking; water is for fighting over." The United States (U.S.) had and continues to experience its fair share of geopolitical water wars, such as the disputes between Arizona and California over the Colorado River, dating as far back as 1931.[1] Globally, water wars are expected to further complicate the existing tensions in conflict regions in the Middle East and South Asia [4]. For instance, following the September 2016 terrorist attack on an Indian army camp, India's Prime Minister, Narendra Modi averred, "blood and water and cannot flow together," a remark made to threaten the 1960s Indus Waters Treaty between India and Pakistan

---

[1] The U.S. Supreme Court decided on a number of water disputes between Arizona and California over the Colorado River from 1930 to 2000. These cases are together referred to as Arizona v. California: 283 U.S. 423 (1931); 292 U.S. 341 (1934); 298 U.S. 558 (1936); 373 U.S. 546 (1963); 376 U.S. 340 (1964); 383 U.S. 268 (1966); 439 U.S. 419 (1979); 460 U.S. 605 (1983); 466 U.S. 144 (1984); 531 U.S. 1 (2000).

E. Frimpong Boamah (✉)
University at Buffalo, State University of New York (SUNY), Buffalo, NY, USA
e-mail: efrimpon@buffalo.edu

© Springer Nature Switzerland AG 2020
K. H. Smith, P. K. Ram (eds.), *Transforming Global Health*,
https://doi.org/10.1007/978-3-030-32112-3_1

1

[5]. Understanding the threats facing our common water future is an imperative now more than ever.

Water has shaped the rise and evolution of civilizations. To understand water is to understand life: the origin, well-being, and evolution of humans, society, and ecology. Throughout ancient and modern histories, how we have managed, engineered, and thought about water has impacted the health, politics, and the socioeconomic and architectural progress of societies. From agricultural development during the Neolithic Revolution to transportation, commerce, and the impressive water and sanitation infrastructures in medieval times (e.g., aqueducts and fountains), human interactions with water have created and shaped great cities and empires.

Along with water-related engineering and architectural infrastructure, other disciplines such as public health, political science, and economics have also shaped how we think and interact with water. For instance, in the first half of the nineteenth century, John Snow, considered the father of modern epidemiology, discovered that contaminated water supplies spread cholera. His germ theory of cholera was unpopular among physicians at the time because it challenged the long-held scientific view that polluted air (miasmas theory), rather than polluted water, caused diseases like cholera. He proved what ancient Greek scholars like Hippocrates and Plato had long observed: water is a source of good or bad health outcomes depending on how well we manage our water resources.

Prominent thinkers have also used economic and political economy theories (e.g., bargaining theory, polis model, rational choice model) to illustrate the economic and political consequences of our decisions about water [6–8]. To explain the paradox of value, economist Adam Smith compared water and diamonds to distinguish market price and economic value:

> Nothing is more useful than water; but it will purchase scarce anything; scarce anything can be had in exchange for it. A diamond, on the contrary, has scarce any value in use; but a very great quantity of other goods may frequently be had in exchange for it. [9]

The idea of economic valuation (e.g., cost-benefit analysis), which was developed in the U.S. in the early twentieth century, popularized an economic approach to viewing the ways in which humans interact with water for drinking, irrigation, transportation, and other uses [10, 11].

The emerging consensus, especially after Garret Hardin's 1968 seminal piece "Tragedy of the Commons," suggested that how we manage and interact with water is rooted in complex governance questions. Wiel had already reminded us to appreciate water for its physical dimension, as a resource used to improve people's health and well-being, and for its governance dimension, given the self-evident truth that water is "publici juris," a resource of the people that must be managed by the people and for the people [12]. Yet, past and recent water crises suggest that we have yet to fully appreciate water's governance dimension, partly because this dimension is "wicked." (see [13]) That is, a simple decision to provide clean water is not all that simple; depending on local contexts, this decision must also address issues related to gender, culture, poverty, sanitation, technology, and disability. How water decisions relate to issues such as public health, engineering, economic development, and climate change is complex. Recent water crises such as "Day Zero" in Cape Town, South Africa; lead-poisoned water in Flint, Michigan; and the slow but gradual "death" of America's second-longest river, the Rio Grande, exemplify this complexity.

This chapter sheds light on the meaning of water governance and its significance for students, professionals, and scholars in disciplines such as public health, public policy, geography, urban planning, and engineering. It first introduces the meaning of water governance and its wicked nature, showing that global health outcomes mirror water-governance outcomes; that is, these governance outcomes are the effects of water and non-water related decisions aggregated at the household, community, and global levels. These decisions and their outcomes transcend disciplinary, professional, and geographic boundaries, showing that water governance is a trans-disciplinary global health subject. We next discuss different typologies of governing or delivering water. Whether they concern preventing communicable water diseases or engineering solutions for flood control, these typologies remind us that if appropriate water-governance arrangements are in place, how we as citizens and professionals interact with water could positively affect local and global water resources and human health outcomes.

Governing the Water Commons: A Framework for Analysis

The Global Water Partnership defines water governance as "the range of political, social, economic and administrative systems that are in place to regulate development and management of water resources and provisions of water services at different levels of society." [14] Some critics have found this definition, despite its wide acceptance, to be too narrow and lacking the analytical rigor necessary to capture the everyday "messiness" of water politics and outcomes (see [15, 16]). Therefore, this chapter defines water governance by drawing from Castrol and Lasswell:

> Water governance is an evolving process of shaping and sustaining authority and power arrangements within which decisions are made regarding (1) what should be the ends and values of water policy and essential water services, (2) who decides on and benefits or suffers from these ends and values, and (3) how will these be achieved. (see [17, 18])

This definition, from the broader environmental governance literature, includes the processes that establish, legitimize, and change institutions (formal and informal rules in use) to guide the behavior of actors in the use of environmental resources [19]. Figure 1.1 provides a framework useful for understanding this definition of water governance and its complexity. This framework communicates three key messages: water decisions are made at multiple scales, or levels, of society (from the household to the global scale)[2]; each level's decisions often involve questions

---

[2] The term *scale* is used loosely here. See rigorous analytical discussions on scale in works such as [20, 21].

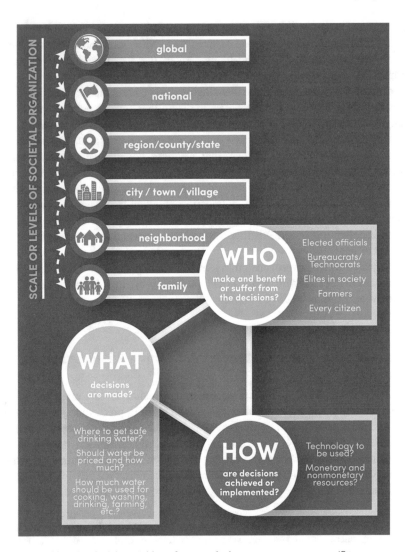

Fig. 1.1 Multi-scale decision-making framework in water governance. (Source: concept: Emmanuel Frimpong Boamah; graphic design: Nicole C. Little)

concerning what, who, and how; and there are inherent conflicts in framing and answering these questions (see [20, 21]).

First, water issues transcend scale and sectors of societal organization. Decisions made at the family level can have consequences at the neighborhood, city, and global levels. Similarly, decisions made at the global level can have consequences for multiple nations, neighborhoods, and even one or a few families. Moreover, water decisions at one level could have non-water consequences at another level, and vice versa. Familial and organizational acts related to water have had multinational ramifications. For example, the Mediterranean Action Plan (Med Plan), involving 21 countries bordering the Mediterranean Sea and from the European community, was enacted to protect marine and coastal communities. It was developed in response to observations of pollution in the Mediterranean Sea and other river basins flowing into it, which had formed an aggregation of pollutants sourced from and interacting at multiple scales.[3] These included unregulated pollutants from industries, direct municipal sewage, runoff from farmlands, and waste from more than 200 million summertime tourists [22]. Another example is the Great Lakes-St. Lawrence River Basin Water Resources Compact, which is a legally binding compact among eight states in the U.S., with inputs from two Canadian provinces. The compact recognizes how water decisions by different jurisdictions (from the city to the state levels) within the Great Lakes basin affect everyone. Hence, it regulates the use and protection of the water in the basin, which comprises about 20% of the earth's fresh surface water [23, 24]. How individuals, households, industries, neighborhoods, cities, and countries decide to use water or dispose of their waste has aggregate effects on our water resources; a trickle of bad water decisions creates a mighty river of water problems. Resulting from poor water and water-related decisions at the individual, household, city, national, and global scales, water crisis is a governance crisis.

Second, water governance is about decisions made; who makes, benefits, and suffers from these decisions; and how the decisions are implemented. We can think about decisions concerning water and water-related issues from the family to the global scales. At the family level, for instance, we make decisions to use water for activities such as cooking and bathing, and we *should* also make conscious decisions about how much water we should use for these activities. At the city, regional, or national levels, authorities make decisions on issues such as water allocation and pricing, and they *should* also consider the equity dimensions of these decisions: should water allocation and/or pricing be different for households based on need, income, disability, and so forth?

Often, decisions about or affecting water are not equitable because of *who* makes the decisions. Who decides how much water should be used for cooking, washing, and so forth at the family level? Who decides how much water in cities and regions should be used for farming and drinking? These questions seem to have obvious answers: parents decide at the family level, and city, regional, and national authorities decide at the city, regional, and national levels, respectively. In reality, these questions are very difficult to address because of issues such as information and power asymmetries as well as gender and disability biases in decision-making. Ostrom et al. discussed the question of who decides when they noted that usually, those affected by decisions are not the ones who make the decisions [25]. This is known as a scale-mismatch problem and is common in water governance.[4] The challenge is to figure out how to effectively gather inputs from everyone in the

---

[3]In 1995, the Med Plan was replaced with the Action Plan for the Protection of the Marine Environment and the Sustainable Development of the Coastal Areas of the Mediterranean (MAP Phase II).

[4]See detailed discussion on scale mismatch and fit in works such as [26, 27].

making of water decisions in a community or among multiple communities sharing a common water resource (see [26, 27]). Again, since water affects and is also affected by other decisions about things like sanitation and land use, other voices must also be present at the decision-making table, not only those directly affected by and/or benefiting from water decisions. In most developing countries in Latin America and Africa, the voices of women and other minority groups, such as the disabled, are often not represented in water decisions because most of these groups are not empowered (e.g., trained in decision-making process, resourced) to fully participate in decision-making about water and other issues (see [28–30]).

The challenges of leaving out voices in water decisions become apparent in the processes of *how* to implement or achieve the decisions. During implementation, non-inclusive decisions reflect what Imbroscio calls technical and political infeasibility: failing to consider voices of technical expertise, including sound economic and other technical analyses (technically feasible), and voices of residents and other stakeholders (politically feasible) [31]. Implementation is challenging when decisions are technically feasible but politically infeasible, and vice versa, or when they are technically and politically infeasible.

The Flint water crisis, involving lead poisoning and Legionnaires' disease, shows how a series of water decisions failed to include other voices. It started when the Flint City Council in 2013, with approval from the city's emergency manager, decided to switch the city's water source from Lake Huron and the Detroit River to the Flint River, without duly considering technical and residents' voices and those affected by the decision. The implementation of this decision did not go well because the council soon had to deal with residents' protests and complaints of health issues caused by the new water source, including residents bringing bottles of discolored tap water to meetings. This poor initial decision-making led to other poor decisions, including the city's emergency manager, Jerry Ambrose, deciding to overrule the city authority's vote to reconnect the water source to the Detroit River. Similarly, protests about South Africa's "water apartheid" problem are one of the many actions happening all over the world in response to water decisions that are not inclusive, open, equitable, and responsive to the needs of a community's many voices [32]. In contrast, the implementation of bulk water pricing in the Paraíba do Sul River Basin, Brazil, was successful partly because it included and was open to different voices (e.g., users, businesses, civic organizations, and government agency representatives) whose actions affect and are affected by water [33].

Case Study 1: The Paradox of Water Abundance and Scarcity in Accra, Ghana

Urban Accra, Ghana illustrates how the abundance of water can become as much a problem as its scarcity when poor water and non-water decisions occur at multiple scales. The city experiences extensive flooding each year with devastating consequences, including the loss of lives and property [34]. For instance, during the city's June 2015 flood, a fuel station exploded, reportedly killing over 250 people, many of whom were taking shelter at the station from the flooding and heavy rain. Outbreaks of diseases such as cholera are also especially common during these floods. The irony is that even though water runs through the city "abundantly" as floods, several communities experience water scarcity. In fact, most communities are served by water tankers, which are private water retailers who fetch water in tanks to sell to households [35]. Accra's flooding has resulted from multiple water and nonwater decisions at multiple scales, for example, by households not properly disposing of waste, building in flood-prone areas, and dumping in water resources such as the Odaw River; and by city authorities not enforcing zoning laws. Accra's water-scarcity challenge is inextricably linked to its flooding and sanitation problems. Dealing with the city's perennial flood problem requires developing measures to harvest rainwater for drinking and farming. It also implies developing sanitation facilities and programs to deal with waste. These seem like common-sense solutions that should have been implemented long ago. Interviews with urban planning practitioners and other stakeholders reveal that the city and national governments, through numerous studies and planning documents, already know what to do, but there is minimum political will to implement the studies' recommendations. The Accra case demonstrates how water governance becomes increasingly wicked due to the gap between knowing what decisions to make (from the household to the national levels) and having the will to implement these decisions.

Conflicts in Water Decisions

The discussion on protests leads to the third and final point about the framework shown in Fig. 1.1: inherent conflicts in framing and answering questions concerning what, who, and how in water governance. Different conflicts can emerge when decisions do not align well with who makes, benefits, or suffers from the decisions and how such decisions will be achieved. Figure 1.2 below illustrates three possible conflicts (but there could be more). The first conflict is the *decision-consequences conflict*: the misalignment between what decisions are made and who makes, benefits, or suffers from them. Water decisions have different impacts on different populations based on their gender, income, disability, and so forth. For instance, to avert Cape Town's Day Zero crisis, city authorities imposed limits on water consumption: 87 l per person per day, which later was reduced to 50 l per person per day. However, this one-size-fits-all decision failed to account for the water needs, especially for sanitation purposes, for groups such as girls and women, including pregnant and lactating mothers. Ostrom long ago warned of the dangers of panacea solutions because they exclude other voices, resulting from privileging the voices of a few powerful actors who often have limited information but can determine the fate of everyone else [36]. Protests are symptomatic of a decision-consequences conflict. They represent attempts to remedy the asymmetry between decisions made and who makes, benefits, or suffers from the decisions.

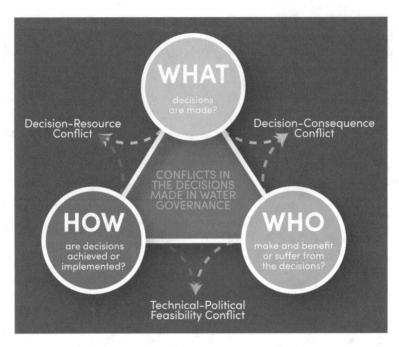

Fig. 1.2 Conflicts in the decisions (what, who, and how) made in water governance. (Source: concept: Emmanuel Frimpong Boamah; graphic design: Nicole C. Little)

The decision-resource and feasibility conflicts are interrelated and often represent the gap between decisions and whether they can be implemented. Resources needed to implement decisions become a challenge when there is misalignment between the scope of the decision and the approach or strategies outlined to implement it. Similarly, and as discussed above, technical and political feasibility becomes a challenge when implementation approaches do not consider those benefiting or suffering from the decision. The implementation of the Korle Lagoon Ecological Restoration Project (KLERP) in Accra, Ghana, embodies all three forms of conflict. The restoration of the coastal wetland Korle Lagoon began in the early 1990s, when the Government of Ghana (GoG) secured $73 million and $48 million for the first and second phases of the project, respectively (see [37, 38]). As of 2017, Ghana's Environmental Pollution Agency had declared the lagoon dead due to its volume of pollutants, despite efforts and resources invested in its restoration. These three conflicts all played a role in the failure to restore this water resource. The decision to restore the lagoon involved evicting nearby residents in two communities: Old Fadama (an informal settlement) and Agbogbloshie. The decision-consequences conflict emerged because the voices of these residents were not duly considered, which resulted in protests and legal suits by citizens with the help of international groups such as the Shack/Slum Dwellers International. These protests and lawsuits called into question the political feasibility of the restoration project, but they also affected the financial resources budgeted for the project. The project stalled because of delays in evicting residents, which added interest to the loans secured for the project [37].

Governance Arrangements/Typologies

Addressing these decisions and their conflicts at each scale requires appropriate governance arrangements. The literature discusses many arrangements, but we will focus on four typologies mentioned by Huitema et al.: polycentric governance, public participation (the focus of this section), experimentation, and a bioregional approach [39].

Scholars address public participation from many disciplinary perspectives. Huitema et al. noted that the word "public" can mean different things (e.g., citizens, water users), and their "participation" in decisions can also mean different things, such as simply showing up at a meeting or providing feedback (see [40]). In urban planning, public participation is discussed as a "ladder" or a "wheel" and focuses on the degree to which citizens have power in decisions (see [41–43]). In social psychology, public participation considers the extent to which both the engagement process and final decisions match the expectations and behavioral patterns of stakeholders involved [44]. In the context of water governance, public participation should ensure that (1) participatory processes and mechanisms are fair and transparent in involving the stakeholders (beneficiaries and victims) of decisions and that (2) final decisions reflect the views of these stakeholders. Some case studies show that public participation supports better decisions in water governance, such as in a case of water reuse projects in Georgia, Texas, and California [45]. However, public participation is not a panacea for the challenges of water governance. For instance, Jacobs et al., in their study of the Yaqui Valley in Mexico; the Upper San Pedro Basin in Arizona; Ceará, Brazil; and the Upper Ping River in Thailand, demonstrate that stakeholders' engagement works well in short-term, low-stakes decisions, such as water allocation, but does not work well in longer-term high-stakes decisions, such as those about water infrastructure [46]. The transaction cost of engaging stakeholders over the long term can be prohibitive. Newig and Fritsch also analyze several water- and non-water-related case studies to highlight the conditions in which public participation is beneficial in environmental-resource governance [47].

Experimentation is either a water-management strategy or a research method [48]. As a research strategy, experimentation helps stakeholders evaluate whether predetermined goals were achieved through decisions implemented [49]. As a water-management strategy, experimentation allows stakeholders to consider the incomplete and uncertain nature of information; hence, decisions are made and adjusted continuously or incrementally as information changes. This strategy traces back to Lindblom's argument that decision makers "muddle through" or rely on incremental information to analyze and compare decisions (see [50, 51]). This critique of rational comprehensive planning posits that decisions should be scientifically made by considering all possible alternative decisions and their causes and effects [52]. As Simon argues, human beings are not *rational* because they do not maximize utility by considering all information; they are *satisficers, or boundedly rational*, because they make incremental decisions based on their cognitive capability and available information [53]. Experimentation is fundamentally grounded in the understanding that decisions and the processes used to make them must allow for "trial and error learning" to meet the changing conditions of reality (see [54, 55]). We see this idea at work not only in water governance but also in water engineering cases. For example, the mosaic of water infrastructure (i.e., dams, pumping stations, canals, ditches) in the Netherlands is engineered as a control panel, which allows the adjustment of water resources, such as the Rhine and Meuse, based on changing information about climate, flooding, drought, and community water needs [56]. Kirby et al. used their integrated hydrology-economics model to demonstrate how experimentation (i.e., using simulation models) can support trial and error, water engineering, and governance decisions in the Murray Darling Basin in Australia [57].

The bioregional or river-basin approach addresses the question of the appropriate scale needed to make cross-boundary water decisions [39]. Above, we discussed how water decisions are made at multiple scales (see Fig. 1.1). For instance, when a river crosses two or more communities, this suggests that decisions about the river (i.e., allocation decisions, pollution, etc.) are appropriately made at the watershed or river-basin level. This would ensure that all decisions incorporate the views of both communities. According to Schlager and Blomquist, this type of decision making can be achieved through two means: (1) the two communities can collaborate in their decisions (i.e., a weak arrangement) or (2) a unitary river-basin authority could be established (a strong arrangement) [58].[5] The Soil and Water Conservation Districts in the U.S. is an example of a strong river-basin arrangement for water decisions [59]. Huitema et al. note that there is the need for more empirical evidence to understand and support the effectiveness of this approach [39]. In their analysis of ten cases on river-basin management in Europe, Mostert et al. found that the river-basin approach is not an "unrealistic ideal," especially in cases where there is social learning, such as experimentation, and stakeholder involvement, such as public participation) [60]. This finding shows that the typologies discussed above are not mutually exclusive. The next section, on the polycentric typology, describes how they function together.

Polycentric Governance in Water Delivery

The meaning of polycentricity varies by discipline. In urban or spatial planning, for instance, polycentricity means the functional connections (e.g., commuting patterns) among multiple city centers in a given region, which are sometimes called

---

[5] Two arrangements, Type I arrangement (hierarchically ordered multi-purpose organizations with no overlapping jurisdictional functions and/or powers) and Type II arrangement (special purpose organizations with cross-jurisdictional functions and/or powers), are presented by [59].

polycentric urban regions (see [61, 62]). In governance terms, polycentricity refers to multiple decision-making centers that are formally autonomous and operate under certain sets of rules [25, 63]. Many scholarly studies on polycentric governance exist, especially in environmental governance. Rather than repeating findings from the literature, this section will introduce a framework to simplify the meaning of polycentric governance. A case study will then offer empirical evidence showing how polycentric governance works in water delivery.

Two interrelated governance debates began in the late 1950s. The first was about Hardin's seminal piece "Tragedy of the Commons," which asserted that humans are "trapped" because, as rational beings, we keep exploiting and destroying the environment (i.e., the commons) to maximize our individual gains without thinking about our collective or societal gains [64]. To address this tragedy, scholars debated whether state (public) or market (private) arrangements are the appropriate governance mechanism to protect us from destroying the commons (See arguments for state or market [65, 66]). The second debate was primarily about the choice between small governments (decentralization) and large government (centralization). In the U.S., some reformers argued that having several municipal governments and special-purpose jurisdictions (e.g., school districts) leads to inefficiencies because functions overlap and are duplicated. In fact, Lind referred to this as a "horde of Lilliputian governments," portraying metropolitan governance as an "organized chaos" or a "crazy-quilt pattern." [67]

These two debates can be summarized as two questions. The first debate asks, is the state or market the appropriate governance mechanism? The second asks, what should the scale of governance be (multiple local, decentralized governments versus one large centralized government, at the regional, national, and global scales)? Drawing on the work of Pahl-Wostl and Knieper, we can use Fig. 1.3 to illustrate the two debates [68]. On one hand, we must decide whether government or market entities should make decisions, on water for instance. On the other hand, we also must decide whether decisions should be made by private or public entities and local or non-local entities.

Polycentric governance, however, helps us understand that the two debates force us into a false dilemma. That is, we must falsely choose between (1) either state or market and (2) either centralized or decentralized institutional arrangements. Two explanations help us to resist this false dilemma as we think through water governance. First, in both abstract and empirical terms, this false dilemma does not adequately capture how society is organized. From an abstract standpoint, Fig. 1.4 uses combination and probability principles to demonstrate a four-by-four symmetrical matrix showing public (local and non-local) and private (local and non-local) institutional arrangements. As the matrix shows, the likelihood of observing cases (e.g., societies) with *purely* centralized and decentralized markets or governments (shown as green) is lower (4 out of 16, or 25%) than the likelihood of observing societies with *mixed (polycentric)* centralized-decentralized, state-market arrangements (75%). In other words, even in federal societies like the U.S., there are instances of monopolistic markets (e.g., telecommunication giants like T-Mobile, Verizon), local markets (e.g., car dealerships), centralized government functions (e.g., federal regulations such as the

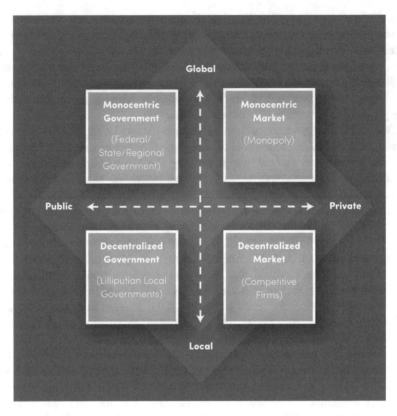

Fig. 1.3 Centralized-decentralized, state-market governance arrangements. (Source: concept: Emmanuel Frimpong Boamah; graphic design: Nicole C. Little)

Clean Water Act), and decentralized government functions (e.g., city zoning ordinances). Such polycentric societies reflect public participation (zoning decisions require public input), experimentation (regulations evolve to meet changing conditions), and bioregional approaches (there are formal organizations to manage transboundary water resources such as the Great Lakes-St. Lawrence River Basin Water Resources Council).

Finally, this false dilemma often frames policy arguments about water privatization. Privatizing water, which seems to support efficiency, often implies the use of market principles, such as pricing, to determine who is included and excluded in water delivery. Making water public, which seems to support equality (water as a human right), often implies that everyone benefits (no one can be excluded) from water. Furthermore, we must decide whether to centralize or decentralize water decisions. In these ways, the false dilemma forces us to decide between *water efficiency* (market-led water governance) and *water equality* (state-led water governance) organized through either centralized or decentralized systems. Polycentric governance teaches us that water equity, in which everyone benefits according to their need and pays according to their ability, is a middle way that can be organized through both (not either or) centralized and decentralized systems. Scholars have conducted several case studies on polycentric police services and river-basin governance, to demonstrate the

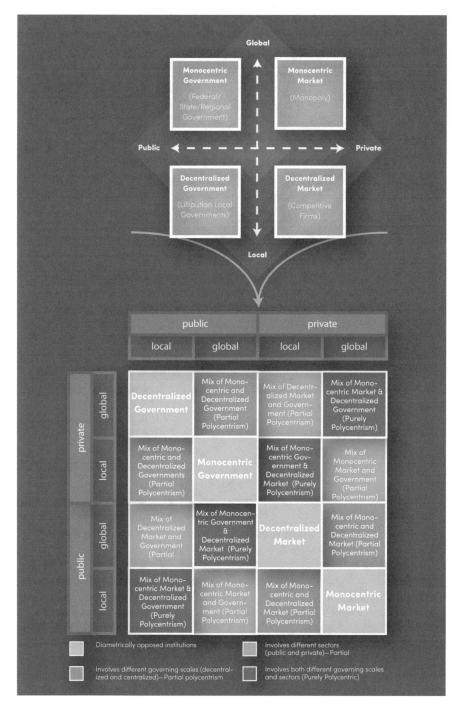

Fig. 1.4 Thinking through polycentric governance by recognizing how governance involves different actors who operate at multiple scales (local and non-local) and within public and private institutional arrangements. (Source: concept: Emmanuel Frimpong Boamah; graphic design: Nicole C. Little)

Case Study 2: Polycentric Water Delivery System in Kenya

Kenya's water-governance reforms since 2002 illustrate polycentric governance in water delivery. The Kenyan Government enacted the Water Act of 2012, which was later replaced by the Water Act of 2002, to end decades-long centralized arrangement of water resources by the state-owned National Water Conservation and Pipeline Corporation. In these reforms, we observe a polycentric governance arrangement in water delivery, which involves state and market organizations at both the local and national levels operating according to multiple rules [72, 73]. This system also involves public participation, experimentation, and bioregional approaches to water governance. There are three main state-owned centralized organizations: the Ministry of Water and Irrigation (MWI), the Water Resource Management Authority (WRMA), and the Water Services Regulatory Board (WASREB). The MWI is in charge of the overall national water policy; the WRMA manages water, including issuing water permits to water service providers (WSPs); and the WSRB selects and regulates these WSPs, including setting and monitoring equitable water tariffs and quality performance standards among the WSPs [74]. These WSPs include a mix of local government-owned water supply companies and private-sector-led water supply companies (e.g., community water projects, water bottling companies), all operating at the local, regional, and national levels. In particular, the community water projects (CWPs) are member-based irrigation projects within different water basins and comprise individual riparian-right owners, farmers, and others [72]. The activities of CWPs within each basin are managed by a basin-level organization, Water Resource Users Associations (WRUAs), comprising representatives from the CWPs. The WRUAs, in consultation with the WRMA, issue collective water-use permits to members of each of the CWPs.

Kenya's water reforms reflect a polycentric system in several respects. First, we observe a mix of public (government) and private (market) organizations operating at the national, regional, and local levels to deliver water to households and non-households (e.g., farmers). Second, this mixture of organizations operates within multiple rules. For instance, while local organizations can make their internal rules (e.g., CWPs can make their own decentralized rules on water use), they must also work with the centralized rules from the WRMA [74]. These rules change as new conditions emerge, which reflects the experimental nature of Kenya's water-governance arrangement. In fact, national reforms in water legislation and organization are examples of experimentation. Finally, the WRUAs exemplify the bioregional approach to water governance, undergirded by an appreciation for the need to involve members of CWPs in basin-level decisions.

effectiveness and benefits of polycentric governance systems [69–71]. A case study on water delivery in Kenya demonstrates the potential of a polycentric governance system to ensure equity in water-delivery decisions, especially in low socioeconomic communities.

Conclusion

Global and local water crises are governance crises. This chapter discussed the wicked nature of the governance crises, which involves making water and non-water-related decisions (regarding what, who, and how) at multiple scales. These decisions are conflict-laden, involving decision-consequences conflicts, decision-resource conflicts, and technical and political feasibility conflicts. Whether at the household or global level, decision makers should be aware of and address these conflicts in their water decisions. Water decisions do not occur in a vacuum; they exist within different governance forms or typologies, discussed in this chapter as public participation, experimentation, the bioregional approach, and polycentric governance. Polycentric governance embodies all these typologies by capturing the mix of centralized and decentralized as well as public and private organizations that work to ensure the delivery of services such as water. This polycentricity manifests prominently in Kenya's reformed water governance, which, according to some studies has achieved benefits such as better coordination in water-use activities, especially during the dry season, alleviation of water disparities among upstream and downstream users, and awareness of how water and non-water decisions affect water and non-water activities [72, 74].

References

1. Peet J. Priceless: a survey of water. The Economist. 2003.
2. McDonald RI, Weber K, Padowski J, Flörke M, Schneider C, Green PA, Gleeson T, et al. Water on an urban planet: urbanization and the reach of urban water infrastructure. Glob Environ Chang. 2014;27:96–105. https://doi.org/10.1016/j.gloenvcha.2014.04.022.
3. UNEP. Options for decoupling economic growth from water use and water pollution: a report of the Water Working Group of the International Resource Panel: United Nations Environment Programme. 2015.
4. U NBC News. Water scarcity fuels tensions across the Middle East; 2018.
5. Bloomberg. A water crisis is brewing between South Asia's arch-rivals; 2019.
6. Maass A. Muddy waters: the army engineers and the nation's rivers. Cambridge, MA: Harvard University Press; 1951.
7. Rogers P. A game theory approach to the problems of international river basins. Water Resour Res. 1969;5(4):749–60.
8. Stone DA. Policy paradox: the art of political decision making. Revised Aufl. Cambridge, MA and New York, NY: Norton.

9. Smith A. cited in Hanemann WM. The economic conception of water. In: Peter Rogers, Ramon Llamas, and Luis Martinez Cortina, editors. Water crisis: myth or reality; 2006, 1776. p. 61–91 CRC Press.
10. Viessmann W Jr. A history of the United States water resources planning and development. In: Russell C, Baumann DD, editors. The evolution of water resource planning and decision making. Northampton, MA: Edward Elgar Publishing; 2009.
11. Boland JJ, Flores N, Howe CW. The theory and practice of benefit-cost analysis. In: Russell C, Baumann DD, editors. The evolution of water resource planning and decision making. Northampton, MA: Edward Elgar Publishing; 2009.
12. Wiel SC. Theories of water law. Harvard Law Rev. 1914;27(6):530–44.
13. Rittel HWJ, Webber MM. Dilemmas in a general theory of planning. Policy Sci. 1973;4(2): 155–69.
14. Rogers P, Hall AW. Effective water governance. Stockholm, Sweden: Global Water Partnership Committee (TEC), 2003.
15. Franks T, Cleaver F. Water governance and poverty: a framework for analysis. Prog Dev Stud. 2007;7(4):291–306.
16. Pahl-Wostl C. Water governance in the face of global change: from understanding to transformation. New York: Springer; 2015.
17. Castro JE. Water governance in the twentieth-first century. Ambient Soc. 2007;10(2):97–118.
18. Lasswell H. Politics: who gets what, when, how. New York: Whittlesey House; 1936.
19. Pahl-Wostl C. A conceptual framework for analysing adaptive capacity and multi-level learning processes in resource governance regimes. Glob Environ Change. 2009;19(3):354–65.
20. Biro A. Water politics and the construction of scale. Stud Pol Econ. 2007;80(1):9–30.
21. Swyngedouw E. Technonatural revolutions: the scalar politics of Franco's hydro social dream for Spain, 1939–1975. Trans Inst Br Geogr. 2007;32(1):9–28.
22. Haas PM. Do regimes matter? Epistemic communities and Mediterranean pollution control. Int Organ. 1989;43(3):377–403.
23. Fisher KR. Foreword: The Great Lakes-St. Lawrence River basin compact and agreement: International Law & (and) Policy Crossroads. Mich St Law Rev. 2006:1085.
24. Olson JM. Navigating the Great Lakes compact: water, public trust, and international trade agreements. Mich St Law Rev. 2006:1103.
25. Ostrom V, Tiebout C, Warren R. The organization of government in metropolitan areas: a theoretical inquiry. Am Pol Sci Rev. 1961;55(4):831–42.
26. Young OR. The institutional dimensions of environmental change. Cambridge, MA: MIT Press; 2002.
27. Cumming GS, Cumming DHM, Redman CL. Scale mismatches in social-ecological systems: causes, consequences, and solutions. Ecol Soc. 2006;11(1):14.
28. Harris LM. Gender and emergent water governance: comparative overview of neoliberalized natures and gender dimensions of privatization, devolution and marketization. Gend Place Cult. 2009;16(4):387–408.
29. Ahlers R. Gender dimensions of neoliberal water policy in Mexico and Bolivia: empowering or disempowering? In: Bennett V, Davila-Poblete S, Nieves Rico M, editors. In opposing currents: the politics of water and gender in Latin America. Pittsburgh, PA: University of Pittsburgh Press; 2005. p. 53–71.
30. Zwarteveen MZ. Water: from basic need to commodity: a discussion on gender and water rights in the context of irrigation. World Dev. 1997;25(8):1335–49.
31. Imbroscio D. Urban America reconsidered: alternatives for governance and policy. Ithaca, NY: Cornell University Press; 2010.
32. Bond P. Water commodification and decommodification narratives: pricing and policy debates from Johannesburg to Kyoto to Cancun and back. Capital Nat Social. 2004;15(1):7–25.
33. Formiga-Johnsson RM, Kumler L, Lemos MC. The politics of bulk water pricing in Brazil: lessons from the Paraíba do Sul basin. Water Policy. 2007;9(1):87–104.

34. Amoako C, Frimpong Boamah E. The three-dimensional causes of flooding in Accra, Ghana. Int J Urban Sustain Dev. 2015;7(1):109–29.
35. Stoler J, Weeks JR, Fink G. Sachet drinking water in Ghana's Accra-Tema metropolitan area: past, present, and future. J Water Sanit Hyg Dev. 2012;2(4):223–40.
36. Ostrom E. The danger of self-evident truths. Polit Sci Polit. 2000;33(01):33–46.
37. Grant R. Out of place? Global citizens in local places: a study of informal settlements in the Korle Lagoon Environs in Accra, Ghana. Urban Forum. 2006;17(1):1–24.
38. Boadi KO, Kuitunen M. Urban waste pollution in the Korle Lagoon, Accra, Ghana. Environmentalist. 2002;22(4):301–9.
39. Huitema D, Mostert E, Egas W, Moellenkamp S, Pahl-Wostl C, Yalcin R. Adaptive water governance: assessing the institutional prescriptions of adaptive (co-) management from a governance perspective and defining a research agenda. Ecol Soc. 2009;14(1):26.
40. Reed MS. Stakeholder participation for environmental management: a literature review. Biol Conserv. 2008;141(10):2417–31.
41. Arnstein SR. A ladder of citizen participation. J Am Inst Plann. 1969;35(4):216–24.
42. Davidson S. Spinning the wheel of empowerment. Planning. 1998;1262(3):14–5.
43. Healey P. Collaborative planning: shaping spaces in fragmented societies. Vancouver, BC: University of British Columbia Press; 1997.
44. DeCaro D, Stokes MK. Public participation and institutional fit: a social–psychological perspective. Ecol Soc. 2013;18(4) https://doi.org/10.5751/ES-05837-180440.
45. Birkhoff J. Community conflict over water reuse. Perspectives from conflict analysis and resolution. In: Hartley TW, editor. Understanding public perception and participation. Alexandria, VA: Water Environment Research Foundation; 2003.
46. Jacobs K, Lebel L, Buizer J, Addams L, Matson P, McCullough E, Garden P, Saliba G, Finan T. Linking knowledge with action in the pursuit of sustainable water-resources management. Proc Natl Acad Sci. 2016;113(17):4591–6.
47. Newig J, Fritsch O. Environmental governance: participatory, multi-level–and effective? Environ Policy Governance. 2009;19(3):197–214.
48. Greenberg D, Linksz D, Mandell M. Social experimentation and public policymaking. Washington, DC: The Urban Institute Press; 2003.
49. Pahl-Wostl C. The importance of social learning in restoring the multifunctionality of rivers and floodplains. Ecol Soc. 2006;11(1):10.
50. Lindblom CE. The science of "muddling through". Pub Admin Rev. 1959;19:79–88.
51. Popper K. Piecemeal social engineering. Princeton, NJ: Princeton University Press; 1944.
52. Boyne GA, Gould-Williams JS, Law J, Walker RM. Problems of rational planning in public organizations: an empirical assessment of the conventional wisdom. Admin Soc. 2004;36(3):328–50.
53. Simon HA. Theories of bounded rationality. Decis Org. 1972;1(1):161–76.
54. Collingridge D. The management of scale: big organizations, big decisions, big mistakes. London, UK: Routledge; 1992.
55. Ostrom E. Understanding institutional diversity. Princeton, NJ: Princeton University Press; 2005.
56. Rijkswaterstaat. Rijkswaterstaat. Amsterdam: Ministry of Infrastructure and Environment; 2011.
57. Kirby JM, Connor J, Ahmad MD, Gao L, Mainuddin M. Climate change and environmental water reallocation in the Murray–Darling Basin: impacts on flows, diversions and economic returns to irrigation. J Hydrol. 2014;518:120–9.
58. Schlager E, Blomquist W. Local communities, policy prescriptions, and watershed management in Arizona, California, and Colorado. Paper presented at the Eighth Conference of the International Association for the Study of Common Property, Bloomington, Indiana, USA; 2000 31 May–4 June.
59. Liesbet H, Garry M. Unraveling the central state, but how? Types of multi-level governance. Am Pol Sci Rev. 2003;97(2):233–43.

60. Mostert E, Pahl-Wostl C, Rees Y, Searle B, Tàbara D, Tippett J. Social learning in European river-basin management: barriers and fostering mechanisms from 10 river basins. Ecol Soc. 2007;12(1):19.
61. Green N. Functional polycentricity: a formal definition in terms of social network analysis. Urban Stud. 2007;44(11):2077–103.
62. Hall P, Pain K. The polycentric metropolis: learning from mega-city regions in Europe. London: Earthscan; 2006.
63. Aligica P, Tarko V. Polycentricity: from Polanyi to Ostrom, and beyond. Governance. 2012; 25(2):237–62.
64. Hardin G. The tragedy of the commons. Science. 1968;162(3859):1243–8. https://doi.org/10. 1126/science.162.3859.1243.
65. Welch WP. The political feasibility of full ownership property rights: the cases of pollution and fisheries. Policy Sci. 1983;16(2):165–80.
66. Ophuls W. Leviathan or oblivion. In: Daly H, editor. Toward a steady-state economy; 1973. p. 214–9.
67. Lind M. A Horde of Lilliputian governments. The New Leader. 1997.
68. Pahl-Wostl C, Knieper C. The capacity of water governance to deal with the climate change adaptation challenge: using fuzzy set qualitative comparative analysis to distinguish between polycentric, fragmented and centralized regimes. Glob Environ Chang. 2014;29:139–54.
69. McGinnis M. Polycentricity and local public economies: readings from the workshop in political theory and policy analysis. Ann Arbor, MI: University of Michigan Press; 1999.
70. Frimpong Boamah E. Polycentricity of urban watershed governance: towards a methodological approach. Urban Stud. 2018;55(16):3525–44.
71. Pahl-Wostl C, Knieper C. The capacity of water governance to deal with the climate change adaptation challenge: using fuzzy set qualitative comparative analysis to distinguish between polycentric, fragmented and centralized regimes. Glob Environ Change. 2014;29:139–54.
72. McCord P, Dell'Angelo J, Baldwin E, Evans T. Polycentric transformation in Kenyan water governance: a dynamic analysis of institutional and social ecological change. Policy Stud J. 2017;45(4):633–58.
73. Baldwin E, Ottombre CW, Dell'Angelo J, Cole D, Evans T. Polycentric governance and irrigation reform in Kenya. Governance. 2016;29(2):207–25.
74. Were E, Roy J, Swallow B. Local organisation and gender in water management: a case study from the Kenya highlands. J Int Dev. 2008;20(1):69–81.

Further Reading

Bakker K. Privatizing water: governance failure and the world's urban water crisis: Cornell University Press; 2010.
Barlow M, Clarke T. Blue gold: the fight to stop the corporate theft of the world's water. New York, NY: The New Press and W. W. Norton & Company; 2002.
Feldman D. Water politics—governing our most precious resource. Cambridge, UK: Polity; 2017.
Ostrom E. Governing the commons: the evolution of institutions for collective action. Cambridge: Cambridge University Press; 1990.
Pahl-Wostl C. Water governance in the face of global change: from understanding to transformation. Cham: Springer; 2015.

Chapter 2
Measuring the Lifetime Environment in LMICs: Perspectives from Epidemiology, Environmental Health, and Anthropology

Katarzyna Kordas, Sera L. Young, and Jean Golding

So Many Exposures!

Imagine the life of a woman who fled her conflict-ravaged home in Burma, spent time in a refugee camp, and finally settled in the United States or of a young malnourished child in Kenya who has frequent infections. Imagine everything their bodies and minds experience in just 1 day. What about other days? How about reaching all the way back to the day they were born? Or even to when they were conceived? How much do their behaviors, experiences, and exposures matter to their development, or their physical or mental health? Is one type of exposure enough to "tip the scales" or is it the average or the totality of experiences that counts? Don't you wish you could measure or record all those exposures and understand their consequences, and eventually, figure out ways to tackle that complex stew?

Across the lifetime, each person will experience and accumulate countless exposures to the natural, built, social, and chemical components of the environment. These exposures occur in many ways, through people's behaviors and lifestyle choices (for example, smoking, diet, or level of activity), physical contact (intentional or not) with chemicals, structures, or physical forces (for example, heat waves or hurricanes), or interactions with people, institutions, policies/laws, and geopolitical forces (for example, war). Whatever a person's lifetime exposures may be,

K. Kordas (✉)
University at Buffalo, State University of New York (SUNY), Buffalo, NY, USA
e-mail: kkordas@buffalo.edu

S. L. Young
Northwestern University, Evanston, IL, USA

J. Golding
University of Bristol, Bristol, UK

scientists believe that they affect human health in important ways, contributing either to the development of disease or to maintenance of good health. In the field of public health, the exposure are called "risk factors" if they contribute to disease and "protective factors" if they contribute to health.

Although scientists recognize the importance of the environment to human health—70–90% of disease risk is thought to be due to differences in environmental factors—they also recognize the difficulty in studying the totality of factors due to their sheer number [1]. To complicate matters, people's exposures produce physiologic responses (for example, increased stress response to unexpected events), which themselves may be risk factors for disease. Many types of environmental factors may also potentiate or antagonize one another, increasing risk of disease when multiple risk factors are present or reducing risk when protective factors are present. Similarly, environmental factors might interact with our innate susceptibility to disease produced by our genes. Inside the human body, metabolic pathways and feedback loops interact with both gene expression and one another, further increasing the complexity of teasing out what causes disease and what is the product of the disease process [2]. Finally, during a person's life, some periods of development are more critical for the initiation of long-term disease processes. Exposures during those critical windows may more powerfully shape later health than exposure during times that are less sensitive or susceptible, like old age [3]. For example, research in the field of developmental origins of health and disease has provided convincing evidence that in utero exposures to malnutrition, stress, and other environmental factors are capable of developmental programming of later disease [4]. Because of our knowledge of critical windows, the first 1000 days of life (from conception to about 2 years) are the focus of many interventions, particularly in low- and middle-income countries (LMICs) [5]. For the same reason, children under 5 years of age, like Maria from Kenya are considered a vulnerable group and may receive targeted interventions in LMICs or high-income countries (for example, the Head Start Program in the U.S.).

When we recognize that environmental factors and experiences shape human health, it is not surprising that many disciplines have put considerable effort into drawing conceptual frameworks to help them understand the totality of factors, to guide research and interventions. The social determinants of health (inequalities) framework, originating in the field of public health recognizes the importance of such factors as gender, income and social status, education, physical environment (transport, housing, waste, energy, industry, water, etc.), social support networks, among others [6]. The total environment framework partitions out the various environments, from chemical to natural, to understand how they affect human development [7]. Taking a related perspective, the socio-ecological framework discusses the multiple levels of influence—intrapersonal, interpersonal, organizational, community, and public policy—that act to promote health [8]. The concept of the exposome accounts for all micro and macro environments and puts them in the context of a time-frame, the entire lifetime [9]. This chapter introduces the concept of the exposome as a framework for measuring environmental factors and studying how they affect human health. The chapter explores some of the debates taking place in this area of science with respect to measurement and conceptualization of the exposome, the challenges of conducting exposome research in LMICs, and the challenge of translating research into effective solutions.

The Exposome Explained

Many branches of science measure human behaviors, experiences, and exposures, from social psychology to toxicology, but they seldom cross disciplinary boundaries. In environmental health, for example, the branch of *exposure science* is dedicated to measuring human exposure to the physical, chemical, or biological environment [10]. This branch of science does not typically measure the social environment or psychosocial stressors. Until recently, exposure scientists focused on isolated points during a person's lifetime and considered each environmental factor separately. For example, scientists studied exposure to single metals or perhaps groups of metals, but this was not combined with other classes of chemicals, like pesticides. Exposure science has been criticized for lacking the tools to characterize human exposures accurately and precisely [9].

The term *exposome* caught the attention of the scientific community around the year 2005, when researchers began to express their frustration with the fact that they knew very little about all the things people are exposed to [9]. The definition of the exposome is "the totality of environmental exposures over a person's lifetime" [11]. A decade after the term was first introduced, it is a growing branch of science and is surrounded by a great deal of excitement [12]. It is also praised for its agnostic nature in the search for connections between exposures to environmental factors and diseases, as opposed to the traditional hypothesis-driven approaches, which can only test the effects of those factors scientists know about or are interested in [12].

Three levels of environmental factors are encompassed by the exposome concept (Fig. 2.1): *general external* (social capital, education, psychological and mental stress, climate, etc.), *specific external* (radiation, infectious agents, chemical contaminants and environmental pollutants, diet, lifestyle factors, medical interventions, etc.), and *internal* (metabolism, endogenous hormones, gut microflora, oxidative stress, inflammation, etc.) [9]. All or any of these factors may interact across a person's lifetime to contribute to health or disease. Scientists continue to refine the concept of the exposome, particularly as they face practical challenges of studying it. The simple definition of the exposome disguises real complexity. Some of the big questions that scientists are grappling with include: how to measure exposures and when, is it necessary to measure every exposure, and how to deal with all the information that is generated?

Fig. 2.1 Illustrating the concept of the exposome for the LMIC context (Source: concept: Katarzyna Kordas; graphic design: Nicole C. Little). Adapted with permission from Vrijheid M. The exposome: a new paradigm to study the impact of environment on health. Thorax. 2014;69:876–8

Measuring the Exposome

There are two main approaches to measuring the exposome [13]. The *top-down* approach follows the recognition that environmental factors (diets, vitamins, drugs, pathogens, pollution, lifestyle factors, etc.) produce their effects by altering metabolite levels or physiological processes in the body [1]. It involves the measurement of chemicals (biomarkers[1] of exposure, susceptibility, or effect) in biological tissues, preferably blood. Exposome science has thus far focused on measuring this internal environment. The other approach is *bottom-up*—measuring people's behaviors, contact, or interaction with the external environment. For example, this may mean measuring individual chemicals in water, air, and diet—the sources of people's environmental exposures [13]. Wearable sensor technologies and apps are becoming increasingly available to measure an individual's external environment.

In the top-down approach, the measurement of biomarkers of exposure to known chemicals such as metals, pesticides, flame retardants in urine, blood, etc., continue to be important. But, the development of new technologies and refinement of laboratory techniques are driving scientific advances [1]. A notable example is the use of high throughput *omics*[2] technologies, such as metabolomics (to measure the molecules that are intermediates or products of the body's metabolic pathways), proteomics (protein), and others. Omics biomarkers often do not reflect exposure to

---

[1] A biomarker is a substance that can be measured in the body, most often in blood, serum, or urine, but also in hair, feces, or even nails, that indicates the presence of an exposure or disease, or the body's physiological state/status.

[2] Omics is a term that refers to the characterization and quantification of biological molecules that are thought to represent the structure, function and dynamics of an organism.

environmental pollutants directly, but rather, the effect of the pollutants on some metabolic pathway or the harm that was done to the body upon exposure. The point of studying omics biomarkers is that they may be more precisely measured than external environmental factors and they include *untargeted* chemicals—things that scientists are not aware of or not hypothesizing as important, and therefore, not targeting specifically. Such untargeted analytics may lead to discoveries of exposure factors or metabolic pathways that play an important, but to-date unknown, role in the disease process. It is important to point out that the proponents of the biomarker approach do not view it antithetical to the bottom-up approach. Both have value—the latter for formulating prevention or intervention strategies, the former for elucidating causes or mechanisms of disease [14].

It is striking that both top-down and bottom-up approaches are influenced by the exposure science tradition, considering chemical exposures or the body's metabolic reaction to the environment. Yet, scientists are encouraged to "embrace the complexity" of environmental exposures and to resist limiting themselves to chemicals [12]. Some commentators point to the importance of directly assessing the external environment, such as air pollution, noise, etc., arguing that some environmental exposures do not leave a measurable trace (biomarker) in the body (for example, heat), whereas some internal biomarkers are not specific to a single exposure (for example, inflammation could be produced by infections as well as obesity) [15]. Wearable physical activity sensors and the use of mobile devices to collect data with apps (for example, on diet, risky behaviors, or mental health) or to track people's movements have also received a great deal of attention, but there is a need to address issues of privacy and to integrate multiple sensors into smaller "packages" to reduce burden on the people who wear them. There are also concerns that the measurement of the social environment is not keeping pace with biomarker discovery. Ultimately, this may shape how the exposome is conceptualized and which disciplines will drive the scientific advances.

Do We Need Exposome Science in LMICs?

The exposome is an exciting area of science, particularly for those wishing to delve into the biological relationships between environmental factors and the health-disease continuum. We may not need exposome science to do meaningful research on environmental determinants of health or develop useful interventions to address the most pressing problems in global health. But the exposome concept suggests important lessons: (1) environment is critical to health, (2) there are usually multiple factors (including social determinants) that contribute to disease, which is why simplistic interventions rarely address the totality of the problem, (3) it is important to take a long view of health, as diseases develop over time, (4) researchers and practitioners wishing to understand and solve global health problems need to consider multiple perspectives, and (5) complex problems require interdisciplinary thinking. It is becoming common for teams representing different areas of expertise (from anthropology to public health to planning to engineering) to tackle problems together. With each discipline contributing its knowledge and perspective, the solutions are more likely to fit the problem and be sustainable.

We also need to recognize the unique challenges related to exposome science in LMICs—in terms of defining environments, conducting exposome-related research, and translating findings into action. These challenges are highlighted below.

The Challenge of Putting It All Together

It is likely that as a concept, the exposome will be increasingly applied to health problems facing populations from LMICs. But collecting and integrating a comprehensive set of risk and protective factors to represent the totality of exposures across the human lifetime is complex. Take for example, the first 1000 days of life, a brief yet very important period. We know that nutrient deficiencies and malnutrition in utero and for the first few years of children's lives affect physical growth and neurobehavioral development [16–18]. We know that access to clean water and hygiene are equally important because they safeguard young bodies from recurring infections and death, and promote better development and growth [19–21]. We also know that exposure to chemicals, like aflatoxin produced by fungi that attack maize or peanuts, metals, such as lead, pesticides, indoor and outdoor air pollution, and many others affects child growth and development during the fetal and early postnatal period and into later childhood [22–32]. There is evidence that ancestral exposures such as famine or smoking may impact the health of later generations [33]. Children in LMICs are very frequently exposed to more than one of these risk factors at a time.

There has yet to be a study that combines all the exposures mentioned above, particularly in LMICs. While we understand that the *social environment* (interactions that occur in the neighborhood, daycare, school, and the larger community) affect child growth, health, and development, as does the *built environment* (physical characteristics of homes and communities), few studies have comprehensively measured these aspects of the environment in LMICs or their effects on health [34–51]. Very few studies have combined chemical exposures with social determinants of health. We know that, just like in high-income countries, the social, built and natural environment to which people are exposed will be underpinned by their socioeconomic status (SES), yet we still struggle to define which aspects of SES are most critical to ensuring optimal health outcomes—assets, employment, income, etc. We also need to recognize that individual experiences occur in the context of geopolitical forces that shape the market, labor, political environments in which people live. It is very challenging to measure and quantify policies or political structures, which occur at regional or country-level and put them together with individual-level exposures, particularly biomarkers representing the internal environment. Such studies would need to involve multiple countries.

The Challenge of Translation

Most scientists do research for the good of society—they want their findings to be used to improve human lives. In LMICs, knowledge is increasingly translated into intervention "packages" that address multiple risk factors [52]. Concepts like the exposome that account for complex contributions of multiple risk factors to health or disease could be very useful for informing interventions. But the translation of findings from exposome science is still an aspiration and will depend on how the exposome is conceptualized. The concept itself is very broad but some writers have criticized it as being too narrowly focused. If we only focus on "all biologically active chemicals inside a person during life," paying less attention to social and policy factors, there is a danger of reducing the exposome to the internal environment [13, 53]. If that happens, it may be difficult to understand the external factors that produce the biomarker profiles detected by powerful machines, and consequently, to make the changes that will protect people's health from environmental hazards and other risk factors [15]. In the same vein, there is fear that exposome science will "take a radically individualized turn," where the individual, molecular-level differences are examined at the exclusion of the social phenomena that contribute to them [53]. The result could be a focus on individual-level interventions like behavior change that already abound in public health and that are criticized as ineffectual in isolation of system-level changes [8].

Many environmental factors, from exposure to pollutants to impacts of natural disasters, are underpinned by deep social, economic, or racial/ethnic disparities [54–59]. Thinking of the refugee woman fleeing the conflict and violence ravaging her country or the little boy suffering from malnutrition and multiple infections, the danger of reducing the expsome to biomakers is particularly great. It could dehumanize their experiences and shift the focus from the geopolitical, systemic changes that would support population health to the small changes they as individuals can make to incrementally improve their own health. It is important that the measurement of the exposome in LMICs account for the broad variety of economic and societal factors that affect health. To counteract the molecular view, the socio-exposome[3] and the public health exposome[4] have been recently put forward [53, 60].

---

[3] Socio-exposome is an extension of the exposome framework and places special emphasis on the measurement of social experiences and the recognition of prior research on broader, social determinants of human health across the lifecycle.

[4] Public health exposome is an answer to the scientific paradigms, which it considers unsatisfactory, for measuring the relationship between the environment, human health, and the disparities in health at the population level. It builds on the socio-ecological framework and extends the original concept of the exposome, with the aim of tracking and integrating complex relationships among exogenous and endogenous exposures across the life span.

The Challenge of Research Conduct and Data Analysis

While technology is propelling exposome science forward, studies designed to measure the exposome often fail to address key questions with respect to the timing of exposures. Longitudinal studies need to be conducted, with data obtained from before the birth of the individual, and at various stages throughout childhood [61, 62]. While some exposures are relatively easy to quantify, there are important logistical and ethical challenges of collecting repeated samples from very young children or vulnerable (low SES, marginalized, indigenous, etc.) populations. People lose interest in studies, and either drop out completely or have long periods of non-participation. Researchers need to find efficient ways to assess the external environment, make research participation enjoyable, and the purpose of research studies clearly relevant to people's lives. Finally, it may be difficult to find (long-term) participants in countries where research has historically not been conducted or is not valued. In some countries (for example, with documented human rights violations or government encroachment on individuals' privacy) people may be more reticent to participate or more likely to drop out of studies.

Assuming the challenges of collecting data over time have been met, there remains the question of how to analyze it. This is a complex question requiring a separate and lengthy discussion. We highlight it here because it is so critical to the future of exposome science—advances are needed not only in measuring environments more accurately and precisely, but also in sifting through, categorizing, combining, and pulling meaning from millions of data points. Environmental factors are closely intertwined, both at any given time point and across time; consequently, new statistical approaches are needed. A few examples of attempts to analyze exposome data are given here, with the encouragement to students interested in "big data" to get involved and contribute to developing new methodologies [63–66].

The Challenge of Ethics

Research in exposome science represents special challenges with respect to ethics in terms of privacy and confidentiality, participant burden, the consent process, and sharing of results. Taking these issues in turn, privacy and confidentiality of data are an important concern, particularly when potentially very sensitive data (sexual risk taking, exact location, mental health, etc.) are collected via apps, stored in digital repositories, and either manipulated in the cloud or transmitted electronically between participants and researchers. The use of commercial apps or devices is fraught with pitfalls because issues of data ownership are murky. When apps are commercial property, who really owns the data and how much agency over the data do the researchers or the participants have?

Participant burden is a concern for research ethics. Because there are currently very few integrated data collection devices (one device collecting all necessary information), people participating in exposome studies are likely to be asked to provide multiple biological samples at multiple time points, wear at least one personal monitor or sampler over several days or even weeks, receive multiple text prompts or app reminders asking them to log information on their mood, diet, activity, etc., and update researchers with changes in address so that any external monitoring data can be linked to them accurately. Participants can perceive this as overly burdensome. Imagine a scenario in which researchers invite into a study a low-resource community in a LMIC setting or refugees who distrust doctors, government officials, or university personnel. Imagine a community so poor that they agree to participate to supplement their income. Because of the underlying distrust or power dynamics, long-term studies that are burdensome or invasive may meet with low acceptance and high level of non-compliance. In some cases, longitudinal studies invite participants to serve on ethics committees or advisory boards, and thus help inform study design, but this is still rare in LMICs [67, 68].

Finally, the practice of reporting back individual or group-level results to study participants is becoming more acceptable, but it is challenging to communicate uncertainty around scientific findings to build accurate expectations in participants of the value and power of the research (as opposed to clinical diagnosis) [69–71]. The use of accurate, understandable, and sensitive language is important. The report-back of results to study participants from LMICs deserves special attention because issues of participant vulnerability, power differentials between researchers and participants, and literacy are likely to be greater than in higher-income countries. Many research studies conducted in LMICs currently are headed by principal investigators from high-income country institutions. Even if very well-intentioned, those investigators may have little knowledge of the local context in terms of perceptions of and comfort with research studies, health belief systems, cultural norms, acceptable behaviors, or linguistic nuances. In those cases, working with community advisory boards will be particularly important. The field of exposome science needs ethics experts to guide the design and implementation of studies; this is yet another area where the involvement and energy of young minds will be essential.

The Role of Students in Measuring Lifetime Environments

Broad thinking is needed to consider how technological advances in "omics" do not overshadow the societal determinants of health and disease. These issues are already being raised and are timely for joint thinking across the disciplines. One example is a very thoughtful consideration for how geographers should engage with environmental health scientists and epigeneticists to understand how space and health relate [72]. They astutely point out that "bodies... are socionatures in which the line between the biological and social is erased." Furthermore, the "permeability of the body to ubiquitous and subtly acting chemicals should put to rest the idea of the body as a citadel that can be protected by intentional individual behaviors" [72]. This realization is a rallying point for a dialogue and collaboration across disciplines; it also opens up space for involvement by students and young researchers.

This dialogue can take on many forms, beginning in the classroom and the halls of academia, and moving to conferences or scientific journals that give a wide range of disciplines the platform to speak of their experiences and contribute to refining the exposome concept and the tools for its measurement. These are just some ideas but the point is that through collaboration and cross-pollination of ideas among many disciplines, solutions to the challenges facing the exposome science can be found. The translation of the molecular exposome into real interventions or meaningful policy solutions is yet to come, but is an exciting future opportunity for researchers willing to engage with other disciplines. Despite its many challenges, particularly when "exporting" the concept to LMICs, exposome science is an exciting area of research, one that should continue to attract bright young minds ready to think of solutions and chart new pathways forward [64] (Fig. 2.2).

The Story of Thiri, A Refugee Woman from Burma Who Settled in Buffalo, NY[5]

By Seth Frndak, PhD Student, University at Buffalo

When her community was destroyed in 2000, Thiri Aung and her family sought shelter at the Mae La refugee camp in Thailand. Thiri was assigned to the kitchen and daily breathed in smoke due to the poor ventilation. Sanitation was a problem in Mae La, so Thiri and her youngest son contracted cholera. Thiri recovered but her son did not. When cholera vaccines were finally distributed throughout the camp, Thiri insisted that her family also take traditional Burmese medicine. Unbeknownst to Thiri, her medicine contained high levels of lead.

After 5 years in the camp, 45-year-old Thiri, her husband, and three children applied for and were permitted entry into the United States and settled in Buffalo, NY. During intake, all of her children screened high for blood lead, possibly due to foods and traditional medicine used in the camp. Thiri was encouraged to re-visit the doctor to monitor her children's lead levels. Because the children continued to have elevated lead levels, the county government inspected Thiri's apartment and found chipping lead paint. They provided education concerning lead exposure and worked with the landlord to repair some of the chipping walls.

After settling into their new community, Thiri started cooking for a small, Burmese restaurant. Her husband struggled to find work. He would often drink and smoke in the house with other unemployed Burmese men. He became irritable, so Thiri let him smoke in the living room in the evenings.

---

[5]The names and other details in this story are fictional, but are based on the lived experiences of refugees in the United States and elsewhere.

Thiri loved cooking, but her children complained about her traditional dishes, so she would treat them to hot dogs and pizza from the corner store—the grocery store was much too far to leave her kids alone at home. Happy to see her children gaining weight, and finding herself eight pounds heavier than in Mae La, Thiri walked to the grocery store less and less.

Thirteen years after resettling in the United States, Thiri proudly saw her eldest child, Hlaing, graduate from high school. Hlaing still lived at home and knew that his mother was often unable to sleep, waking and gasping for air. Hlaing had encouraged her to visit the doctor after his father died of a heart attack a few years prior. While resistant, Thiri finally allowed her son to schedule an appointment. Thiri's blood pressure was 160/110 with a BMI of 30. She had obstructive sleep apnea and stage 2 hypertension. She was at high risk for stroke and heart failure. The doctor recommended changes in diet and activity and prescribed medication to control her blood pressure. Thiri listened to the advice and began slowly losing weight, reducing her blood pressure. Today she enjoys gardening with other Burmese women.

A Day in the Life of Maria, A 5-Year Old from Kisumu County, Kenya[6]
By Patrick Mbullo, PhD Student, Northwestern University

During household visits, a community health volunteer (CHV) is required to report the health of children, especially those under 5 years of age. A CHV will record the presence of ailments such as cough, diarrhea, and dysentery, as well as immunization status. When a team of CHVs and caseworkers from *Pamoja Community Based Organization* visited Mabungo, a small village in Nyanza region, the disturbing report about the health condition of two young children was not unexpected.

When we arrived at the household, Mercy's 5-year-old daughter, Maria, lay on an old papyrus mat on the floor, immobile; her frail body—feet, fingers, and buttocks—were all covered with raw sores inflicted by jiggers (*tungiasis*), a type of flea that affects humans. Some sores were fresh while others were re-infected. On Mercy's lap was her emaciated 3-year-old-daughter, who seemed severely malnourished. "My children have had persistent diarrhea for more than three days now, I am worried because they can't drink or eat. Maria can't play or go to school. She spends most of her time sleeping on this mat," Mercy narrated, her teary eyes revealing the state of her pain. "Yesterday, I saw long worms come out as she defecated. At first, I thought the intestines were coming out. Because I was scared, I called the community

---

[6] This case study is based on a family in Kenya, but depicts the lived experiences of families in this country and elsewhere. For privacy reasons, names and other personal details have been changed.

health worker," she continued. While we were busy documenting the case, we observed Mercy retrieve clothes from a pile of dirty laundry for the kids. She had neither water nor soap to wash the clothes.

With the household sanitation situation, it was obvious the diarrhea would only get worse if nothing was done. Mercy's house (made of earthen floor and mud wall) was infested with jiggers; she had no pit latrine, and most importantly, lacked food. A more comprehensive intervention was needed. The CHV prepared the case for referral to the nearest health facility, as the rest of us zeroed on addressing the household hygiene condition. A standard requirement for a CHV during home visits is to give health education and other needed household services.

Mercy's household mirrors the plight of many households in Mabungo village, where children are exposed to numerous environmental challenges preventing them from attaining good health and wellbeing. Everyday in this village, many children miss school because of bacterial or parasitic infections such as jiggers or tapeworms. Change of seasonality exacerbates these environmental exposures, with rainy season bringing bouts of cholera, schistosomiasis, and other waterborne diseases, including malaria, a leading cause of death among children under five in this region.

Despite the government's push for universal health care, the households' inability to access basic needs such as clean water and sanitation, food and shelter will make health care a mirage dream.

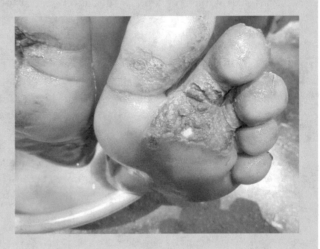

Fig. 2.2 Tungiasis-infested feet (Source: Diana Awuor, Social Worker at Pamoja Community Based Organization, in Kisumu Kenya)

Acknowledgements Seth Frndak, a Ph.D. student at the University at Buffalo, wrote the story of Thiri. Patrick Mbullo, a Ph.D. student at Northwestern University, wrote the story of Maria.

References

1. Rappaport SM, Smith MT. Environment and disease risks. Science. 2011;330:460–1.
2. Rappaport SM. Biomarkers intersect with the exposome. Biomarkers. 2012;17:483–9.
3. Rice D, Barone S Jr. Critical periods of vulnerability for the developing nervous system: evidence from humans and animal models. Environ Health Perspect. 2000;108:511S–33S.
4. Dover GJ. The Barker hypothesis: how pediatricians will diagnose and prevent common adult-onset diseases. Trans Am Clin Climatol Assoc. 2009;120:199.
5. Martorell R. Improved nutrition in the first 1000 days and adult human capital and health. Am J Hum Biol. 2017;29:e22952.
6. Marmot M, Friel S, Bell B, Houweling TAJ, Taylor S. Closing the gap in a generation: health equity through action on the social determinants of health. Lancet. 2008;372:1661–9.
7. Tulve NS, Ruiz JDC, Lichtveld K, Darney SP, Quackenboss JJ. Development of a conceptual framework depicting a child's total (built, natural, social) environment in order to optimize health and well-being. J Environ Health Sci. 2016;2:1–8.
8. McLeroy KR, Bibeau D, Steckler A, Glanz K. An ecological perspective on health promotion programs. Health Educ Q. 1988;15:351–77.
9. Wild CP. Complementing the genome with an "exposome": the outstanding challenge of environmental exposure measurement in molecular epidemiology. Cancer Epidemiol Biomark Prev. 2005;14:1847–50.
10. National Research Council (NRC). Exposure science in the 21st century: a vision and a strategy. Washington, DC: National Academies of Science; 2012.
11. Wild CP. The expsome: from concept to utility. Int J Epidemiol. 2012;41:24–32.
12. Cui Y, Balshaw DM, Kwok RK, Thompson CL, Collman GW, Birnbaum LS. The exposome: embracing the complexity for discovery in environmental health. Environ Health Perspect. 2016;124:A137–40.
13. Rappaport SM. Implications of the exposome for exposure science. J Expo Sci Environ Epidemiol. 2011;21:5–9.
14. Lioy PJ, Rappaport SM. Exposure science and the exposome: an opportunity for cohoerence in the environmental health sciences. Environ Health Perspect. 2011;119:A466–7.
15. Peters A, Hoek G, Katsouyanni K. Understanding the link between environmental exposures and health: does the exposome promise too much? Br Med J. 2012;66:103–5.
16. Argaw A, Hanley-Cook G, De Cock N, Kolsteren P, Hyubregts L, Lachat C. Drivers of under-five stunting trend in 14 low- and middle-income countries since the turn of the millennium: a multilevel pooled analysis of 50 demographic and health surveys. Nutrients. 2019;11:2485.
17. Georgieff MK, Ramel SE, Cusick SE. Nutritional influences on brain development. Acta Paediatr. 2018;107:1310–21.
18. Leroy JL, Ruel M, Habicht JP, Frongillo EA. Linear growth deficit continues to accumulate beyond the first 1000 days in low- and middle-income countries: global evidence from 51 national surveys. J Nutr. 2014;144:1460–6.
19. Prüs-Ustün A, Bartram J, Clasen T, Colford JM Jr, Cumming O, Curtis V, Bonjour S, et al. Burden of disease from inadequate water, sanitation and hygiene in low- and middle-income settings: a retrospective analysis of data from 145 countries. Tropical Med Int Health. 2014; 19:894–905.
20. Dillingham R, Guerrant RL. Childhood stunting: measuring and stemming the staggering costs of inadequante water and sanitation. Lancet. 2004;363:94–5.

21. Lima AAM, Guerrant RL. Strategies to reduce the devastating costs of early childhood diarrhea and its potential long-term impact: imperatives that we can no longer afford to ignore. Clin Infect Dis. 2004;38:1552–4.
22. Smith LE, Prendergast AJ, Turner PC, Humphrey JH, Stoltzfus RJ. Aflatoxin exposure during pregnancy, maternal anemia, and adverse birth outcomes. Am J Trop Med Hyg. 2017;96:770–6.
23. Hu H, Téllez-Rojo MM, Bellinger D, Smith D, Ettinger AS, Lamadrid-Figueroa H, Schwartz J, Schnaas L, Mercado-García A, Hernández-Avila M. Fetal lead exposure at each stage of pregnancy as a predictor of infant mental development. Environ Health Perspect. 2006;114:1730–5.
24. Cantoral A, Téllez-Rojo MM, Levy TS, Hernández-Ávila M, Schnaas L, Hu H, Peterson KE, Ettinger AS. Differential association of lead on length by zinc status in two-year old Mexican children. Environ Health. 2015;14:95.
25. Plusquellec P, Muckle G, Dewailly E, Ayotte P, Jacobson SW, Jacobson JL. The relation of low-level prenatal lead exposure to behavioral indicators of attention in Inuit infants in Arctic Quebec. Neurotoxicol Teratol. 2007;29:527–37.
26. Suarez-Lopez JR, Checkoway H, Jacobs DR Jr, Al-Delaimy WK, Gahagan S. Potential short-term neurobehavioral alterations in children associated with a peak pesticide spray season: The Mother's Day flower harvest in Ecuador. Neurotoxicol. 2017;60:125–33.
27. Rosas LG, Eskenazi B. Pesticides and child neurodevelopment. Curr Opin Pediatr. 2008;20: 191–7.
28. Torres-Sánchez L, Schnaas L, Rothenberg SJ, Cebrián ME, Osorio-Valencia E, Hernández MC, García-Hernández RM, López-Carrillo L. Prenatal p,p'-DDE exposure and neurodevelopment among children 3.5–5 years of age. Environ Health Perspect. 2013;121:263–8.
29. Rana J, Uddin J, Peltier R, Oulhote Y. Associations between indoor air pollution and acute respiratory infections among under-five children in Afghanistan: do SES and sex matter? Int J Environ Res Public Health. 2019;16:e2910.
30. Khan KM, Weigel MM, Yonts S, Rohlman D, Armijos R. Residential exposure to urban traffic is associated with the poorer neurobehavioral health of Ecuadorian children. Neurotoxicol. 2019;73:31–9.
31. Patel SK, Patel S, Kumar A. Effects of cooking fuel sources on the respiratory health of children: evidence from the Annual Health Survey, Uttar Pradesh, India. Public Health. 2019;169:59–68.
32. Winneke G. Developmental aspects of environmental neurotoxicology: lessons from lead and polychlorinated biphenyls. J Neurol Sci. 2011;308:9–15.
33. Pembrey ME. Does cross-generational epigenetic inheritance contribute to cultural continuity? Environ Epigenet. 2018;4:1–8.
34. Slykerman RF, Thompson JM, Pryor JE, Becroft DM, Robinson E, Clark PM, Wild CJ, Mitchell EA. Maternal stress, social support and preschool children's intelligence. Early Hum Dev. 2005;81(10):815–21. https://doi.org/10.1016/j.earlhumdev.2005.05.005.
35. Baldassare M. The effects of household density on subgroups. Am Sociol Rev. 1981;46(1):110–8. https://doi.org/10.2307/2095030.
36. Bartlett S. Does inadequate housing perpetuate children's poverty? Childhood. 1998;5(4):403–20. https://doi.org/10.1177/0907568298005004004.
37. Fuller TD, Edwards JN, Sermsri S, Vorakitphokatorn S. Housing, stress, and physical Well-being - evidence from Thailand. Soc Sci Med. 1993;36(11):1417–28. https://doi.org/10.1016/0277-9536(93)90384-G.
38. Loo C, Ong P. Crowding perceptions, attitudes, and consequences among the Chinese. Environ Behav. 1984;16(1):55–87. https://doi.org/10.1177/0013916584161003.
39. Youssef RM, Attia MSED, Kamel ML. Children experiencing violence I: parental use of corporal punishment. Child Abuse Negl. 1998;22(10):959–73. https://doi.org/10.1016/S0145-2134(98)00077-5.
40. Bradley RH, Caldwell BM. The home inventory and family demographics. Dev Psychol. 1984;20(2):315–20. https://doi.org/10.1037//0012-1649.20.2.315.

41. Bradley RH, Whiteside L, Mundfrom DJ, Casey PH, Kelleher KJ, Pope SK. Early indications of resilience and their relation to experiences in the home environments of low-birth-weight, premature children living in poverty. Child Dev. 1994;65(2):346–60.
42. Wachs TD, Camli O. Do ecological or individual characteristics mediate the influence of the physical-environment upon maternal-behavior. J Environ Psychol. 1991;11(3):249–64. https://doi.org/10.1016/S0272-4944(05)80186-0.
43. Evans GW, Bullinger M, Hygge S. Chronic noise exposure and physiological response: a prospective study of children living under environmental stress. Psychol Sci. 1998;9(1):75–7. https://doi.org/10.1111/1467-9280.00014.
44. Evans GW. Child development and the physical environment. Annu Rev Psychol. 2006;57:423–51. https://doi.org/10.1146/annurev.psych.57.102904.190057.
45. Ruiz-Casares M, Nazif-Muñoz JI, Iwo R, Oulhote Y. Nonadult supervision of children in low- and middle-income countries: results from 61 national population-based surveys. Int J Environ Res Public Health. 2018;15:e1564.
46. Gascon M, Vrijheid M, Nieuwenhuijsen MJ. The built environment and child health: an overview of current evidence. Curr Environ Health Rep. 2016;3:250–7.
47. Foster H, Brooks-Gunn J. Children's exposure to community and war violence and mental health in four African countries. Soc Sci Med. 2015;146:292–9.
48. Parchment TM, Small L, Osuji H, McKay M, Bhana A. Familial and contextual influences on children's prosocial behavior: South African caregivers as adult protective shields in enhancing child mental health. Glob Soc Welf. 2016;3:1–10.
49. Stansfeld SA, Berglund B, Clark C, Lopez-Barrio I, Fischer P, Ohrstrom E, Haines MM, et al. Aircraft and road traffic noise and children's cognition and health: a cross-national study. Lancet. 2005;365(9475):1942–9. https://doi.org/10.1016/S0140-6736(05)66660-3.
50. Sullivan SM, Broyles ST, Barreira TV, Chaput JP, Fogelholm M, Hu G, Kuriyan R, Kurpad A, Lambert EV, Maher C, Maia J, Matsudo V, Olds T, Onywera V, Sarmiento OL, Standage M, Tremblay MS, Tudor-Locke C, Zhao P, Katzmarzyk PT, ISCOLE Research Group. Associations of neighborhood social environment attributes and physical activity among 9-11 year old children from 12 countries. Health Place. 2017;46:183–91.
51. Evans GW, Gonnella C, Marcynyszyn LA, Gentile L, Salpekar N. The role of chaos in poverty and children's socioemotional adjustment. Psychol Sci. 2005;16(7):560–5. https://doi.org/10.1111/j.0956-7976.2005.01575.x.
52. WHO. Packages of interventions for family planning, safe abortion care, maternal, new born and child health. Geneva: World Health organization; 2010.
53. Senier L, Brown P, Shostak S, Hanna B. The socio-exposome: advancing exposure science and environmental justice in post-genomic era. Environ Sociol. 2017;3:107–21.
54. Hernandez M, Collins TW, Grineski SE. Immigration, mobility, and environmental justice: a comparative study of Hispanic people's residential decision-making and exposure to hazardous air pollutants in Greater Houston, Texas. Geoforum. 2015;60:83–94.
55. Schwartz NA, von Glascoe CA, Torres V, Ramos L, Soria-Delgado C. "Where they (live, work and) spray": pesticide exposure, childhood asthma and environmental justice among Mexican-American farmworkers. Health Place. 2015;32:83–92.
56. Voelkel J, Hellman D, Sakuma R, Shandas V. Assessing vulnerability to urban heat: a study of disproportionate heat exposure and access to refuge by socio-demographic status in Portland, Oregon. Int J Environ Res Public Health. 2018;15:E640.
57. Moody H, Grady SC. Lead emissions and population vulnerability in the Detroit (Michigan, USA) Metropolitan Area, 2006-13: a spatial and temporal analysis. Int J Environ Res Public Health. 2017;14:E1445.
58. Dowling R, Ericson B, Caravanos J, Grigsby P, Amoyaw-Osei Y. Spatial associations between contaminated land and sociodemographics in Ghana. Int J Environ Res Public Health. 2015;12:13587–601.
59. Laurent É. Issues in environmental justice within the European Union. Ecol Econ. 2011;70:1846–53.

60. Juarez PD, Matthews-Juarez P, Hood DB, Im W, Levine RS, Kilbourne BJ, Langston MA, et al. The public health exposome: a population-based, exposure science approach to health disparities research. Int J Environ Res Public Health. 2014;11:12866–95.
61. Golding J. Who should be studied and when in a longitudinal birth cohort? Paediatr Perinat Epidemiol. 2009;23(Supp 1):15–22.
62. Golding J, Jones R, Bruné MN, Pronczuk J. Why carry out a longitudinal birth survey? Paediatr Perinat Epidemiol. 2009;23(Supp 1):1–14.
63. Steer CD, Bolton P, Golding J. Preconception and prenatal environmental factors associated with communication impairments in 9 year old children using an exposome-wide approach. PLoS One. 2015;10:e0118701.
64. Cifuentes P, Reichard J, Im W, Smith S, Colen C, Giurgescu C, Williams KP, Gillespie S, Juarez PD, Hood DB. Application of the public health exposome framework to estimate phenotypes of resilience in a model of Ohio African-American women's cohort. J Urban Health. 2019;96:57–71.
65. Golding J, Gregory S, Iles-Caven Y, Lingam R, Davis JM, Emmett P, Steer CD, Hibbeln JR. Parental, prenatal, and neonatal associations with ball skills at age 8 using an exposome approach. J Child Neurol. 2014;29:1390–8.
66. Agier L, Basagaña X, Maitre L, Granum B, Bird PK, Casas M, Oftedal B, et al. Early-life exposome and lung function in children in Europe: an analysis of data from the longitudinal, population-based HELIX cohort. Lancet Planet Health. 2019;3:e81–92.
67. Birmingham K. Pioneering ethics in a longitudinal study. Bristol: Policy Press; 2018.
68. Kordas K, O'Hare D, Jacobs-Pearson M. Longitudinal studies: engaged cohort good for science. Nature. 2014;516:170.
69. Morello-Frosch R, Brody JG, Brown P, Altman RG, Rudel RA, Pérez C. Toxic ignorance and right-to-know in biomonitoring results communication: a survey of scientists and study participants. Environ Health. 2009;8:6.
70. Ohayon JL, Cousins E, Brown P, Morello-Frosch R, Brody JG. Researcher and institutional review board perspectives on the benefits and challenges of reporting back biomonitoring and environmental exposure results. Environ Res. 2017;153:140–9.
71. Perovich LJ, Ohayon JL, Cousins EM, Morello-Frosch R, Brown P, Adamkiewicz G, Brody JG. Reporting to parents on children's exposures to asthma triggers in low-income and public housing, an interview-based case study of ethics, environmental literacy, individual action, and public health benefits. Environ Health. 2018;17:48.
72. Guthman J, Mansfield B. The implications of environmental epigenetics: a new direction for geographic inquiry on health, space, and nature-society relations. Prog Hum Geogr. 2012;37:486–504.

Chapter 3
Transforming Well-Being for Refugees and Their Communities: Perspectives from Medicine, Nursing, Education, and Social Work

Kim Griswold, Jessica Scates, and Ali Kadhum

Introduction

Imagine—You have traveled far away from the country of your birth, you will probably never go home. You have left some family and many friends behind. You may have suffered trauma before, and during your journey. Now you must settle in a new place. You must integrate into a new culture, while preserving and honoring your own. You are known as a refugee, but you will integrate as a citizen into a new country—arriving with your unique culture and history, and striving for safety and a prosperous life for you and your children.

Forces driving migration and socio-political ideologies that contribute to refugee and immigrant health are defining our global twenty-first century. Complex immigration policies are often in conflict with principles of human rights, social justice, economic theory, and autonomy—of individuals and societies. This chapter will outline the complicated nature of displacement, and the impact it, as well as resettlement and *emplacement* have on the health and well-being of refugees. We will also present a perspective from Buffalo, NY, a community with long experience in refugee resettlement, and with expertise in meeting the needs of asylum seekers.

K. Griswold (✉)
Primary Care Research Institute, Buffalo, NY, USA
e-mail: griswol@buffalo.edu

J. Scates
University at Buffalo, State University of New York (SUNY), Buffalo, NY, USA

A. Kadhum
BestSelf Behavioral Health, Buffalo, NY, USA

© Springer Nature Switzerland AG 2020

K. H. Smith, P. K. Ram (eds.), *Transforming Global Health*,
https://doi.org/10.1007/978-3-030-32112-3_3

Forcibly Displaced Populations

In 2018, wars, violence, and persecution uprooted and forcibly displaced a record 68.5 million people around the world [1]. Of these, 40 million internally displaced persons sought safety within the borders of their home country, 3.1 million asylees crossed borders to request refugee status, or legal protection in another country, and 25.4 million were classified as refugees; they received the legal status and right to seek shelter, safety, and make a home in another country [1]. Although refugees, asylees, and internally displaced persons are on the move, they cannot be equated with migrants. Migrants cross international borders to join family, search for livelihood, escape natural disasters, or for other purposes. Forcibly displaced peoples have fled their homes because of persecution, war, or violence for reasons of race, religion, nationality, political opinion, or membership in a particular social group [1]. International laws require nation states to protect refugees, yet at borders, each country's asylum laws are considered matters of national sovereignty [2] (Fig. 3.1).

History and Policies of Migration and Displacement

In the half century after WWII, civil conflict and military aggression in Africa and Asia caused further forced migration, which stabilized to some extent in 2005; however, increasing conflict in the Middle East, Syria, and Sudan culminated in the 2015 estimate of over 60 million men, women, and children displaced and seeking safety [3, 4].

Countries in Eastern Europe that border conflict zones receive disproportionate numbers of refugees, and may be under-resourced for their resident population, thus incurring greater economic burden. Recent waves of nationalism sweeping through Europe complicate the integration of refugees and often feed xenophobia.

Extreme poverty in Mexico and many Latin American nations has fed the migration route of economic migrants primarily to the Unites States (U.S.). However, gang violence and persecution against citizens within many of these countries is responsible for increasing numbers of political refugees [5]. Vagaries of the U.S. immigration laws (specifically as they relate to refugees and asylum seekers) and their arbitrary interpretation result in controversial treatment of migrants and refugees at U.S. borders and reflect the current political climate. The U.S. does have a history of isolationist policies after world conflicts [6]. For example, refusal to accept Jewish refugees into the U.S. following WW II, and the separation of children from their parents between 2017 and 2018, are examples of repressive immigration stances.

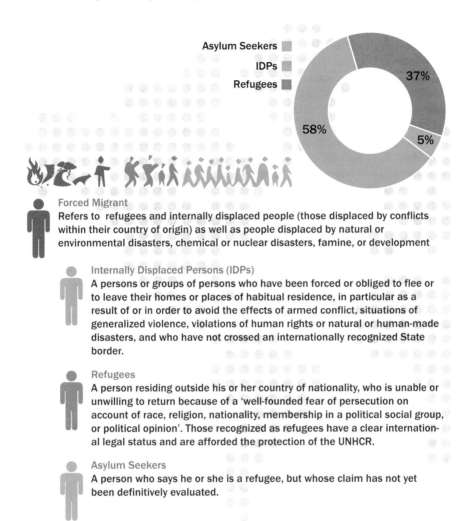

Fig. 3.1 Forced migration statistics. (Source: design: Ingrid Calderon, Mrunmayee Pathak, Verónica Yuqui, prepared for "Rethinking Resettlement," graduate architecture studio taught by Erkin Özay, University at Buffalo, Spring 2016; modified by Nicole C. Little)

Refugee Resettlement in the U.S.

Refugee resettlement is the legal relocation of refugees to a safe country. Resettlement provides refugees with legal and physical protection, including access to civil, political, economic, social, and cultural rights similar to those enjoyed by nationals. However, resettlement is a lengthy and complicated process that includes registration with the United Nations High Commissioner for Refugees (UNHCR), sponsorship from international resettlement organizations, and many levels of background checks and medical appointments. Less than 1% of all refugees resettle in an accepting country [3].

Within the U.S., the Department of State oversees refugee resettlement. UNHCR and U.S. Embassies around the world refer refugee applicants to nine Resettlement Support Centers external to the U.S. that process their claims. There, biographical and historical data are collected, and representatives from the Departments of State and Homeland Security complete security screening. Officers from the U.S. Citizen and Immigration Services (USCIS) then review all information, and conduct personal interviews with each applicant. Finally, all accepted refugees undergo a complete physical screening and exam before entry into the U.S [7].

Once accepted into the U.S., overseen by the Office of Refugee Resettlement, refugees resettle to cities throughout the nation that have established resettlement organizations and where there may already be communities of similar culture and background.

Historically, the U.S. has been the top resettlement country in the world, capping refugee resettlement at around 75,000 people per year. Despite a lack of evidence, citing fears that terrorists might infiltrate the program, the U.S. refugee resettlement program was temporarily suspended in 2017 [8–10]. Immigration politics and a rise in nationalism resulted in the lowest historical numbers of refugees admitted to the U.S., 45,000 in 2018 and an expected 30,000 in 2019 [11].

Forced Migration, Health, and Resilience

Global conflicts and violations of human rights (for example, ethnic cleansing, citizens as human shields in war, religious persecution, rape as a weapon of war) force individuals and families to flee as refugees [12]. Theoretical constructs seek to understand how the trauma of flight and pre- and post-migration factors affect individuals' overall health and welfare, mental health, and capacity for successful resettlement in safe countries. The losses that refugees endure (family, support systems, social identity, place) put them at risk for health problems (such as chronic pain, depression, anxiety, post-traumatic stress symptomatology) necessitating equitable access to primary care and behavioral health systems.

Assessment of an individual's mental health should take into consideration premigration, migration, and post-migration factors [13]. Mental health screening for recently resettled refugees may be an effective strategy, providing there are adequate follow up resources [14]. However, it is important to remember that individual responses to trauma and violence are mediated through a cultural lens, and that all people may experience and display distress differently [15].

Martin-Baró, in studying violence on society, stated, "Salvadorian displaced farmers manifested hypervigilance, distrust of outsiders and paranoid thinking but, rather than concluding that these were symptoms of poor mental health, he understood them as realistic responses to life conditions and manifestations of damaged social relationships. Abnormality and poor health are not with the individual but in the social environment" [16, p. 7, 17]. Thus, models of trauma-informed care may not always be the most effective means of achieving optimal health and wellness. Dona suggests that focusing on individual trauma and its treatment presents a negative view of refugee health—versus the need for a positive public health approach toward wellbeing and welfare [16].

Systems working to improve health and well-being for resettled refugees focus on individual health problems while disregarding discussions about quality of life, meaningful existence, and structural barriers [16]. Additionally, prescription medications or personal therapy that focuses on the individual, rather than the collective social/religious/cultural network, may lead to further isolation and create further oppression for already marginalized populations "under the guise of professional help" [16, p. 5].

Recovery must be grounded in "social recovery shaped by rights and collective justice, and personal healing needs to be situated within social processes that aim to re-establish collective safety, communalization and bereavement" [18]. In this context, the five pillars of the "ADAPT" (adaptation and development after persecution and trauma) model reinforce a more comprehensive view of healing: provision of security, building and reinforcing networks, achieving justice, finding roles and identities, and integrating existential meaning; repairing these domains is necessary for restoring communal mental, and physical, health [19].

Voices from the Field: Resettlement

During a breast cancer education workshop, May Shogan encouraged an Iraqi woman in her 40s to receive a mammogram. Her screening showed abnormalities. Because the woman's mother previously had ovary surgery, her doctors wanted to check her ovaries as well. She was scared. She did not know enough about her mother's health history to answer the doctors' questions. Most importantly, though, she was unmarried and a virgin. She was concerned that a gynecologist screening her ovaries might affect her virginity. Despite the doctor's assertion that an internal ultrasound would not be invasive, she refused to go any further. The doctor was concerned about saving her life; her concerns—the ramifications of losing her virginity from an invasive procedure—originated from her religious and cultural beliefs.

May says, "Culture is not a fashion statement that changes depending on the time and place." Practices going back centuries, many based on tradition and religion, are very hard to break. Women from diverse cultures refuse cancer screening—they do not believe they can control God's will.

Providers caring for patients who observe diverse religious and cultural traditions must consider that treating the patient goes beyond treating their physical health. Patients who feel a provider is undermining their beliefs may discontinue their care or treatment and endanger their health and well-being. To ensure a patient receives care that improves health and well-being, providers must take time to understand religious and cultural expectations. Patients that deeply trust their providers may be more open to sharing information and comply with treatment plans.

– May Shogan, director of education and international visitors at the International Institute of Buffalo led a multi-year Susan G. Komen grant to educate refugee women about the causes and early detection of breast cancer and to provide free breast cancer screenings [20].

Fundamental components of culturally effective clinical care include appropriate use of trained interpreters, a focus on health literacy, and knowledge and understanding of each refugee patient's cultural, religious, and/or traditional beliefs [21, 22]. Evidence based guidelines and culturally aware and competent care can help to ensure that the delivery and receipt of health care is optimal, and result in positive health outcomes. Comprehensive guidelines published by the Canadian Collaborative for Immigrant and Refugee Health suggest care providers should pay particular attention to and become knowledgeable about the culture from the country of origin, type of migration (forced versus voluntary), degree of health literacy, and vaccination history [23]. Mental health screening and care should be individualized and culturally congruent to improve diagnosis, treatment, and outcome. Trained interpreters and bi-cultural "brokers" or advisers are essential components of culturally informed care [13].

Voices from the Field: Clinical Care
Caring for refugee populations has made me a better physician, partly because I have had to learn to listen very well—usually through the use of trained interpreters. Their stories are spellbinding and heartbreaking. Their ability to teach my colleagues, students, and me is masterful. They are focused on creating a better life for their families, obtaining education, and offering their unique talents to a new place. Refugee patients and colleagues now help us in the University at Buffalo's Jacobs School of Medicine and Biomedical Sciences by teaching interprofessional health trainees how to use interpreters appropriately [24]. Part of this training emphasizes the importance of cultural awareness and sensitivity, patients as the best teachers, and the importance of humility on the part of the clinician. It also brings to me a continual sense of joy to be engaged in the communities that we care with, and for.

– Kim Griswold, Professor, Department of Family Medicine, Jacobs School of Medicine and Biomedical Sciences, University at Buffalo.

Making Places: The Journey from Displacement to Emplacement

The study of displacement has been historically rooted in humanitarian assistance despite its ties to economics, political science, or history, and refugees have been grouped according to their supposed common conditions or nature [25]. However, refugees reflect diverse political, religious, socio-economic, and cultural backgrounds. Wrought with horror and suffering, their displacements differ from geography and context. For many, theirs is a forced re-placement, resulting in a scattering of families and communities around the world [26]. They courageously adjust "home" to reflect the physical spaces in which they feel at ease and safe, along each point in their journey—whether displacement within their home countries, residence in informal border towns, lengthy stays in resettlement camps, or third country resettlement.

A radical shift in displacement literature has been occurring over several decades. As globalization increases, scholars have rejected the notion that geographical displacement equates with loss of identity and culture [16, 25, 27–29]. From "stable, rooted and mappable identities to fluid, transitory and migratory phenomena... people are now thought of moving continuously through flexible, open-ended and contested space" [30, p. 317]. However, increased globalization has led to increasingly restrictive immigration policies [31]. Although the rich and poor live in the same mobile world, they experience it differently and are impacted by it inequitably [32]. "The way people experience movement to a new place, and the extent to which this is a shocking and disruptive experience, is determined by the conditions under which they move" [32, p. 275]. For displaced peoples, movement into new areas is often stigmatized.

Turton argues there is a need for a theory of place that gives equal credence to self-identify; one that is as dependent on location as it is to homelessness [32]. People on the move are not moving through indifferent space, they move through places that are shaping them and being shaped by them [30]. Displacement is not only about the loss of a place, but also the struggle to *make a place* in the world [32, 33]. Although we live in a fluid world, *emplacement*, the act of making a place a home or linking together space, place, and power, creates a sense of belonging and meaning, something for which displaced persons yearn [30, 34].

Individuals can develop a sense of place through...deliberate acts like storytelling, re-creating features of a lost environment, keeping mementos safe, and establishing and maintaining social links [27, 32, 35–37]. Many refugees bring strength, adaptability, a sense of belonging and purpose to new environments [38]. However, the communities and neighborhoods in which they resettle contribute as much to their resilience and ability to make place. Emplacement can only be possible through the building of relationships, whether they be negotiations between refugees and those who assist them, connections with neighbors, partnerships with policy makers, camaraderie with teachers and peers, or friendships with local ethnic communities [16, 27, 32, 36, 38]. Positive social interactions lead to economic and educational opportunities, increased civic engagement, legal and resettlement support, and celebrations of diversity of cultural expressions through art, dance, and music [39]. For emplacement to occur, hosts and newcomers must actively engage, understand, and learn from one another [36].

Displaced: "In" But Not "Of"

> Because of diminishing resources, income-generating opportunities, and problems of unemployment, host populations are becoming increasingly hostile to refugees. As a result, most refugee communities see no future for themselves or their offspring in their country of asylum…Instead of working to develop roots in the new place, [refugees] aim becomes return to the country of origin from where they were forcibly uprooted. [31, p. 389].

Refugees often exist within a physical residence yet do not belong; they are "in" but not "of" a space [28, p. 344]. Power structures, politics, and socio-economic, racial, and educational discrimination are the backdrop to their struggles. Local and national policies that limit access to social services and sources of livelihood as well as freedom of movement reduce a newcomer's ability to make a place a home [31]. Refugees must contend with segregated camps and neighborhoods, bias and hatred, stigma of the other, nationalism and anti-refugee sentiment, and even conflict between personal, cultural, and host country norms. Differences in language, culture, climate, geography, food, and community complicate the act of emplacement. Take Amira, who resettled in Buffalo, NY after leaving Iraq:

Voices from the Field: Displaced New Americans
Life in Iraq was very different for Amira and her family. As a homemaker, Amira prided herself in taking good care of her husband, kids, and family. After fleeing Iraq and resettling in the U.S., her new reality was very different. Amira had to navigate unfamiliar systems like health care and education for her children. To help her family, Amira found a job. The initial separation from her children was very difficult—she experienced constant fear and worry. However, Amira prevailed and began to earn more money, at almost the same rate as her husband. Accustomed to traditional gender roles throughout her life—the husband worked while the wife cared for the children—the couple experienced conflict over money and power dynamics. These conflicts led to the couple's divorce.

In Amira's culture, when divorce occurs women do not remarry. While living in Iraq, Amira and her husband's families would have helped the couple solve their problems. From Amira's perspective—and with limited social supports—in the U.S. the only resource they have is the police. Today, Amira struggles raising her children alone—the children's father is not involved in their lives. Thinking back on the past and their time in a refugee camp, Amira notes that even facing fear and hunger, their family was at peace. In the refugee camp, she knew that there was no future for her children, but they were together. Now she feels that she is losing her kids, has no husband, and is all alone. She wished she had stayed in the camp instead of coming to the U.S.

Literature is full of stories like Amira's. Farah, a refugee from Somalia who traveled through Kenya to Sweden, Denmark and then Finland, suffered abuse and attacks from racists living in Finland. Post-displacement, he sought "human life" in Kenya, where there was peace and where he could raise his family [40, p. 9]. Children face marginalization by peers and even teachers resulting in poor academic achievement and increased likelihood to join gangs [41, 42]. Sudanese refugees living in Australia note a major cause of misconduct and suicide among youth is idleness and inadequate mid- to long-term settlement opportunities [37].

Resettlement in a community can be just as, if not more, traumatic and destructive than initial displacement. Bronstein's 2011 systematic review on psychological distress in refugee children surveyed 22 studies covering over 3000 children from over 40 countries [43]. Trauma and pre-migration were factors for psychological distress but post-migration factors like uncertainty regarding migration status, the process of immigration, discrimination, a lack of personal and structural support, and greater restrictions to living conditions had direct relationship to higher scores of depression and PTSD. Bogic, studying factors associated with mental disorders in long-settled refugees assessed 854 refugees living in Germany, Italy, and the UK [44]. The "stressful social and material conditions in resettlement independently predicted mental health status as well as, or better than, actual exposure to war trauma" [44]. Post-migration stressors, specifically not feeling accepted by the host country, having a temporary residence status, unemployment, and substance use disorders explained the most significant variance in rates of mood and anxiety disorders [44].

For many refugees, resettlement results in a lack of financial, social, and psychological support required to ensure their health and well-being [37]. In many places, refugees face discrimination and criminalization, deficit-based paradigms that blame and marginalize, and recovery programs and longer-term community opportunities that are only accessible for a portion of the population [37, 42, 45]. "Refugee and humanitarian entrants are...asked to fend for themselves before they have had time to acquire the necessary cultural competencies and alternative means of life support" [37].

Emplaced: Leaving Our Imprints

Daily routines and habits "leave our imprints on places – and…places leave their imprints on us" [30, p. 328]. Emplacement practices like rituals, re-establishment of supportive familiar systems, relationships with community and family, and story-telling make sense of one's being-in-the-world and play an important role in *rehumanizing* impermanent living environments [30, 37, p. 45, 46]. Take Jubair's family who left Iraq and resettled in Buffalo, NY:

Voices from the Field: Emplaced New Americans
Jubair came to the U.S. from Iraq eager to support his children. He enlisted their help as leaders within the home, knowing that they would be windows into the U.S. culture and traditions. In Iraq, this type of arrangement would be unheard of; the father advises the children. Thinking about their new relationship, Jubair notes that in Iraq, work took him from his family; he spent most of his time with the community and never spent time with his children. After coming to the U.S., he began to enjoy spending time with his children. Although he lost the social connections he had overseas, he was content knowing his children.

Jubair's wife worried about the well-being of her family and questioned why they were there. She wanted to go back to Iraq, back to her mother. To ease her stress, Jubair encouraged her to work. Her work led to friendships and she began to create meaning in the place she lived. Today, Jubair encourages other families from Iraq to encourage their wives to work.

Jubair admits life in America was not easy at first. He had to reflect on certain aspects of American culture for a while before being OK with them. In Iraq, when Jubair and his wife had problems, their parents would solve them. In the U.S., they needed to solve those problems together. They have worked to understand one another, agreeing that a problem cannot go for more than a day. Jubair and his wife note that the key to success in their relationship is that they talk about their problems together. Their children are also succeeding in the community—one son was recently selected as the best student in the entire school and another, who attends a local college, has begun to receive job offers from IT companies.

Emplaced people cultivate their unique cultures and traditions while embracing and being embraced by the host communities in which they live. Jubair's family created a space that was unique to them and their culture while establishing new traditions and nurturing relationships within their resettled community. Similarly, Halima, a refugee from Somalia, was confident that she survived a near death experience to do something meaningful with her life—that work with the community was the medicine that helped her to live [30]. In creating a sewing group to dialogue with other Somali women, volunteering in a multicultural children's playgroup, and running a local school lunchroom, Halima combatted feelings of estrangement "by actively carving out her own paths. In doing so, she was literally placing herself back into the world" [30, p. 326]. Stories of refugee resilience are common. All 4 One[2], a group of Sudanese hip-hop artists, constructed a transnational virtual network of artists and fans [37]. Karen refugees created a sense of belonging and meaning by erecting a monastery that locals and refugees frequent [29]. And, a Somali Bantu community maintained gardens that provided a place for cooperation and collaboration, physical, mental, and emotional therapy, and memories of home [27].

Evidence suggests that refugees can thrive and benefit the host nation economies in which they settle [47, 48]. In releasing the labor market impact of refugees in 2017, the U.S. Department of State published "robust causal evidence that there is no adverse long-term impact of refugees on the U.S. labor market" [49]. Similarly, an article published in *Nature* analyzed 30 years of data from Western Europe refuting the suggestion that refugees pose a financial burden on a host country economy [50]. Alexander Betts Refugee Economies explores how refugees, when provided basic economic freedoms, live complex and varied economic lives, are highly entrepreneurial, and connected to the global economy [51]. "Far from being an inevitable burden, refugees have the capacity to help themselves and contribute to their host societies—if we let them" [51].

Both qualitative and quantitative data suggest that an individual alone cannot succeed (or fail) to cope in a place and be healthy; to be a part "of" these spaces requires acceptance and collaboration [16]. Both newcomers and host communities play equal role in ensuring a refugee's successful emplacement (Fig. 3.2).

Fig. 3.2 The concept of emplacement. (Source: concept: Jessica Scates and Kim Griswold; graphic design: Nicole C. Little)

A Case Study: Buffalo, NY, the City of Good Neighbors

Buffalo, NY has a century old tradition of welcoming immigrants, refugees, and asylum seekers. Four local resettlement agencies help legally accepted refugees acclimate to their new environment by providing assistance with housing, employment, and health care, enrolling children in school, linking with social services, and navigating the city. Buffalo also hosts a large shelter for asylum seekers, and a Center for Survivors of Torture (CST) through Jewish Family Services of Buffalo and Erie County [52]. The CST serves individuals (resettled refugees as well as asylum seekers) who suffered persecution, violence, and torture.

Buffalo's refugee populations collaborate with local municipalities, businesses, and higher education institutions, contributing to the revitalization of the economy. Places like the West Side Bazaar, Buffalo String Works, and Stitch Buffalo create jobs while celebrating diverse arts and culture [53–55].

Community based organizations are serving their neighbor's needs, reducing stigma through annual events like World Refugee Day [56]. Local government established an Office for New Americans, giving refugees a political voice [57]. Partnerships with the local police department developed a language access plan, ensuring refugees can communicate their needs and understand their rights [58]. Established clinics serve the most vulnerable populations including victims of trauma and torture, asylees, pregnant women, and children. Many hire refugees— valuable employees who provide cultural and linguistic support in delivery of health care. Resettlement organizations tirelessly seek funding to establish community gardens, legal services, and trainings in small business development, interpreting, and cultural competency. The Human Rights Initiative, a medical student-led group at the University at Buffalo Jacobs School of Medicine and Biomedical Sciences, arranges and provides pro bono services to conduct forensic exams for asylum seekers to assist in cases alleging torture and ill treatment in their countries of origin [59]. Researchers spanning planning to public health, social work to nursing, management to medicine seek to understand the barriers to good health and well-being for refugees, offering programmatic and policy change solutions [60]. They are also training the next generation of medical doctors, nurses, pharmacists, social workers, public health professionals, and the like through multi-disciplinary courses, workshops, and annual summits focused on improving refugee health and well-being.

Because of Buffalo's long history of caring for individuals from diverse cultures, health and medical care have developed to meet the distinct needs of individuals and families. The Jericho Road FQHC provides a best practice primary care model to care for local communities, and offers services that focus on identified needs [61]. For example, the Priscilla Project supports new refugee expectant mothers by linking them with medical students for health education and support; and VIVE La Casa is a large residential shelter for individuals and families seeking asylum. Buffalo also supports two additional Federally Qualified Health Centers, and an integrated primary care/behavioral health clinic, each providing care to sizeable refugee populations.

Conclusion

Motivated by atrocities committed during the previous world wars, the General Assembly of the United Nations adopted the Universal Declaration of Human Rights on December 10, 1948. The declaration expresses values shared by all members of the international community and has given rise to legally binding international agreements like the Convention Relating to the Status of Refugees. Article 14(1) guarantees the right for persecuted people to seek and enjoy asylum in other countries [62]. However, world leadership still struggles to balance the provision of humanitarian aid with maintaining border and national security. Globalization has resulted in the increase of movement of goods and services across nation states, but has also led to an increase in nationalistic ideologies that breed fear and persecution of the "other" [31]. Some fear what they do not know.

Buffalo NY is an example of a city seeking to emplace and embrace refugees and asylees by involving them in the local social, economic, and political fabric.

Beyond the debate of whether refugee resettlement is ethically and morally right, or whether it is a humanitarian obligation, it is certain that for the health and well-being of all, community members new and old, we must create spaces that allow for safe emplacement.

> To emphasize the horror and pain of the loss of home...and to say nothing—or little—about the work of producing home or neighborhood...is to see them—as virtually everyone who writes about refugees urges us *not* to see them—as a category of 'passive victims' who exist to be assisted, managed, regimented and controlled—and for their own good. Above all, it makes it more difficult for us to identify with the suffering stranger, to see him or her as an ordinary person, a person like us, and therefore as a potential neighbor in *our* neighborhood [32, p. 278].

The positive and negative stories of emplacement should cause us to reflect. What is missing from this discourse? What is in the best interest of the physical and mental health of populations around the world? How are we, as nations, cities, and neighborhoods, welcoming and empowering our new neighbors?

References

1. Edwards A. Forced displacement at record 68.5 million. 2018. https://www.unhcr.org/news/stories/2018/6/5b222c494/forced-displacement-record-685-million.html. Accessed 16 Dec 2018.
2. Dragostinova T. On 'strategic frontiers': debating the borders of the post-second world war Balkans. Contemp Eur Hist. 2018;27(3):387–411. https://doi.org/10.1017/S0960777318000243.
3. El-Gamal A. Protecting Syrian refugees: short term solution to unsustainable burden and the necessity of eliciting aid from the global north. Georget Immigr Law J. 2017;31(3):597.
4. UNHCR. Figures at a Glance. 2001–2009. UNHCR.org. Accessed 12 Nov 2019.
5. Hyland S. Hacer América and the American dream: global migration and the Americas. Origins. 2015. 8(8).

6. United States Holocaust Memorial Museum. https://www.ushmm.org/. Accessed 6 Feb 2019.
7. U.S. Refugee Admissions Program (USRAP). Consultation and worldwide processing priorities. USCIS.gov. Last updated 03 May 2019.
8. International Rescue Committee. How the U.S. vetting and resettlement process really works. 2019. Rescue.org. Accessed 12 Nov 2019.
9. Cowger S, Bolter J, Pierce S. The first 100 days: summary of major immigration actions taken by the trump administration. Washington, DC: Migration Policy Institute; 2017.
10. Karasapan O. Refugees, migrants, and the politics of fear. 2017. https://www.brookings.edu/blog/future-development/2017/04/12/refugees-migrants-and-the-politics-of-fear/. Accessed 6 Feb 2019.
11. U.S. Department of State. Proposed refugee admissions for fiscal year 2019. 2018. https://www.state.gov/j/prm/releases/docsforcongress/286157.htm. Accessed 12 Dec 2018.
12. Amnesty International. Amnesty international report 2017/18: state of the world's human rights. London: Amnesty International; 2018.
13. Kirmayer LJ, Narasiah L, Munoz M, Rashid M, Ryder AG, Guzder J, Hassan G, et al. Common mental health problems in immigrants and refugees: general approach in primary care. CMAJ. 2011;183(12):E959–67. https://doi.org/10.1503/cmaj.090292.
14. Polcher K, Calloway S. Addressing the need for mental health screening of newly resettled refugees: a pilot project. J Prim Care Community Health. 2016;7(3):199–203. https://doi.org/10.1177/2150131916636630.
15. Shannon PJ, Wieling E, McCleary JS, Becher E. Exploring the mental health effects of political trauma with newly arrived refugees. Qual Health Res. 2015;25(4):443–57. https://doi.org/10.1177/1049732314549475.
16. Giorgia D. Rethinking Well-being: from contexts to processes. Int J Migr Health Soc Care. 2010;6(2):3–14. https://doi.org/10.5042/ijmhsc.2010.0606.
17. Martin-Baro I. Political violence and war as causes of psychosocial trauma in El Salvador. Int J Ment Health. 1989;18(1):3–20. https://doi.org/10.1080/00207411.1989.11449115.
18. Maynard KA. Rebuilding community: psychosocial healing, reintegration, and reconciliation at the grassroots level. In: Rebuilding societies after civil war: critical roles for international assistance. Boulder: Lynne Reinner; 1997.
19. Silove D. The ADAPT model: a conceptual framework for mental health and psychosocial programming in post conflict settings. Intervention. 2013;11(3):237–48. https://doi.org/10.1097/WTF.0000000000000005.
20. International Institute of Buffalo. https://iibuffalo.org/. Accessed 4 Sept 2018.
21. George M. A theoretical understanding of refugee trauma. Clin Soc Work J. 2010;38(4):379–87. https://doi.org/10.1007/s10615-009-0252-y.
22. Griswold KS, Pottie K, Kim I, Kim W, Lin L. Strengthening effective preventive services for refugee populations: toward communities of solution. Public Health Rev. 2018;39(1):3. https://doi.org/10.1186/s40985-018-0082-y.
23. Pottie K, Greenaway C, Feightner J, Welch V, Swinkels H, Rashid M, Narasiah L, et al. Evidence-based clinical guidelines for immigrants and refugees. CMAJ. 2011;183(12):E824–925. https://doi.org/10.1503/cmaj.090313.
24. University at Buffalo Global Health Equity. The value of refugees and interpreters as standardized patients for interprofessional education. https://www.buffalo.edu/globalhealthequity/global-projects/refugeehealthandwellbeing/buffalo/the-value-of-refugees-and-interpreters-as-standardized-patients-for-interprofessional-education.html. Accessed 4 Sept 2018.
25. Malkki LH. Refugees and exile: from "refugee studies" to the national order of things. Annu Rev Anthropol. 1995;24(1):495–523. https://doi.org/10.1146/annurev.an.24.100195.002431.
26. Sampson R, Gifford SM. Place-making, settlement and well-being: the therapeutic landscapes of recently arrived youth with refugee backgrounds. Health Place. 2010;16(1):116–31. https://doi.org/10.1016/j.healthplace.2009.09.004.

27. Coughlan R, Hermes SE. The palliative role of green space for Somali bantu women refugees in displacement and resettlement. J Immigr Refug Stud. 2016;14(2):141–55. https://doi.org/1 0.1080/15562948.2015.1039157.
28. Bauman Z. In the lowly nowherevilles of liquid modernity: comments on and around Agier. Ethnography. 2002;3(3):343–9. https://doi.org/10.1177/146613802401092788.
29. Rangkla P. Refuge and emplacement through Buddhism: Karen refugees and religious practices in a northwestern border town of Thailand. Asia Pac J Anthropol. 2013;14(1):8–22. https://doi.org/10.1080/14442213.2012.743581.
30. Lems A. Placing displacement: place-making in a world of movement. Ethnos. 2016;81(2): 315–37. https://doi.org/10.1080/00141844.2014.931328.
31. Kibreab G. Revisiting the debate on people, place, identity and displacement. J Refug Stud. 1999;12(4):384–410. https://doi.org/10.1093/jrs/12.4.384.
32. Turton D. The meaning of place in a world of movement: lessons from long-term field research in southern Ethiopia. J Refug Stud. 2005;18(3):258–80. https://doi.org/10.1093/refuge/fei031.
33. Xenos N. Refugees: the modern political condition. Altern Glob Local Politi. 1993;18(4):419–30. https://doi.org/10.1177/030437549301800401.
34. Glick Schiller N, Çağlar A. Displacement, emplacement and migrant newcomers: rethinking urban sociabilities within multiscalar power. Identities. 2016;23(1):17–34. https://doi.org/10.1 080/1070289X.2015.1016520.
35. Easthope H. A place called home. Hous Theory Soc. 2004;21(3):128–38. https://doi.org/ 10.1080/14036090410021360.
36. Brun C. Reterritorializing the relationship between people and place in refugee studies. Geogr Ann Ser B Hum Geogr. 2001;83(1):15–25.
37. Wilson MJ. 'Making space, pushing time': a Sudanese hip-hop group and their wardrobe-recording studio. Int J Cult Stud. 2012;15(1):47–64.
38. Resilience and coping. 2011. https://refugeehealthta.org/physical-mental-health/mental-health/adult-mental-health/resilience-and-coping/. Accessed 4 Sept 2018.
39. Griswold K, Kim I, Scates J. Building a community of solution with resettled refugees. J Community Med Health. 2016;6:404. https://doi.org/10.4172/2161-0711.1000404.
40. Hautaniemi P. Fugitive memories. How young Somali men recall displacements and emplacements in their childhood. Anthropol Yearb Eur Cult. 2006;15:77–91.
41. Dryden-Peterson S. The educational experiences of refugee children in countries of first asylum. Washington, DC: Migration Policy Intsitute; 2015.
42. Roy L, Roxas K. Whose deficit is this anyhow? Exploring counter-stories of Somali bantu refugees' experiences in "doing school". Harv Educ Rev. 2011;81(3):521–42. https://doi.org/10.17763/haer.81.3.w441553876k24413.
43. Bronstein I, Montgomery P. Psychological distress in refugee children: a systematic review. Clin Child Fam Psychol Rev. 2011;14(1):44–56. https://doi.org/10.1007/s10567-010-0081-0.
44. Bogic M, Njoku A, Priebe S. Long-term mental health of war-refugees: a systematic literature review. BMC Int Health Hum Rights. 2015;15(1):29. https://doi.org/10.1186/s12914-015-0064-9.
45. Muir J, Gannon K. Belongings beyond borders: reflections of young refugees on their relationships with location. J Community Appl Soc Psychol. 2016;26(4):279–90. https://doi.org/10.1002/casp.2260.
46. Ager A. Feedback. Dev Pract. 1997;7(4):402. https://doi.org/10.1080/09614529754198.
47. Taylor JE, Filipski MJ, Alloush M, Gupta A, Rojas Valdes RI, Gonzalez-Estrada E. Economic impact of refugees. Proc Natl Acad Sci. 2016;113(27):7449–53. https://doi.org/10.1073/pnas.1604566113.
48. d'Albis H, Boubtane E, Coulibaly D. Macroeconomic evidence suggests that asylum seekers are not a "burden" for Western European countries. Sci Adv. 2018;4(6):eaaq0883.
49. Mayda AM. The labor market impact of refugees: evidence from the U.S. resettlement program. Washington, DC: United States of America Department of State; 2017.

50. Maxmen A. Migrants and refugees are good for economies. Nature. 2018. https://doi.org/10.1038/d41586-018-05507-0.
51. Betts A, Bloom L, Kaplan JD, Omata N. Refugee economies: forced displacement and development. 1st ed. Oxford: Oxford University Press; 2016.
52. Jewish Family Service of Buffalo & Erie County. Western New York Center for Survivors of Torture. http://www.jfsbuffalo.org/wny-survivors-center-. Accessed 4 Sept 2018.
53. West Side Bazaar. https://www.westsidebazaar.com/. Accessed 4 Sept 2018.
54. Buffalo String Works. https://www.buffalostringworks.org/. Accessed 4 Sept 2018.
55. Stitch Buffalo. https://www.stitchbuffalo.org/. Accessed 4 Sept 2018.
56. Facebook. World refugee day in western New York. https://www.facebook.com/WorldRefugeeDayBuffalo/. Accessed 4 Sept 2018.
57. City of Buffalo. Office of new Americans. https://www.buffalony.gov/422/Office-of-New-Americans. Accessed 4 Sept 2018.
58. Partnership for the Public Good. Collaboration, communication and community-building: a new model of policing for 21st century Buffalo.
59. University at Buffalo Jacobs School of Medicine and Biomedical Sciences. Human rights initiative. http://medicine.buffalo.edu/about/community_outreach/human-rights-initiative.html. Accessed 4 Sept 2018.
60. University at Buffalo Global Health Equity. Refugee health and wellbeing in Buffalo. https://www.buffalo.edu/globalhealthequity/global-projects/refugeehealthandwellbeing/buffalo.html. Accessed 4 Sept 2018.
61. Jericho Road Community Health Center. https://www.jrchc.org/. Accessed 4 Sept 2018.
62. International Justice Resource Center. Asylum and the rights of refugees. https://ijrcenter.org/refugee-law/. Accessed 4 Sept 2018.

Chapter 4
Interactive Systems in Nutrition: Perspectives from Epidemiology, Veterinary Science, Nutrition, Anthropology, and Community Health

Laura E. Smith, Roseanne C. Schuster, Sarah E. Dumas, and Brendan T. Kerr

Introduction

Nutrition is a complex and multifaceted influencer of health. Everything we consume, from a sip of water, to those green beans from a community garden, to that egg you had for breakfast are components of nutrition. But what are the factors that influence your nutrition, and, in turn, your health? Consider every choice you make around what you eat. You probably consider personal preference and nutritional value of the food, but what determines the food available to eat? And what determines your ability to access that food? It turns out that nutrition is influenced by a constellation of factors that exist within a dynamic and interacting system. In this chapter, we consider three broad components of this system, land, food, and the body, in the context of child malnutrition (Fig. 4.1).

Malnutrition and Stunting: A Global Problem

Child malnutrition underlies half of all global deaths in children, causes life-long negative health effects, and hinders the economic potentials of both individuals and entire societies. Malnourished children face higher mortality rates, are more

L. E. Smith (✉) · B. T. Kerr
University at Buffalo, State University of New York (SUNY), Buffalo, NY, USA
e-mail: lesmith6@buffalo.edu

R. C. Schuster
Arizona State University, Tempe, AZ, USA

S. E. Dumas
New York City Department of Health and Mental Hygiene, New York, NY, USA

© Springer Nature Switzerland AG 2020
K. H. Smith, P. K. Ram (eds.), *Transforming Global Health*,
https://doi.org/10.1007/978-3-030-32112-3_4

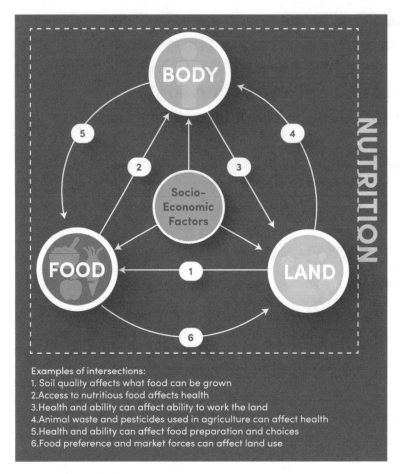

Fig. 4.1 Conceptual framework and examples among body, food, and land. (Source: concept: Laura Smith; graphic design: Nicole C. Little)

vulnerable to and more likely to die from infection, and have delayed recovery when they get sick.

One indicator of malnutrition is *stunting*, or low height-for-age. Stunted children are of a height below two standard deviations from the average height among children of the same age and sex. This means stunted children are below the normal height range, and are suffering from delays in their physical, and subsequently their neurocognitive, development. Stunting is caused by a variety of socio-economic, environmental, and nutritional factors, with nutrition long thought to be chief among them. The majority of stunting occurs during the first 1000 days of life from conception to age two, a period of rapid growth and development. Insufficient nutrients and disease during this period can lead to growth faltering that is largely irreversible, and can influence the entire life-course of a child, from increased mortality and reduced school achievement in childhood, to lower economic productivity in

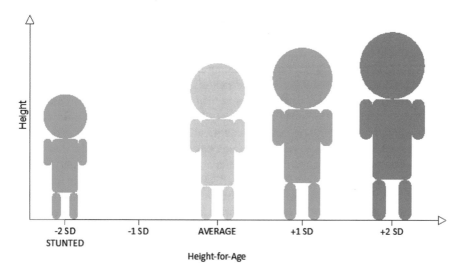

Fig. 4.2 Stunting in children under five globally in 2017. Stunted children are more likely to have poor academic performance, motor development, immune function, cognitive ability, and economic outcomes in adulthood. (Source: Laura Smith)

adulthood [1]. Once established, stunting is largely irreversible and prevention is therefore critical.

The Food and Agriculture Organization (FAO) estimated that 149 million children under 5 years old globally are stunted, as of 2018 [2]. Stunting disproportionately affects children in South Asia, Eastern and Southern Africa, and West and Central Africa, where approximately one in three children are stunted. While stunting and malnutrition are still large threats to the health of children globally, the percentage of children under 5 who are stunted globally has dropped from one-third in 2000 to one-fourth in 2017. Although the improvement is profound, we need to be mindful that still 21.9% of children remain stunted today [3] and there are still many factors that need to be considered reduce the prevalence of stunting, globally (Fig. 4.2).

Current Approaches and Challenges of Nutritional Interventions

Historically, malnutrition has been thought of as solely a problem of insufficient nutrient intake. The majority of interventions that have been enacted in global communities to prevent malnutrition and stunting have been *nutrition-specific interventions*, which specifically target the immediate determinants of malnutrition, by supplementing the nutrition of participants. Nutrition-specific interventions can include promotion and support of breastfeeding, fortification of foods, provision of micronutrient supplements in individuals, and treating those with severe malnutrition [4]. This approach is effective for treating micronutrient deficiencies, but often fails to influence more complex and multifactorial outcomes, like chronic malnutrition or stunting. For these broader issues *nutrition-sensitive interventions* are more effective, considering the broad influencers of nutrition and the dynamic system surrounding it by focusing on factors like agriculture, social safety nets, early child development, schooling, and economic empowerment, alongside nutrition. Unfortunately, nutrition-sensitive interventions are still relatively novel and require greater resources to carry out than the relatively simple nutrition-specific interventions, so the effects of nutrient-sensitive interventions remain understudied [5].

Recently, researchers who study childhood malnutrition estimate that if existing nutrition-specific interventions were to expand their coverage to 90% coverage in the 34 countries with 90% of the global stunting burden, there would only be a 20% decline in stunting. The other 80% of the stunted population would remain stunted, due to factors other than those addressed by the nutrient-specific interventions. This reduction would only cover half of the United Nation's Sustainable Development Goal to reduce stunting globally by 40% by 2025, and end malnutrition globally by 2030. This "impact gap" highlights the need for nutrition-sensitive interventions which focus on outside factors and the dynamic system influencing nutrition [6].

Body: Health and Well-Being

When we consider the body, we are largely considering the health and well-being of an individual. The World Health Organization (WHO) defines *health* as a "state of complete physical, mental, and social wellbeing and not merely the absence of disease or infirmity." *Well-being* focuses on the biological state, specifically illness and mortality, but also refers to an individual's means or ability to lead an active, healthy life [7]. A broader use of well-being will consider well-being in the context of social and cultural environments, or necessitate that the state of well-being be sustainable over time [8]. Different forms of well-being are within the broader concept of health [9].

Health and well-being shape how people interact with concepts of food and land. For example, a farmer who is ambulatorily disabled from a childhood illness could have difficulty working the land, gaining access to different resources, or maintaining a stable income for adequate nutritional security. The farmer would suffer from additional health problems stemming from reduced access to nutritious foods, additional resources, and economic prosperity. Health and well-being affect economic productivity and ability to get out of poverty trap.

It is important to consider other factors on multiple levels affecting the body as well. Health and well-being are influenced by individual (e.g., vaccination status, gender, age, education), household (e.g., socio-economic status, caregiving behaviors), health service (e.g., availability, access, quality, ability to identify illness), environmental (e.g., pollutants), and societal/governmental (e.g., political infrastructure, governance) factors, along with innumerable other factors that influence health. These distal factors feed into the complex and dynamic web, and make the task of addressing malnutrition difficult to do without considering numerous factors and effects.

Food: Food System, Decision-Making, and Food Security

When we conceptualize food in relation to nutrition we cannot solely consider diet, we must also consider broader factors like the overall food system, nutritional security, and the decisions families make surrounding food. In this section, we will break down these concepts and describe the relationships food has with land and body.

A *food system* is a network of processes including food production, processing, distribution, storage, to its consumption and waste management. This also includes intangible elements that influence the system, including government policies and socio-cultural constructs. The food system has a large degree of influence on what foods are accessible and practical to prepare, store, and dispose of. Inadequate access can lead to nutritional deficits, especially if the accessible foods alone are not satisfactory sources of micronutrients. The food system can have a disproportionate effect on nutritional status in regions with poor infrastructure or economic outcomes, and contribute to security of food. *Food security*, as defined at the World Food Summit in 1996, exists when "when all people, at all times, have physical and economic access to sufficient, safe and nutritious food that meets their dietary needs and food preferences for an active and healthy life" [10]. An unstable and insecure food system could prevent a household from maintaining a secure supply of nutritious food and contribute to malnutrition and stunting in children.

Food decision-making on a household level involves the choices a household makes surrounding food, eating environments, family food policies, and strategies for mobilizing family food resources. Food decision-making at the household level is considered the most proximal system affecting child nutrition. Decisions made regarding food are affected by the food system and food security, along with food preparation knowledge and skills, available technology, household income, breastfeeding practices, and availability of different foods throughout time.

Daily, individuals and households make thousands of small choices that go into major decisions, weighing a number of factors. Understanding this complex system requires an understanding of the individual or household within the complex system. Households make decisions based on their objectives, endowments, and incentives. To identify effective points to intervene in a system, a clear understanding of participants' motivations is necessary. This includes reasons for joining groups, like sharing food or cooperative farming and hunting, and motivations to adopt new technologies or behaviors. In each of the observations, we must consider broader socio-economic issues as well.

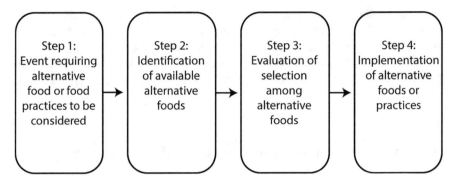

Fig. 4.3 Stages to the decision-making process. (Source: Laura Smith)

There are several stages to the decision-making process, each listed in Fig. 4.3 [11]. The event spurring the need for this process, or the availability of alternate foods, could stem from the changes in the food system or the household's nutritional security. Additionally, nutritional security and food systems affect the decision-making surrounding food in a household, influencing availability, access, and economics surrounding food, in turn affecting cost–benefit analyses the family intrinsically performs.

Land: Land Use, Resource Allocation, and Agricultural Food Production

Much like food, land cannot be conceptualized as just the physical environment. Land is the summation of the natural and developed environment, the policies and choices surrounding the usage of land, allocation of resources, and in the context of nutrition, agricultural food production.

In a special report, the intergovernmental panel on climate change defined *land use* as "the total of arrangements, activities, and inputs that people undertake in a certain land cover type" [12]. Alternatively, land use is the designation of space to a specific use by humans, like agriculture, waste management, housing, or resource extraction. Usually, the same plot of land cannot be allocated to separate and conflicting roles, leading to choices that need to be made on the allocation of land. *Resource allocation*, the designation of natural resources to a specific use by humans, is similarly limited in that an allotment of resources cannot be designated to separate and conflicting uses and also necessitates choices on use of limited resources.

The combination of decisions regarding of land use and resource allocation are important for *agricultural food production*, the cultivation of food for human or animal consumption. What foods are produced where, to what degree, and in what time frame are a function of the availability of land and natural resources for agriculture, and a variety of other factors including agronomic skills and knowledge, availability of technology and fertilizer, animal ownership, susceptibility to pests and disease, and socio-cultural values. In turn, agricultural food production is often the first stage in the food system and contributes to the food stability in a region, exemplifying the complexity of this dynamic system that influences child malnutrition and other aspects of health.

Multilevel Systems: Household, Community, National/Global

Achieving adequate nutrition and preventing child stunting requires cooperation among three complex and dynamic systems, on multiple ecological levels. The most proximal level to individual child nutrition would be the *household level*, the level which contains factors within a family or household unit and would include food production and soil quality on a household plot, household policies on food consumption, or ability of household members to produce or prepare food. The *community level*, which consists of a group of households bound by geography, identity, or both, is the next most proximal. At this level, factors like community access and production of foods, local infrastructure, concentration of poverty, and community networks and safety nets are to be considered. The most distal level is the national/global level, which encompasses national, international, and worldwide aspects of the food system. This could include federal decisions on food policy, international trade, gross domestic product, climate change, and agroecological zones. These factors within one level often feed into factors on another, influencing the status of nutrition in entire populations and in specific households.

Livestock Ownership: Considering Land, Food, and Body

What Is Livestock?

To illustrate the dynamic relationships between land, food, and the body surrounding nutrition we will discuss one component of this system that affects the lives of many small, rural farmers: *livestock*. Livestock are animals domesticated to produce food (e.g., meat, milk, eggs), materials (e.g., fiber, leather), and manure. Livestock contribute greatly to the livelihood of small farmers around the world, by serving as financial instruments for savings and risk protection, enhancing social status, and helping with manual labor (e.g., ploughing, transporting goods) [13]. Consequently, livestock cannot be treated simply as a commodity, because they serve multiple functions in and are integral to the farm household.

Challenges Facing Livestock Production and Cultivation

While the production of livestock can be lucrative and positive for small famers, livestock production can be challenging. Many considerations need to be made to successfully cultivate livestock. In tropical and arid regions, livestock face nutritional constraints from the native grasses they consume, which generally have lower nutritive values, and from the limited amount of nutritious feed farmers can grow for their livestock, due to constraints on land ownership. Additionally, animal disease is a major impediment to successful livestock production. Trypanosomiasis, rinderpest, foot and mouth disease, contagious bovine pleuropneumonia, and bovine tuberculosis, for example, are all major diseases affecting cattle in sub-Saharan Africa and contribute to a significant loss in animal productivity. Moreover, disease is often associated with drought, putting farmers at significant risk of total asset loss, due to the combination of losses in crops and livestock. Animal loss is one of the key reasons behind small farmers falling into poverty, and animal loss prevention needs to be considered for effective poverty prevention. An index-based livestock insurance program can help reduce the risk of falling into poverty from animal loss, and could alleviate the risks associated with livestock ownership [14].

Successful animal breeding is an important component of livestock production. Animals are often bred for specific traits (e.g., to produce more food, to be more efficient farm tools, to be more disease-, heat-, or drought-tolerant), but this often comes as a tradeoff for other desirable traits. For example, exotic cattle in the global south bred may be bred to produce more milk, but have a lower tolerance for endemic infectious disease. Additionally, focusing on a single trait may be efficient on a larger farm focused on producing a single commodity and has sufficient resources, but often fails on small farms which need animals to serve multiple purposes. To illustrate, a breeding program for yaks in central Asia may focus on higher fiber production, but neglects to breed for traits that make the yaks ideal pack animals. This could lead the animals to be less useful to small farmers or pastoralists who need the yaks to produce wool *and* carry heavy loads.

Land and Resources: A Network of Decisions Surrounding Livestock

Any livestock production effort requires a certain amount of land and natural resources to be successful and allocation of this limited resource creates a need to make tradeoffs. Does a farmer choose to own chickens, which could take up less land but are more vulnerable to predators? Does she choose to own a donkey who could need considerable amounts of food but could plough the fields? Or does she choose none at all? Considerations like land, food, and water availability, disease prevalence and susceptibility, specific needs of the farm (e.g., ploughing), or market forces all influence how the farmer will choose how to allocate her resources.

Yet, these decisions are not always based on environmental or economic factors. As with all choices made by humans, social and cultural beliefs influence the decisions made around land and livestock, which makes understanding farmer's decisions about livestock complicated because farmers are not solely maximizing income. Rather, farmers make decisions in a greater context of beliefs and the other uses of livestock which may often seem irrational to an economist. For example, a household may keep a hen long past the time when they can no longer produce eggs in order to slaughter it for a special occasion, like welcoming an honored guest or for a marriage. At this point, the chicken has an important social value and their economic value is difficult to quantify. In addition to social value, farmers have to consider the value of farm products (e.g., crop residuals) needed to support livestock and the relative importance of the work the animals do apart from material production (e.g., ploughing, pest control, nutrient cycling).

Some farms attempt to combine crop and livestock production in what is known as an integrated crop-livestock system. In this system, livestock provides "nutrient cycling" by producing manure or urine, which is used to fertilize crops, some of which in turn is used to feed the livestock. Nutrient cycling is such an important component of this system, that farmers rank manure as highly as milk as a desired product of raising livestock [15]. Often, livestock accelerate the process of nutrient cycling via digestion and excretion, returning nutrients to the soil quicker than if plants were allowed to decompose naturally. Nutrient cycling is also a means to import nutrients from other areas to the farm, because livestock often are fed feed from elsewhere or are allowed to graze on uncultivated land.

Climate Change and Livestock

Increasing environmental stresses from climate change will limit the productivity of livestock. As resources become scarce, livestock will have to focus more energy on maintaining body weight and their health, rather than producing food or material for farmers. For example, an area which is transitioning to a semi-arid climate may lose nutritious grasses needed to support cattle, and thus smallholder farmers will lose access to the nutrient cycling provided by cattle, along with the milk and meat that would be produced by well-fed cattle.

Nomadic pastoralists also will not escape the effects of climate change. As droughts become more common, inherently unstable arid and semi-arid areas will not be able to support livestock and force pastoralists to search for suitable land. Pastoralists will migrate away from their traditional areas, and may spread deforestation and desertification via overgrazing, compounding upon the already substantial environmental damage. Alternatively, pastoralists could become sedentary as they lose animals, and exhaust resources like water or suitable soil needed for agriculture.

Livestock and the Body: Health and Well-Being

Livestock can have significant impact on human nutrition and health through multiple pathways. Livestock can directly improve human nutrition by providing smallholder farmers and rural communities with access to animal source foods (ASF) without the need to hunt. ASF provide a dense collection of nutrients hard to find in plant-based diets, including highly bioavailable amino acids, heme iron, vitamin B12, and zinc [16]. Furthermore, animal production can provide income and long-term financial security, allowing farmers the ability to purchase a more well-rounded and nutritious diet, education for their children, and adequate healthcare. In turn, access to healthcare can prevent health problems, one of the key reasons for falling into and keeping people in poverty [13, 17]. Finally, livestock ownership can empower women, especially the ownership of poultry and milk-producing animals. This allows women to have greater control over resources and decision-making linked to child nutritional outcomes. In these ways, livestock production can serve as a buffer against poverty and child malnutrition, in both direct and indirect ways.

Conversely, livestock can negatively influence health and well-being. Ownership of livestock can expose children to zoonotic pathogens, affecting their nutritional well-being via enteropathy and diarrheal disease. Maternal livestock ownership could also increase maternal time burden and energy demands, and influence the amount of time and energy they have to care for children. While these are important considerations about health and livestock ownership, the evidence for negative relationships is mixed and it is generally agreed that livestock is a positive influencer of health and well-being [13, 18].

Water Insecurity: Considering Land, Food, and the Body

Global Importance of Water

Water is essential to life. The average adult human body is composed of approximately 50-65% water in mass [19]. Adult women are estimated to meet needs for bodily function with an intake of 2.7 L of water from any source per day, with adult men requiring 3.7 L [20]. However, the human need for water is more expansive than merely physiological intake. The World Health Organization recommends a minimum of 20 L of safe water per person per day to meet basic personal and food hygiene needs, including sanitation and cooking [21]. Beyond these personal uses, water is required for household agriculture and income-generating activities, as well as within the broader systems for agriculture, manufacturing, transportation, energy, and other activities that are critical for food systems to support household and community nutritional needs.

Water and Land: Disparity and Conflict

Water is a major input in agriculture, thus water availability and allocation are critical influencers of the global food supply. Of the global water supply, only 2.5% is considered freshwater and <0.01% of all water is fresh and available for human use, in lakes, rivers, reservoirs, and easily accessible aquifers [22]. Along with other climatic and environmental factors, this limits what crops can be grown in which region and at what time of year [23]. Industrial agriculture and irrigation have mitigated this to a degree, however, freshwater scarcity and abundance continue to vary regionally and disparities are expected to be exacerbated by climate change.

Water access and allocation are major drivers of conflict worldwide, often driven by agricultural needs. Across sub-Saharan Africa, conflicts continue to emerge between pastoralists, who move livestock in search of good grazing land and water, and sedentary agrarians. This tension is recognized as one of the several drivers of the war in Darfur, Sudan. In many cases, conflicts over water use and rights are exacerbated by changing or unclear land rights.

To further illustrate, the Colorado River in the American Southwest supports nearly two million acres of farmland across an arid landscape. Demand for water in this region continues to grow, while supply is expected to diminish, in the context of projected drier and hotter climates. Thus, management of the Colorado River is a source of policy conflict among the Colorado River Basin States (Arizona, California, Colorado, New Mexico, Nevada, Utah, Wyoming) and northwestern Mexico. This has led to the Colorado River being one of the most "litigious" rivers globally [24].

Water and Food: Decisions at the Household Level

With four billion people facing severe water scarcity for at least 1 month every year, households make many daily decisions to prioritize their water supply and meet their physiological, sanitation, and cooking water needs [25]. Similar to food decision-making, these choices a household makes—such as how and from where to obtain water and how to prioritize finite quantities to support all household needs—often include weighing important tradeoffs. For example, in contexts where household members—often women and children—travel long distances to a water source or must wait in long lines at water pumps, time spent acquiring water takes time away from income-generating activities. Potential income lost this way could be used to purchase nutritious food and lost time could have been used for household activities like food preparation.

In situations of water scarcity, families are faced with having to prioritize water for cooking, direct consumption, and other household needs like personal hygiene, livestock, and gardens. Households in Kenya report changing what foods they prepare—such as cooking foods with less water requirements, cooking less nutritious foods, and decreasing diversity of foods consumed—when water is scarce [26]. In the dry season, households may be faced with the decision of whether to allocate money typically used to purchase food, to purchase water. In the context of infant and young child feeding, time spent acquiring water has decreased women's opportunities to breastfeed, and the stressed financial resources force women to provide fewer or less nutritious foods for children [27].

Water and the Body: Health and Well-Being

Water insecurity is defined as "the inability to access and benefit from affordable, adequate, reliable and safe water for wellbeing and a healthy life" [28]. Each year, 1.8 billion people drink unsafe water, including 582 million people relying on unimproved sources and surface water for household drinking, cooking, and hygiene [29]. Drinking or using unclean water for food preparation can directly lead to sickness via infectious diseases (e.g., typhoid, cholera) or parasites (e.g., guinea worm, cryptosporidium) based in or spread through unsafe water [30]. Moreover, households lacking sufficient water to use for food preparation and personal hygiene may suffer from physical illness stemming from environmental contaminants that are not properly removed from food or the body using water.

Water also affects psychosocial health. Globally, women are typically most knowledgeable about water in their household and act as household water managers. Women may also bear the burden of water acquisition and the stress related to procurement of water. Worry and lack of sleep over about running out of water and navigating social relationships, in order to borrow water, have been documented in Latin America and sub-Saharan Africa [26, 31]. Psychosocial stressors are known to affect caregiving behaviors, which may in turn affect child nutrition and health in the household [32].

Conclusion

As we have reviewed, nutrition is influenced by a complex, multilevel, dynamic, and interacting system of factors, which plays an important role in human health and childhood stunting. We considered the roles of "body," "food," and "land," and how they interact with each other within this system. Body and well-being shape how the individual can interact with land and food systems and is a critical link between land use, food systems, and health. Understanding these complex systems highlights the need for multi-sectoral interventions to improve child malnutrition. The systems are interlinked, and to improve child health, it is important to intervene considering the linkages between the systems. Additionally, nutrition policies and systems occur at the household, community, national, and global levels. We have presented two examples, land use and water use to explore how body, food, and land interact. Programs and research programs must incorporate an understanding of these interacting systems to address the challenges surrounding human nutrition, and the factors influencing it.

The Chicken or the Egg? An Intervention in Rural Zambia

Rural Zambia: Background

Inhabitants of Zambia's Luangwa Valley region, most of whom rely on small-scale agriculture to get by, have long faced chronic poverty and food insecurity. This is driven by poor infrastructure and market access, inconsistent rainfall, growing population, resource degradation, and economic reliance on volatile cash crops. Smallholder farmers here are vulnerable to shocks, or events that threaten their economic well-being and stability (e.g., market collapse, drought), and in turn their health. Children in the region have limited access to animal source foods (ASF; meat, milk, and eggs), which are dense in the energy, protein, and micronutrients that children need to grow and develop, and 40% of children under the age of 3 years in the region are stunted [33]. Researchers proposed that improving poultry production in the region could help bolster farmer's resilience and improve child nutrition outcomes by diversifying income, empowering women, and increasing access to ASFs.

Why Poultry?

Poultry, which include chicken, duck, guinea fowl, and other domestic fowl, can provide farming households with a stable source of protein, energy, and micronutrients from the meat and eggs that they produce. They additionally provide rural farmers with a "safety net" to cushion the impact of environmental or economic shocks. They are low-input, meaning that they require little land, labor, or feed investments, making them more accessible to poor farmers than other types of livestock. Interventions to improve poultry productivity may additionally empower women, as women are most commonly the primary caretakers of poultry and often have greater ownership and control over them compared to other livestock [34].

Yet, poultry production is not without its challenges. Flocks face high rates of mortality from endemic infectious diseases, predation, poor husbandry practices, and poor nutrition, which—along with the genetics of indigenous breeds—limits flock size and egg production. However, development programs can address these barriers to increase poultry outputs (e.g., growth, survival, and egg production) and maximize the benefits of poultry ownership for smallholder farming families.

Interventions

Two poultry development programs were established in the Luangwa Valley, and their impact on poultry outputs and family nutrition outcomes was evaluated. The first intervention aimed to build off the existing "backyard" or village poultry production system in the region. This program, involving 5271 farmers, provided community vaccination against Newcastle disease and farmer training on topics such as supplementary feeding, disease prevention, and improved poultry housing. Over 4 years, researchers found a 160% increase in average flock size among participating households, compared with no change in non-participating control communities, and a 138% increase in poultry profitability. However, there was no change in chicken meat or egg consumption, in part due to farmers' preference to hatch eggs and sell birds to address a specific household need [35].

The second intervention established 20 egg producing centers (EPC) in 20 rural communities. At each facility, 4-5 local farmers (86% women) were recruited and trained, and 40 layer hens were provided. Over 1 year, the EPC produced 156,188 eggs that they sold into their communities, generating an average net profit of US$ 313 per EPC [36]. Families living near EPCs were significantly more likely to consume eggs than those in control communities, and women and young children were especially likely to benefit [37]. However, due to production challenges in this first year of the program—including unreliable access to layer feed, excessive ambient temperatures, and suboptimal management practices—egg production declined over time, limiting the long-term impact of the program.

Discussion

The results of these interventions are mixed, but promising. The first intervention demonstrates that targeted interventions promoting improved poultry management practices can improve the resilience of the "backyard" poultry system and contribute to the economic well-being of households. The second intervention explored a feasible, acceptable, and scalable strategy for local egg production in low-income, rural communities. While there are a number of production challenges that should be addressed before scaling up the EPC model, this study is among the first to show a high potential for community-level impact of livestock production on ASF consumption.

Sustainable, nutrition-sensitive, livestock-based strategies to improving human nutrition, health, and well-being will vary according to the local agro-ecological, economic, and socio-cultural context. Despite production limitations, poultry play a vital role in many resource-limited communities throughout the world, often filling multiple roles within the household livelihood strategies. This research demonstrates that challenges in small-scale poultry production can be successfully addressed through a diversity of approaches. Extension and improvement of these approaches will help to maximize the benefits of poultry to improve family diets, increase incomes, and generate long-term economic and nutritional security.

References

1. Black RE, Victora CG, Walker SP, Bhutta ZA, Christian P, De Onis M, Ezzati M, Grantham-McGregor S, Katz J, Martorell R. Maternal and child undernutrition and overweight in low-income and middle-income countries. Lancet. 2013;382(9890):427–51.

2. FAO, et al. The State of Food Security and Nutrition in the World 2018. Building climate resilinece for food security and nutrition. In: The State of Food Security and Nutrition in the World. Rome: FAO; 2018.
3. UNICEF/WHO/World Bank. Joint child malnutrition estimates; March 2019 ed.
4. Kraemer K, JB Cordaro J, Fanzo M, Gibney E, Kennedy A, Labrique JS, Eggersdorfer M. 4 Nutrition-specific and nutrition-sensitive interventions. In: Good nutrition: perspectives for the 21st century. Basel: Karger Publishers; 2016. p. 276–88.
5. Ruel MT, Alderman H, Maternal, and Child Nutrition Study Group. Nutrition-sensitive interventions and programmes: how can they help to accelerate progress in improving maternal and child nutrition? Lancet. 2013;382(9891):536–51.
6. Bhutta ZA, Das JK, Rizvi A, Gaffey MF, Walker N, Horton S, Webb P, Lartey A, Black RE, The Lancet Nutrition Interventions Review Group. Evidence-based interventions for improvement of maternal and child nutrition: what can be done and at what cost? Lancet. 2013;382(9890):452–77.
7. Kracht U, Schulz M. Food security and nutrition: the global challenge. Münster: LIT Verlag; 1999.
8. Douglas M. Fundamental issues in food problems. Curr Anthropol. 1984;25(4):498–9.
9. Bauer JW, Braun B, Olson PD. Welfare to well-being framework for research, education, and outreach. J Consum Aff. 2000;34(1):62–81.
10. World Food Summit. Rome declaration on world food security and world food summit plan of action. Rome: FAO; 1996.
11. Gillespie AH, Gillespie GW Jr. Family food decision-making: an ecological systems framework. J Fam Consum Sci. 2007;99(2):22.
12. Bolin B, Sukumar R, Ciais P, Cramer W, Jarvis P, Kheshgi H, Nobre C, Semenov S, Steffen W. IPCC special report on land use, land-use change and forestry, chapter 1. Cambridge: Cambridge University Press; 2000.
13. Randolph TF, Schelling E, Grace D, Nicholson CF, Leroy JL, Cole DC, Demment MW, Omore A, Zinsstag J, Ruel M. Invited review: role of livestock in human nutrition and health for poverty reduction in developing countries. J Anim Sci. 2007;85(11):2788–800.
14. Chantarat S, Mude AG, Barrett CB, Carter MR. Designing index-based livestock insurance for managing asset risk in northern Kenya. J Risk Insur. 2013;80(1):205–37.
15. Lekasi JK, Tanner JC, Kimani SK, Harris PJC. Manure management in the Kenya highlands: practices and potential. Coventry: Henry Doubleday Research Association; 2001.
16. Allen LH. Global dietary patterns and diets in childhood: implications for health outcomes. Ann Nutr Metab. 2012;61(Suppl. 1):29–37.
17. Kristjanson P, Krishna A, Radeny M, Nindo W. Pathways out of poverty in western Kenya and the role of livestock. Rome: FAO; 2004.
18. Zambrano LD, Levy K, Menezes NP, Freeman MC. Human diarrhea infections associated with domestic animal husbandry: a systematic review and meta-analysis. Trans R Soc Trop Med Hyg. 2014;108(6):313–25.
19. Watson PE, Watson ID, Batt RD. Total body water volumes for adult males and females estimated from simple anthropometric measurements. Am J Clin Nutr. 1980;33(1):27–39.
20. Campbell S. Dietary reference intakes: water, potassium, sodium, chloride, and sulfate. Clin Nutr Insight. 2004;30(6):1–4.
21. World Health Organization. How much water is needed in emergencies. In: Technical notes on drinking-water, sanitation and hygiene in emergencies. Geneva: WHO; 2013. p. 1–4.
22. Igor S. World fresh water resources. In: Gleick PH, editor. Water in crisis. Oxford: Oxford University Press; 1993.
23. Levy BS, Sidel VW. Water rights and water fights: preventing and resolving conflicts before they boil over. Am J Public Health. 2011;101(5):778–80.
24. Ghosh N. Interstate water disputes in the Colorado Basin in western United States. Decision. 2009;36(3):23.
25. Mekonnen MM, Hoekstra AY. Four billion people facing severe water scarcity. Sci Adv. 2016;2(2):e1500323.

26. Collins SM, Owuor PM, Miller JD, Boateng GO, Wekesa P, Onono M, Young SL. 'I know how stressful it is to lack water!' Exploring the lived experiences of household water insecurity among pregnant and postpartum women in western Kenya. Glob Public Health. 2018; 14(5):649–62.
27. Butler MS, Schuster RC, Collins SM, Young SL. "Unbreastfed": the far-reaching consequences of water insecurity for infant feeding practices in low- and middle-income countries. J Hum Lact. 2018;34(3):603–4.
28. Jepson WE, Wutich A, Colllins SM, Boateng GO, Young SL. Progress in household water insecurity metrics: a cross-disciplinary approach. Wiley Interdiscip Rev Water. 2017;4(3):e1214.
29. World Health Organization, UNICEF. Progress on drinking water, sanitation and hygiene: 2017 update and SDG baselines. New York: UNICEF; 2017.
30. Stoler J, Brewis A, Harris LM, Wutich A, Pearson AL, Rosinger AY, Schuster RC, Young SL. Household water sharing: a missing link in international health. Int Health. 2019;11(3): 163–5.
31. Wutich A, Ragsdale K. Water insecurity and emotional distress: coping with supply, access, and seasonal variability of water in a Bolivian squatter settlement. Soc Sci Med. 2008;67(12): 2116–25.
32. Matare CR, Mbuya MNN, Pelto G, Dickin KL, Stoltzfus RJ. Assessing maternal capabilities in the SHINE trial: highlighting a hidden link in the causal pathway to child health. Clin Infect Dis. 2015;61(suppl_7):S745–51.
33. Dumas SE, Kassa L, Young SL, Travis AJ. Examining the association between livestock ownership typologies and child nutrition in the Luangwa Valley, Zambia. PLoS One. 2018; 13(2):e0191339.
34. Dumas SE, Maranga A, Mbullo P, Collins S, Wekesa P, Onono M, Young SL. "Men are in front at eating time, but not when it comes to rearing the chicken": unpacking the gendered benefits and costs of livestock ownership in Kenya. Food Nutr Bull. 2018;39(1):3–27.
35. Dumas SE, Lungu L, Mulambya N, Daka W, McDonald E, Steubing E, Lewis T, Backel K, Jange J, Lucio-Martinez B. Sustainable smallholder poultry interventions to promote food security and social, agricultural, and ecological resilience in the Luangwa Valley, Zambia. Food Secur. 2016;8(3):507–20.
36. Dumas SE. Evaluating the impact of poultry interventions on maternal and child nutrition outcomes in the Luangwa Valley, Zambia [Dissertation]. New York: Cornell University; 2017.
37. Dumas SE, Lewis D, Travis AJ. Small-scale egg production centres increase children's egg consumption in rural Zambia. Matern Child Nutr. 2018;14:e12662.

Chapter 5
Integrating Traditional and Modern Medical Practices: Perspectives from the History of Science and Medicine

James J. Bono

Introduction

> Seventeenth-century [European medical] travelers to China marveled at the astonishing prowess of local healers, and especially at their exquisite feel for the pulse. The uncanny accuracy of their diagnoses bordered on the incredible [1].

As historian of medicine Shigehisa Kuriyama tellingly reminds us, we have always lived in a world replete with multiple, independent, and yet effective, medical systems. Not surprisingly, then, there remains even today much to be learned of value to practitioners and their patients from efforts to integrate traditional and modern medical practices and health perspectives. Over the course of centuries, distinct medical systems have come into contact with one another to varying degrees and with different results. The movement of peoples over time and space—through, for example, trade routes, warfare, the spread of the world's major religions, or the exigencies of sociopolitical, economic, or environmental change leading to forced or voluntary migrations—has continually shaped and reshaped practices, beliefs, goals, expectations, institutions, and the very roles signified by categories such as "patients" and "practitioners" that we associate with the presence and operation of established medical traditions. Few medical systems are "islands" unto themselves; interaction, change, appropriation, and a host of (positive, negative, or neutral) "symbiotic" relations mark historic and emergent exchanges among the world's multiple medical systems and traditions.

Whether working globally or locally, health professionals are continually faced with increasingly complex and variable sets of beliefs, practices, and frameworks among patients they encounter in their own and in different communities and countries. For modern allopathic health and medical practitioners, it is fundamentally

J. J. Bono (✉)
University at Buffalo, State University of New York (SUNY), Buffalo, NY, USA
e-mail: hischaos@buffalo.edu

© Springer Nature Switzerland AG 2020
K. H. Smith, P. K. Ram (eds.), *Transforming Global Health*,
https://doi.org/10.1007/978-3-030-32112-3_5

important to recognize this fact, and to resist uncomprehending and dismissive attitudes with respect to the beliefs, practices, and medical systems of "others." The challenge of recognizing and responding appropriately to differences and different medical systems is not new: most importantly, those who persist may reap the reward of an expanded repertoire of techniques and modes of understanding vital to addressing the needs of all patients.

Whether or not we recognize it, the fact is that patients bring to their interactions with modern medicine a variety of beliefs, practices, and frameworks regarding bodies, health, illness, and healing. Depending upon the histories and circumstances of specific individuals from diverse communities, "traditional" medical systems and practices may play a significant role in shaping individual or collective beliefs, practices, and frameworks. Typically, allopathic professionals know little about traditional medical systems: about how those systems might affect patients with whom they interact professionally, and what to do should they become aware of individual patients' adherence to such systems. While it is neither practical nor necessary for such professionals to become experts in a range of traditional medical systems, nonetheless, misalignments between patient and practitioner beliefs, practices, and systems typically result in one or more problems: miscommunication and misunderstanding; recommendations that are at odds with patient understandings, beliefs, and/or values; mistrust leading possibly to anger and rejection of modern medicine; patient "non-compliance" coupled with deteriorating medical conditions and/or physician frustration, to cite some prominent consequences.

As an issue confronting global health care and equity, the effects of multiple, co-existing medical systems are complex and variable. In high-income, industrialized countries, indigenous systems of traditional medicine still exist. In addition, patterns of immigration—including the resettlement of refugees from across the globe—have led to multiple, pluralized medical cultures co-existing, often silently, barely noticed by modern medical practitioners. Thus, two differential sets of challenges must be addressed: First, in countries with relatively robust "traditional" medical systems including large numbers of adherents among practitioners and patients. Second, in countries like the USA, where a great variety of indigenous and transplanted traditional medical systems exist—often unrecognized—as part of the cultural inheritance of multiple and diverse patient populations. Complicating this picture is the fact that, with different patient populations, knowledge of such systems and adherence to their practices, beliefs, and values vary tremendously from community to community, from family to family, from one generation to the next, and, of course, from one individual to another (Fig. 5.1).

Global Health and Modern Medicine: One Among Many

While a widespread presence, in many areas of the world modern allopathic medicine remains a far from dominant actor, especially among non-elites and extra-urban populations. Where advocates of modern medicine and global health equity seek to

Fig. 5.1 Traditional Chinese Medicine being sold at market. (Source: Xi'an, China—Traditional Medicine Market, Joel, 2007, unmodified, CC BY-ND 2.0. Image: www.flickr.com/photos/jminnick/522693227. License: https://creativecommons.org/licenses/by-nd/2.0/)

improve health outcomes in such settings, the challenges are often daunting. Take, for example, women's and neonatal health. As one study soberly notes, "Maternal and neonatal health has received increasing attention as mortality rates refuse to fall in the poorest populations. More than two-thirds of all newborn deaths occur in just 10 countries, the same countries in which more than 60% of maternal deaths occur." [2] While overall such statistics have improved in recent years, nonetheless, it remains advisable that in addressing and ameliorating such stark conditions, as this same study asserts, we "develop interventions that meet the health needs of disadvantaged women and newborn infants," which, in turn, "requires knowledge of the context in which poorest families make decisions on their health." [2] More explicitly still, a key obstacle to understanding such decision-making contexts and formulating "appropriate interventions" is precisely the "lack of understanding of local beliefs and practices (and the reasons for them)." [2]

Tackling such problems and developing effective strategies and interventions, then, must begin with understanding contexts, practices, and meanings of health and health-related care among specific local populations in question. An important tool for achieving such goals is undertaking a "qualitative study of care practices and beliefs during pregnancy, birth, and postpartum." [2] Importantly, the goal of such studies represents something far more robust than simply developing a general "cultural competence" among advocates and providers of modern Western medicine. Rather than focusing upon some generalized and abstract notion of "culture," such studies aim instead to sort-out and understand the local contexts, practices, and beliefs among individuals and families. That is to say, rather than acting on pre-formed assumptions based upon stereotypes of cultural beliefs and social norms, the key to such qualitative studies is to unearth how specific individuals and communities translate generalized beliefs and norms into localized beliefs, practices, and actions. Acquiring such knowledge and understanding always entails listening, observing, and nuanced interrogation. Rather than racing toward conclusions based upon partial (and presumptive) cultural understanding, the appropriate stance should be one of "hesitation"—of slowing-down in order to understand and before acting [3–5]. We shall return to this issue later in this chapter.

To achieve such goals in a nuanced and rigorous manner, the article cited provides a sample of some exemplary approaches. For one, beyond the views and practices embraced by centralized, and typically modernized, government health care systems, the authors describe and advocate the work of numerous researchers in multiple countries with at-risk populations in establishing "community level" partnerships with village representatives, "local women's groups," and such key actors as "female community health volunteers" and local practitioners, including those considered to be a "traditional healer." [2] As the authors rightfully suggest, such a "participatory approach was also attractive because of its tolerance of diversity. Contextual differences within the study area might affect how an intervention is received. Participatory development practitioners believe that to transform society, local stakeholders should be active participants in problem identification, planning, implementation and evaluation." [2]

In practice, this meant attending carefully to the roles played by a variety of actors—including those noted above—in seeking care for pregnant women and neonates. Traditional healers, of course, bring to those seeking care selective aspects of indigenous medical systems: selective because in each instance the knowledge and practices brought to bear in a given case varies with the level of sophistication achieved by the practitioner in question. In rural Nepal, the study not surprisingly reports that in many cases a "healer who was a friend or relative was called first." In cases where "the condition worsened," those responsible for making decisions would then turn to a person with a reputation as a more sophisticated and knowledgeable healer, who might well hail from a more distant village [2]. Even prior to seeking such professional advice, pregnant women and mothers of newborns would typically depend upon a hierarchy of decision-makers within their family: especially elders (e.g., mothers and mothers-in-law) whose views were informed by experience, by expertise in family remedies, and by their own localized variant of shared traditional knowledge of pregnancy, childbirth and neonatal health, and beliefs, concepts of health, illness, and the body and its vulnerabilities. As the authors note, the existence of multiple actors, of a hierarchy of decision-makers, and the circulation of beliefs and practices—in part traditional and shared; in part localized and variable—point to the "social nature of decision making during illness" in rural Nepal [2].

Of course, at any given moment and in any specific case, any number of cultural beliefs and practices may also come into play, all susceptible to individual and localized variations: concerns over "pollution" or "uncleanliness"; "embarrassment or shame"; the perceived need to protect vulnerable individuals against others (strangers; "evil spirits") [2]. All of these factors form the background to, and precede chronologically, any attempt to introduce modern allopathic medical interventions. Indeed, the authors argue convincingly that any attempt to suggest the introduction of such interventions must take into consideration such backgrounds. Moreover, the implication of their study is that such a methodology can and should apply to other different, but analogous settings in a variety of countries and continents where local customs, beliefs, decision-making hierarchies, and traditional medical systems form the setting within which other actors seek to introduce modern medical interventions.

Key to the results of their study and attendant recommendations is operationalizing the "participatory approach" that they deem vital to successful integration of traditional and modern medical approaches. Such operationalizing begins with seeking appropriate, context-dependent, "answers to four questions: (1) Who should implement the intervention? (2) What should the intervention address? (3) Who should be the major stakeholders in the intervention? and (4) what should the intervention be?" [2]. In discussing these questions in the context of women's and neonatal health in rural Nepal, let us cite a few major points raised by the authors: points that can, and we believe should, be generalized to many other analogous situations faced by global health advocates across the world. First, the authors note the important point that "an intervention should tackle delays in care seeking by addressing the familial context." This seemingly unremarkable suggestion is easily ignored or forgotten in practice, with consequences that can be disastrous. Any number of cultural beliefs and attitudes—not just traditional medical beliefs—can prove to be barriers in seeking interventions beyond those provided by traditional medical practices. Of course, one structural component in care-seeking behavior, and therefore in seeking access to modern allopathic medicine, in many societies across the globe (including urban and industrialized settings) can be the status of women within families and within societies. With respect to their rural Nepalese families, the authors suggest that "Women's low status in the home may affect their ability to act on existing or learned knowledge: it would take great courage and personal strength for her to challenge tradition." The authors' own intervention, then, recognizes and responds to such constraints by attempting to address "women's empowerment" in culturally and socially nuanced ways. More specifically, they observe that "An intervention that directly challenges tradition … might be inappropriate." This is precisely where "understanding of local contexts" as noted above must be combined with "good facilitation skills"—with careful, culturally and structurally competent, forms of communication and intervention strategies [2]. The formation of "groups of women" who collectively identified and discussed problems facing pregnant women and neonates and decided how to respond to possible new interventions from outside traditional beliefs and practices proved decisive and effective in enabling "a community-based, community-responsive intervention to be initiated" [2]. What results ensued? "In intervention areas," the authors in conclusion report, "large reductions in neonatal mortality and maternal mortality were

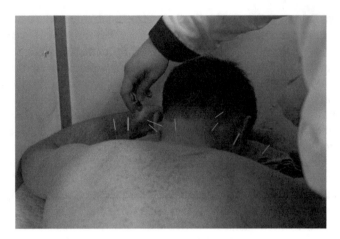

Fig. 5.2 Acupuncture at Taeyoungdang Oriental Medicine, a component of traditional Chinese medicine. (Source: Dongdaemun-gu, 2010; CC BY 2.0, SellyourSeoul. Image: www.flickr.com/photos/sellyourseoul/4273475678. License: https://creativecommons.org/licenses/by/2.0/)

observed, and there was increased care seeking for health problems. Care behaviors … also improved, and groups continued to meet regularly." The key seems to be "building an intervention that is acceptable" [2].

Other studies echo and support many of the approaches and conclusions discussed above. Understanding the contexts of care—the hierarchy of care-givers, including traditional healers—and the dynamics of decision-making within families and localized communities; attention to obstacles affecting care-seeking behaviors and receptivity to modern allopathic medical systems, especially those rooted in cultural and religious beliefs and practices; developing models based on listening, on learning how to communicate effectively with diverse individuals and groups espousing a variety of traditional beliefs and practices; empowering individuals—especially women and new or prospective mothers—by fostering participatory approaches to understand and change complex health-related behaviors through the organization of community-based discussion, support, and planning groups: some combination of these approaches seem to represent a growing wave of new and effective intervention strategies among recent field-based researchers. Indeed, such approaches offer the promise of significant, positive results not only in the case of Nepal highlighted earlier, but more broadly in India, Pakistan, South Asia, Africa, and Central America [6–13] (Fig. 5.2).

Strangers in Strange New Lands? The Ubiquity of Modern Medicine within Pluralistic Urban-Industrial Landscapes

With declining child mortality and rising life expectancy, modern medicine has achieved global recognition in most highly industrialized and urban settings, signaled not least by the sheer presence of massive medical research and clinical complexes; extensive institutional, governmental, and governance structures; and, as a consequence, vast economic resources coupled with privileged social and cultural authority. Taken together such highly visible complexes (hospitals, public and private research complexes) and behind the scenes institutions not readily visible to most ordinary citizens (e.g., government agencies, professional biomedical organizations) testify to the power, prestige, and centrality of modern medicine. Yet, in the very shadows of such complexes and institutions, modern urban-industrial societies harbor diverse, plural populations embracing congeries of distinctive beliefs, practices, values, and stories regarding the body, health, illness, and well-being. Highly technologized practitioners of modern medicine must, again, continually acknowledge and take account of the presence of patients with divergent views and competing systems of care if they are to achieve the goal of global health equity at home as well as within distant lands.

In some urban-industrialized settings, the presence of competing systems is a palpably visible and historically acknowledged fact. Within such settings (for example, in Korea and Taiwan), institutions, health-care practitioners, and patients frequently encounter differences and conflicts requiring negotiation, which may or may not result in accommodation [14–17]. Less visible and often ignored are the continued presence of pluralized, "traditional" beliefs—alternative views of the body, health, and disease; and health-care practices—in urban-industrialized settings where the dominance of modern Western medicine appears virtually hegemonic, most notably in the USA. While both sorts of settings present their own challenges, the second in particular proves especially difficult due to the relative invisibility of alternatives to mainstream modern medicalized understandings of health and illness. In what follows, we shall focus attention on the latter.

Nonetheless, it is worth noting that both kinds of settings offer the prospect of addressing the encounter between "modern" and "traditional" medical systems in two separate, though not mutually exclusive, ways: (1) a chiefly medical response that seeks first and foremost to assess the strengths and weaknesses of mainstream and alternative medical systems, either in a spirit of contestation, or one of emergent cooperation embracing an ideal of "integrative medicine" and (2) a patient-centered approach whose goal is to facilitate the best possible outcomes as mutually and cooperatively defined through careful listening, questioning, and narrative construction of the history and meaning of illness in ways that are capable of describing culturally responsive, and personally resonant (yet possible and pragmatic) trajectories for *this* individual patient and *this* specific illness.

While in some circles there is widespread interest in possibilities for integrating modern and traditional medical systems, such possibilities remain controversial, if not contentious, for much of mainstream Western academic medicine and its many practitioners. Comparative assessment of alternative medicine itself can be controversial; certainly, no consensus has yet emerged, nor have we yet witnessed robust and ubiquitous commitments to such assessments. That said, the fact that the very term, "Integrative Medicine" (referring to actively pursuing pragmatic models for combining Western allopathic with "traditional" medical practices and therapies) is now well established—that there are centers or units designated for the study of integrative medicine or medical therapies at such medical schools as the University of Arizona and Harvard University—points to a growing insistence that plural medical approaches to health, illness, diagnosis, and therapy are desirable, of value to patients, and, indeed, necessary to promote health and well-being. On the one hand, as Dr. Andrew Weil notes, while openness to such efforts in the USA and China may be on the upswing, strong resistance remains, with the very notion of what counts as evidence clearly itself a matter of dispute [18]. Indeed, methodologies (such as Koch's postulates from the late nineteenth century) for establishing causative agents in disease, or contemporary insistence on the need for evidence-based medicine, play precisely into a logic in which only those phenomena—and therapeutic results—for which we have established tools and technologies capable of uncovering underlying "mechanisms" or revealing statistically persuasive relationships can be considered real. What such tried and tested approaches to knowledge-production rule out-of-bounds are exactly alternative forms of knowledge and "expertise" that champion the utility of phenomena, practices, and therapies that are not, or cannot be, studied in such ways.

In addition, Weil cites the case of "Latin America," where "traditional herbal medicine and other forms of folk medicine are widely used." Yet, he notes that, "these are completely separated from conventional medicine. There is no communication, no integration ... in these Latin American countries, academic medicine and conventional medicine are very rigid, very closed to any other ways of thinking" [18]. On the other hand, in much of Europe and Australia efforts are under way to document, describe, criticize, and refine models for integrating traditional and modern medicine, especially in primary care settings—to understand and redefine their relationship as symbiotic [19].

While integrative medicine remains open to multiple systems of traditional medicine, for Weil and many others, attention not surprisingly focuses largely upon traditional Chinese medicine [20, 21]. As a robust system with a long and documented history, Chinese medicine presents modern Western medicine with the testimony of longstanding practices providing experiential evidence of diagnostic and therapeutic successes. Those open to integrative approaches oftentimes see such practices and successes as complementary to modern Western medicine with the weaknesses of one system (e.g., pain management) complemented by the strengths of the other. Such complementarity should not be surprising and seems well worth serious attention by researchers and clinicians. As historian of medicine Shigehisa Kuriyama (cited earlier) has effectively argued, Chinese and Western medicine, respectively, employ differential techniques of seeing, feeling, observing, and describing the body in health and in illness that serve to capture diverse if not divergent aspects—different empirical regularities, irregularities, and interrelationships—of organs and bodies than the other system (Fig. 5.3).

While efforts to assess medically the encounter between traditional and modern medical systems and consequently arrive at consensus regarding their possible integration in theory and in practice remain ongoing and unresolved, if unquestionably important, there should be no question at all that modern Western medical practice must attend to the system of beliefs, conceptions of the body, health, and illness, and

Fig. 5.3 Cupping with bamboo, a component of traditional Chinese medicine. (Source: Randy Adams, 2012, CC BY-ND 2.0. Image: www.flickr.com/photos/7830239@N06/2445098397. License: https://creativecommons.org/licenses/by-nd/2.0/)

the health-care practices associated with traditional medicine. For such beliefs, concepts, and practices continue to circulate among diverse groups within the USA and elsewhere [22]. Despite various degrees of assimilation, we remain a heterogeneous nation comprised of the descendants of indigenous peoples, historic-era settlers, and waves of subsequent immigrants and refugees some more and some less removed from folk and traditional modalities of health beliefs, values, and practices. Whether deeply ingrained within patterns of traditional medical behaviors and beliefs rooted in a shared experience of a still cohesive community, or unselfconsciously and selectively adopting and adapting distantly remembered patterns inherited from one's forebears, each one of us as individuals and patients brings our own distinctive, if unarticulated, understandings of health, illness, and what constitutes appropriate care to our encounters with physicians and health-care providers. Such individualized views and practices shape our decisions and subsequent behaviors. Let us, then, conclude with discussion of how a patient-centered practice of modern medicine ought to respond to such challenges.

Perspectives from the Health and Medical Humanities

Increasingly, the practice of medicine has come to be informed by perspectives championed by the health and medical humanities, with the latter's emphasis upon eliciting the experience and meaning of illness serving as the basis for the mutual negotiation of care between patients and physicians especially in cases of chronic and serious illness [23–25]. Negotiating care builds on the capacity to listen and communicate effectively with patients in order to explore and appreciate the goals of individual patients, the contexts shaping those goals, and the nature and sources of any barriers that stand in the way of effective communication and negotiation of care. Not surprisingly, the meaning and experience of illness vary from patient to patient, affected by a variety of factors, especially the patient's beliefs and understanding, their past history, their social contexts and support networks, and the existence and perception of a host of vulnerabilities. Key to understanding patients and negotiating care, then, is eliciting and learning how to address such factors appropriately. Where patients' beliefs about the body, health, and illness—including the meaning and experience of illness—vary dramatically from that of "mainstream" physicians and patients familiar to them (much more likely to be the case with patients raised within different, traditional medical systems and cultures of care) cultivating such techniques for negotiating care become all the more critical.

Advocates of "Narrative Medicine" appreciate the power of patients' own implicit stories—the still emergent narrative trajectories of their lives together with challenges, threats, and possibilities posed by their illnesses [26–29]. Such emergent life-stories shape patients' experience of illness and work to construct the meanings they have for them. As a result, cultivating the ability to discern patterns in patients' experience of and response to illness, coming to recognize the sources of concern, suffering, fear, hope—in short, the roots of the imagined meanings of their illnesses—prove invaluable to the narratively competent health-care provider. Often, negotiating care becomes a matter of working together with a patient to facilitate the adaptive re-imagining of their life-stories in the face of serious and chronic diseases that threaten to alter their trajectories.

Many illnesses pose such threats to patients. Yet, time and again, we can be surprised both by what individual patients do or do not perceive as a threat, and why. The very existence, nature, and meaning of a threat are not themselves predictable. Two patients faced with the same diagnosis of cancer with similar prognoses are as likely as not to react in remarkably different ways. One patient faced with a chronic condition that limits mobility—or some other non-life threatening, yet consequential, function—may suffer terribly from such an emergent loss, where another may prove immune to suffering, adapting, instead, to changed circumstances. As no less a figure as Dr. Eric J. Cassell has noted, the reasons why patients suffer are multiple, rooted in their most fundamental sense of identity, and require that we talk with patients if we wish to understand and help [30, 31]. Here, then, is where histories and contexts matter. Here, in turn, is where the need arises to cultivate (and critically interrogate) what many have called "cultural competence" among health-care providers.

But, what is cultural competence and what does it demand of the practitioner? We start with a commonplace: individual patients are members of communities, frequently members of multiple communities—an ethnic group or nationality; a religious community; a vocational or professional community; etc. As members of one or more community, each patient has access to entire repertoires of beliefs, concepts, practices, and values. Among these "cultural" elements we may include those associate with the body, health, illness, and well-being: as noted above, the degree to which they are shaped by inherited traditional medical systems depends on many factors. In any case, we have long recognized specific patterns of belief and practices regarding health and illness that have come to be attributed, for example, to members of specific immigrant communities and ethnicities [32–34]. Cultural competence certainly involves the ability to recognize the existence, operation, and effect of such cultural factors—including effects of traditional medical beliefs and practices—*together with* cultivating strategies for assessing their impact on individual patients, while responding appropriately to them in ways that exemplify "narrative competence": how they "fit" into patterns and life-stories unique to *this* patient who stands before us. Failing to do so could result in misunderstandings, misdiagnoses, and mismanagement of patients and their families in ways that may prove harmful, or even tragic.

However, we must also recognize what cultural competence does not, or should not, include. Most certainly, it does not demand that health-care providers be professional anthropologists: that they be experts in one or more cultures, nor that they be experts in one or more traditional medical systems. A truly critical cultural competence should instead hold up as an ideal the ability to recognize when a patient—as an individual, and not as a mere stereotypical member of an ethnicity or "culture"—may adhere to beliefs or practices that diverge from "mainstream" modern medicine, and to assess when such commitments are consequential with respect to negotiating care and creating the conditions for the best possible outcome for the specific patient in question. Of course, the conditions affecting patient care and outcomes go beyond those critically apprehended and addressed by cultural competence alone. Often, structural factors—whether related to cultural factors such as those facing recent immigrant and refugee populations, or not—play a large role in patient access to care, to health-seeking behaviors, and to equity in health-care more generally. Poverty, housing, access to food, class and racial barriers, and a host of other structural factors must not be ignored: cultural competence is inadequate if not conjoined with structural competence [35, 36].

Finally, pursuing the goal of cultural competence should not and must not generate unintended consequences that serve to undermine effective care and genuinely dialogic negotiation between patient and practitioner. As the renowned medical anthropologist and physician, Arthur Kleinman (and co-author, Peter Benson) note:

> One major problem with the idea of cultural competency is that it suggests culture can be reduced to a technical skill for which clinicians can be trained to develop expertise. The problem stems from how culture is defined in medicine, which contrasts strikingly with its current use in anthropology … Culture is often made synonymous with ethnicity, nationality, and language. For example, patients of a certain ethnicity—such as, the "Mexican

patient"—are assumed to have a core set of beliefs about illness owing to fixed ethnic traits. Cultural competency becomes a series of "do's and don'ts" that define how to treat a patient of a given ethnic background. The idea of isolated societies with shared cultural meanings would be rejected by anthropologists, today, since it leads to dangerous stereotyping—such as, "Chinese believe this," "Japanese believe that," and so on—as if entire societies or ethnic groups could be described by these single slogans [37].

As noted at the very beginning of this chapter, few medical systems [if, indeed, any!] are "islands" unto themselves. Medical systems are inherently historical, dynamically changing, and symbiotic: when, by contrast, we treat them as monoliths—as silently, hegemonically, and invariably imposed upon individuals without regard for the multiplicity of possible meanings, interpretations, and practices that shape individuals' understandings of medical knowledge and how they respond to medical advice—we do so at our, and our patients', peril. As always, we must ask, rather than assume; we must listen, rather than pronounce; we must learn to formulate questions appropriately, and attend carefully to the nuances and individuality of what we hear from our patients (Fig. 5.4).

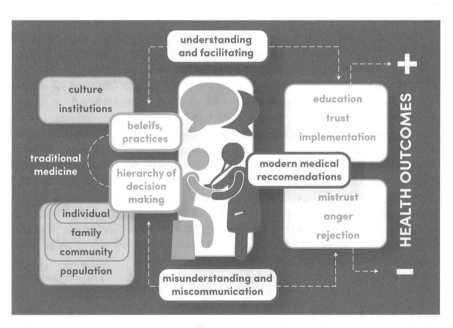

Fig. 5.4 Integration of traditional medicine and modern medicine. (Source: Nicole C. Little and James Bono)

Perspectives from the Field: Cultural Competency or Ethnography?

It is of course legitimate and highly desirable for clinicians to be sensitive to cultural difference, and to attempt to provide care that deals with cultural issues from an anthropological perspective. We believe that the optimal way to do this is to train clinicians in ethnography. "Ethnography" is the technical term used in anthropology for its core methodology... What sets this apart from other methods of social research is the importance placed on understanding the native's point of view. The ethnographer practices an intensive and imaginative empathy for the experience of the natives—appreciating and humanly engaging with their foreignness, and understanding their religion, moral values, and everyday practices.

Ethnography is different than cultural competency. It eschews the 'trait list approach' that understands culture as a set of already-known factors, such as 'Chinese eat pork, Jews don't.' (Millions of Chinese are vegetarians or are Muslims who do not eat pork; some Jews, including the corresponding author of this paper [i.e., Kleinman], love pork.) Ethnography emphasizes engagement with others and with the practices that people undertake in their local worlds. It also emphasizes the ambivalence that many people feel as a result of being between worlds (for example, persons who identify as both African-Americans and Irish, Jewish and Christian, American and French) in a way that cultural competency does not. And ethnography eschews the technical mastery that the term 'competency' suggests. Anthropologists and clinicians share a common belief—i.e., the primacy of experience. The clinician, as an anthropologist of sorts, can empathize with the lived experience of the patient's illness, and try to understand the illness as the patient understands, feels, perceives, and responds to it [37].

Kleinman's Template of Exemplary Questions: Eliciting the Patient's Understanding of Illness
What do you call this problem?
What do you believe is the cause of this problem?
What course do you expect it to take? How serious is it?
What do you think this problem does inside your body?
How does it affect your body and your mind?
What do you most fear about this condition?
What do you most fear about the treatment? [37].

References

1. Kuriyama S. The expressiveness of the body and the divergence of Greek and Chinese medicine. New York: Zone Books; 1999. p. 21.
2. Morrison J, Osrin D, Shrestha B, et al. How did formative research inform the development of a women's group intervention in rural Nepal? J Perinatol. 2008;28:S14–22.
3. Stengers I. The cosmopolitical proposal. In: Latour B, Weibel P, editors. Making things public: atmospheres of democracy. Cambridge: MIT Press; 2005. p. 994–1003.

4. Stengers I. Cosmopolitics I. Minneapolis: University of Minnesota Press; 2010.
5. Stengers I. Cosmopolitics II. Minneapolis: University of Minnesota Press; 2011.
6. Neonatal Mortality Formative Research Working Group. Developing community-based intervention strategies to save newborn lives: lessons learned from formative research in five countries. J Perinatol. 2008;28:S2–8.
7. Syed U, Khadka N, Kahn A, Wall S. Care seeking practices in South Asia: using formative research to design program interventions to save newborn lives. J Perinatol. 2008;28:S9–S13.
8. Iyengar SD, Iyengar K, Martines JC, et al. Childbirth practices in rural Rajasthan, India: implications for neonatal health and survival. J Perinatol. 2008;28:S23–30.
9. Mohan P, Iyengar SD, Agarwal K, et al. Care-seeking practices in rural Rajasthan: barriers and facilitating factors. J Perinatol. 2008;28:S31–7.
10. Darmstadt GL, Kumar V, Yadav R, et al. Community perceptions of birth weight in rural Uttar Pradesh, India: implications for care of low-birth-weight infants. J Perinatol. 2008;28:S53–60.
11. Awasthi S, Srivastava NM, Pant S. Symptom-specific care-seeking behavior for sick neonates among urban poor in Lucknow, Northern India. J Perinatol. 2008;28:S69–75.
12. Hill Z, Manu A, Tawiah-Agyemang C, et al. How did formative research inform the development of a home-based neonatal care intervention in rural Ghana? J Perinatol. 2008;28:S38–45.
13. Schooley J, Mundt C, Wagner P, et al. Factors influencing health care-seeking behaviours among Mayan women in Guatemala. Midwifery. 2009;25:411–21.
14. Cho MO. Health care seeking behavior of Korean women with lymphedema. Nurs Health Sci. 2004;6:149–59.
15. Lee MS, Shin B-C, Choi T-Y, Kim J-I. Randomized clinical trials on eastern-western integrative medicine for health care in Korean Literature: a systematic review. Chin J Integr Med. 2011;17(1):48–51.
16. Hung J-Y, Chiou C-J, Chang H-Y. Relationships between medical beliefs of superiority of Chinese or western medicine, medical behaviours and glycaemic control in diabetic outpatients in Taiwan. Health Soc Care Community. 2012;20(1):80–6.
17. Willsky G, Bussmann RW, Ganoza-Yupanqui ML, Malca-Garcia G, Castro I, Sharon D. Utilizing local practices: integrating traditional medicine and the Peruvian Public Health System in Northern Peru. In: Smith K, Ram P, editors. Transforming global health: interdisciplinary challenges, perspectives, and strategies. Cham: Springer International Publishing AG; 2019.
18. Weil A. The state of the integrative medicine in the U.S. and Western World. Chinese Journal of Integrated Medicine. 2011;17(1):6–10, p. 8, p. 10.
19. Templeman K, Robinson A. Integrative medical models in contemporary primary health care. Complement Ther Med. 2011;19:84–92.
20. Eisenberg D. Reflections on the past and future of integrative medicine from a lifelong student of the integration of Chinese and western medicine. Chin J IntegrMed. 2011;17(1):3–5.
21. Robinson N. Integrative medicine—traditional Chinese medicine, a model? Chin J Integr Med. 2011;17(1):21–5.
22. Barnes P, Powell-Griner E, McFann K, Nahin R. Complementary and Alternative Medicine Use among Adults: United States, 2002. Advance Data Report #343; 2004. p. 1–19. https://nccih.nih.gov/ [National Center for Complementary and Integrative Health (NIH): search under "Complementary and Alternative Medicine Use among Adults"].
23. Jones T, Wear D, Friedman LD, editors. Health humanities reader. New Brunswick: Rutgers University Press; 2014.
24. Cassell EJ. The place of the humanities in medicine. In: Ethics and the life sciences. Hastings-on-Hudson: Institute of Society; 1984.
25. Cassell EJ. Talking with patients. Vol. 1, the theory of doctor-patient communication. Vol. 2, clinical technique. Cambridge: MIT Press; 1985.
26. Kleinman A. The illness narratives: suffering, healing, and the human condition. New York: Basic Books; 1988.
27. Hunter KM. Doctors' stories: the narrative structure of medical knowledge. Princeton: Princeton University Press; 1991.
28. Charon R. Narrative medicine: honoring the stories of illness. New York: Oxford University Press; 2006.

29. Schleifer R, Vannatta JB. The Chief concern of medicine: the integration of the medical humanities and narrative knowledge into medical practice. Ann Arbor: University of Michigan Press; 2013.
30. Cassell EJ. The nature of suffering and the goals of medicine. N Engl J Med. 1982;306(11):639–45.
31. Cassell EJ. The nature of suffering and the goals of medicine. New York: Oxford University Press; 1991.
32. Harwood A. The hot-cold theory of disease: implications for treatment of Puerto-Rican patients. J Am Med Assoc. 1971;216:1153–8.
33. Hatcher E, Whittemore R. Hispanic adults' beliefs about type 2 diabetes: clinical implications. J Am Acad Nurse Pract. 2007;19:536–45.
34. Kang Y, Crogan NL. Social and cultural construction of urinary incontinence among Korean American elderly women. Geriatr Nurs. 2008;29(2):105–11.
35. Metzl JM, Hansen H. Structural competency: theorizing a new medical engagement with stigma and inequality. Soc Sci Med. 2014;103:126–33.
36. Judelsohn A, Orom H, Kim I, et al. Planning the city of good (and New) NEIGHBOURS: Refugees' experiences of the food environment in Buffalo, New York. Built Environ. 2017;43(3):426–40.
37. Kleinman A, Benson P. Anthropology in the clinic: the problem of cultural competency and how to fix it. PLoS Med. 2006;3(10):e294. https://doi.org/10.1371/journal.pmed.0030294.

Chapter 6
Combatting Gender-Based Violence: Perspectives from Social Work, Education, Interdisciplinary Studies, and Medical Anthropology

Nadine S. Murshid, Melinda Lemke, Azfar Hussain, and Shahana Siddiqui

Field Research: Navigating Through Dhaka Medical College Hospital for Sexual Assault Examination and Support

With trepidation, the young woman kept shifting her gaze to the duty doctor at the Forensic Medicine Department, to me, and then to the notes the doctor was writing on the government prescribed medico-legal format. After getting the basic information of the patient/victim's name, address, and current work/education status, the doctor, originally trained in Ob/Gyn, started inquiring into the incident for which the patient was seeking legal recourse in the first place. The patient, a medical student herself, relayed a story of betrayal and manipulation in her marriage leading to unwanted sexual intercourse with her spouse, against whom she had filed the rape case. The husband was caught having an extra-marital affair earlier, leading to long drawn out family feud. He had no intention of leaving either of the women but lied to his wife on wanting to work on the marriage, persuading her to have intercourse with him with the hopes of her getting pregnant yet again (they already have two children together) as a way of keeping her in the marriage.

The duty doctor's eyes widened as she listened on to what sounded like a Hindi-series episode. Reality is truly stranger than fiction. After rechecking all the information with the earlier police report and the patient's narrative, the doctor explained that she would need to carry out a physical examination. Both the patient and the doctor were well aware that no traces of spermatozoa

N. S. Murshid (✉) · M. Lemke
University at Buffalo, State University of New York (SUNY), Buffalo, NY, USA
e-mail: nadinemu@buffalo.edu

A. Hussain
Grand Valley State University, Grand Rapids, MI, USA

S. Siddiqui
University of Amsterdam, Amsterdam, The Netherlands

© Springer Nature Switzerland AG 2020
K. H. Smith, P. K. Ram (eds.), *Transforming Global Health*,
https://doi.org/10.1007/978-3-030-32112-3_6

or abrasion will be found since the case had come to the police far after the 72 h cut off point and even later to the doctors. More so, both of them knew that given that the patient was married, the problematic "two-finger test" will fail this case because according to the established medico-legal examination procedure, the patient will be considered to be "habituated to sex" (as it is stated in the standard Forensic Medicine and Toxicology textbooks ascribed by the medical curriculum board), raising questions on the nature of consent at the time of intercourse.

As part of graduate research on public health responsiveness to violence against women, I carried out an extensive ethnography at the Forensic Department of Dhaka Medical College Hospital which provides a wide number of services to survivors from One Stop Crisis Center support to medico-legal examination at the Forensic Department to post assault treatment at the Women's Clinic. Despite all the efforts on the part of the different ministries in Bangladesh to better respond to high rates of gender-based violence (GBV) incidents, the legal cases primarily hinge on the medico-legal examination that, during my fieldwork, continued to apply the two-finger test in determining the nature of vaginal penetration.

In 2014, the Bangladesh High Court repealed the much contested two-finger test, framed within the Penal Code 1860 and Evidence Act 1872, as the scientific procedure to determine forceful intercourse in the medico-legal examination process. High Court finally banned the two-finger test in April 2018, with the line ministries having done very little work on what a new medico-legal framework would entail for all stakeholders, particularly the victim-patients.[1] In the absence of a rape/sexual assault care guidelines, at the ground level, the two-finger examination process was not only the standard practice that generations of doctors were trained in, but also a point of entry for the forensic doctors to carry out a physical examination to concur with victim's testament. Because consent is difficult to prove medically/biologically, and because doctors understand the limitations of the two-finger test, they combine both the physical examination along with detailed account of the incident from the side of the patient to demonstrate that some form of betrayal of trust or non-consensual incident had taken place.

As seen in this case, the duty doctor, who knew the limitations of the two-finger test and the absence of marital rape in the legal system, went beyond her call of duty and provided unsolicited advice on how to get the husband into trouble with his current job. Where medical science, framed within archaic legal structures and understanding of the female anatomy and sexuality, doctors find creative ways to address the

---

[1] "Victim-patient" is a term taken from Sameena Mulla's work on forensic nurses and their use of photography in sexual assault evidence collection in USA.

unique and complex cases of sexual assault that come into their examination room. In the end, I heard the doctor telling the young woman that your life is not yet over… "You will find someone and have a family in no time. Just concentrate on your studies. That's all that matters." Though the doctors lack a sense of sophistication and/or finesse in their patient–doctor relations, they provide "tough love" every day to the kaleidoscope of women and their unique cases. Importantly, however, the doctors are seemingly unaware of the complex set of uses that potentiate GBV, and in minimizing the problem, they offer few solutions, suggesting the need for support teams that consist of a variety of professionals in addition to doctors who can provide support to women experiencing violence: social workers, lawyers, and activists, and other medical health professionals such as nurses and mental health counselors. The case at hand is a depiction of how structural violence is located on the body, highlighting how institutions and support systems are inherently violence against women, deepening the culture and norms around GBV.

A Compelling Global Problem

Gender-based violence (GBV) is defined by the United Nations as a human rights violation that disproportionately affects girls and women across the world [1]. According to the World Health Organization, GBV affects one in three women and one in five children, and includes physical and sexual violence, among others. The high prevalence of GBV—accompanied by a culture of impunity and normalization of violence—in turn creates a culture of fear for all women and girls. As demonstrated in the previous narrative, GBV encompasses a range of practices—domestic/familial, social, economic, political, cultural, or ideological, even epistemic. Some obvious examples of GBV include genital mutilation in Africa, dowry-related murder in South Asia, sexual harassment on streets, and intimate partner violence at home. This is, however, not to suggest that GBV is germane to what is called the "Third World." In fact, GBV is a global phenomenon with a variety of consequences for victimized, subjugated genders as well as for the oppressed at large. But as those examples highlight, most perpetrators of violence are known to victims, making it particularly complicated to tackle it as a social problem [2, 3].

Gender-based violence is connected to a host of other problems including losing control over one's body; long-term physical and mental health problems; reproductive health issues, including increased risk of unwanted pregnancies and HIV contraction; difficulties in forming and maintaining healthy relationships; substance abuse as a coping mechanism; higher odds of engaging in risky sexual behaviors and self-harm [4–7]. Moreover, factors associated with risk of perpetration as well as victimization include low levels of education, exposure to violence at home as a child, substance abuse, and negative, stereotypical views of women sanctioned and legitimized by the patriarchal culture itself [8–10]. Violence in all these instances also is not merely an individual-level concern; it has implications for families and communities. Indeed, the violence in question is known to be transmitted intergenerationally as a learned behavior, and as a tool of exerting and maintaining control over others [11, 12].

In this chapter, we redefine GBV to accentuate the fact the question of gender ranges beyond the male/female binary. Thus, the continuum of ongoing violence perpetrated on different genders cannot simply be reduced to a few sites identified by liberal government and non-government organizations. We have included a glossary (see Appendix) to assist readers new to this topic with terms used throughout the chapter. Second, we discuss drivers of and interventions against GBV. We do so through a case study of GBV in Bangladesh. Finally, we conclude with a discussion of emerging concepts, theories, and prospective interdisciplinary innovations to this global phenomenon. Specifically, we discuss how feminist approaches to policy development and the specific implementation in the field of education might provide a critically necessary and transformative mechanism for raising awareness about and ameliorating GBV.

Gender-Based Violence

To begin, "gender is embedded in a range of social, political, cultural, and ideological formations" [13, p., 101, 104]. Rather than being binary male-female, it includes trans men and women, as well as intersex and other forms of gender identities. Thus, it might be more accurate to speak of "gender" in the plural—*genders*. Though gender-based violence is documented to disproportionately affect females, GBV is indiscriminate across class, race, ethnicity, nationality, religious, and age lines [14, 15]. We therefore define GBV as a range of overt and covert practices that discriminate against, subjugate, and oppress—in a variety of ways—individuals, based on their gender. It includes not only physical and sexual violence, but also emotional, verbal, intellectual, economic, and various coercive and controlling tactics such as threats, deprivation, surveillance, and degradation.

The core sites of GBV are the family, community, and state, including both private and public spaces of the home, workplace, places of worship, and schools. Gender-based violence occurs in sociopolitical spaces under the interconnected systems of capitalism, imperialism/colonialism, racism, and patriarchy. The black feminist theorist-activist bell hooks and Patricia Hill Collins, among others, calls them the "interlocked" systems of oppression that obtain and operate in—and affect—both micro and macro sites [16]. We identify four areas in which such violence is located: land, labor, language, and the body [17]. In this chapter, we focus primarily on the body, but with the understanding that bodily violence does not exist in isolation, but rather often intersects with the other loci, for instance, by way of commercial exploitation, which can involve violence on the body via lack of access to sanitary conditions, threats to one's life, and withholding of wages.

Though GBV is a worldwide phenomenon crossing all class and culture lines, violence against cis-gendered women has received more attention in the literature than others, even though trans women continue to experience violence in all sorts of ways. Intimate partner violence (IPV) and to an extent non-partner sexual violence continue to be a primary focus in research focused on cis-gendered women [3]. As an example of intersectional oppression by race, sex, and nationality, neoliberal restructuring of the Global South has produced a limitless supply of cheap labor, that when coupled with restrictive immigration policy in the Global North, engenders a market for labor and sex trafficking and other forms of commercialized violence [14]. Other types of violence, such as honor killings and genital mutilation, have received extensive attention as well, but mostly as examples of violence in "exotic" locations in Africa and South Asia, i.e., the Global South where rights of women are argued to be more limited than in the Global North [18].

International organizations such as the United Nations recognize the problem of gender-based violence through a variety of internationally ratified protocols and conventions, often framing it as an issue of gender inequality. This most recently was witnessed when GBV was deemed one of 17 Sustainable Development Goals in 2016, with the understanding that addressing gender inequality would improve the treatment of women, if not all genders, worldwide [19]. But, as Angela Davis rightly points out, the solution to the problem of GBV "must involve a consciousness of capitalism, and racism, colonialism, and post-colonialities, and ability, and more genders that we can imagine, and more sexualities than we ever thought we could have" [13, p., 101, 104].

As the diagram below demonstrates, gender equality and GBV cannot be examined without an examination of its intersection with other forms of oppression. Though invisible and taken-for-granted, everyday cultural norms, mores, and practices are perpetuated by larger systems violence broadly conceptualized as structural violence (Fig. 6.1).

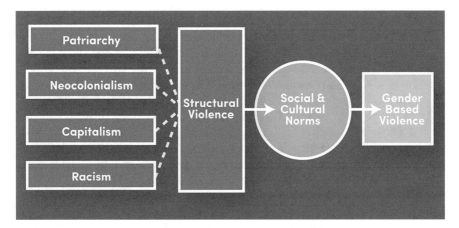

Fig. 6.1 Conceptual model showing how cultural norms are shaped by interlocked systems. (Source: concept: Nadine S. Murshid; graphic design: Nicole C. Little)

Why GBV Persists

Gender-based violence exists for several reasons and chief among these are the interlocked systems of oppression—patriarchy, capitalism, imperialism/(neo)colonialism, and racism. We argue that GBV is integral to each system of oppression. Such systems variously produce and reproduce violence through micro, or individual acts and threats of, and macro or state actions that normalize violence [18]. McClintock, for example, posits how nationalism is framed in familial terms such as motherland and fatherland to create a space of "unity," in the same way that the family is, which then sanctions gender difference and social hierarchy [20].

GBV also exists because of how the problem is framed. GBV typically is discussed as "violence against women." This passive construction of violence holds out men as the sole perpetrator of violence. In framing GBV as a "women's issue," power dynamics that support and maintain GBV are overlooked. In doing so, such violence becomes a male problem or the problem of a couple of "good" men to solve. Thus, individual women, as opposed to larger society, are somehow responsible for the violence they experience as they are the only identified actors in that construction [21]. This framing also often conflates GBV with "intimate partner violence," or "domestic violence" in the public's eye, obscuring different forms of and places within which gendered violence occurs [22].

The "exoticization" of certain kinds of violence as a way to create separation from those problems, as located in "distant" places, is yet another way in which violence is obscured. For instance, imagining violence in foreign lands blatantly erases how patriarchal cultural norms, along with the other systems of oppression, permeate and affect—even if unevenly—the entire world, from the USA to Bangladesh. Such exoticization erases more subtle forms of coercive control that occurs in countries like the USA, for instance, while it ignores the fact that patriarchy is the hegemonic power-relation that is institutionalized to the extent that subordination of women is the norm. Foregrounding certain places, particularly formerly colonized spaces as the only grounds of anti-women violence, makes it easy to erase the violence committed by colonizers and neocolonizers [23].

Indeed, there are even racist ways of talking about patriarchy and about the violence perpetrated on women, which, we argue, cannot really address the pervasive problem of gender-based violence. In the USA, for example, one in four women reports to be assaulted every year, which is only slightly lower than the global estimate of one in three women. However, the racialized reading of patriarchy and women's position in society makes it appear that violence against women is a nonproblem in the USA, which then allows it to persist. In a study on Texas trafficking policy designed to educate teachers, counselors, and administrators about trafficking, Lemke found that some legislative respondents discussed trafficking in ways that perpetuated troubling us-them dichotomies and were indicative of normative bias including that only certain female victims were worthy of law enforcement protection [14]. This finding is significant given legislative views on a specific form of gender-based violence directly affect trafficking legislation, and in this case what educational practitioners might learn about this phenomenon. Further, violence is seen as a humanitarian crisis. Its portrayal in the media is mainly meant to evoke emotional responses in people.

Using the lens of intersectionality, we can also argue that gender-based violence is not only inevitable within the structures of oppression, but that it is exacerbated by other unequal social relations and subjugated identities, based on class, race, and sexuality, for instance—ones that in fact compound the effects of discrimination [24]. Further, in addition to unequal class, gender, and race relations, other asymmetrical relations are also socially constructed and produced by nations. And all such unequal relations—which give rise to different forms of discrimination in society—are often fraught with violence. In other words, violence is structurally produced and then obscured. This obscuring, non-neutral phenomenon allows violence to persist. However, just because women are deemed to be socially and politically subordinate to men, it does not mean all women have "equal shares or stakes in, or experiences of, the system itself," nor are their perspectives and interests homogenous [25, 26].

> **The Case of Bangladesh: Intimate Partner Violence as GBV**
> Bangladesh had to adopt structural adjustment policies imposed by the World Bank and the International Monetary Fund, as financial loans given to the country were profoundly tied with structural change that eventually led to changes in the very structure of the country's economy. With rapid industrialization in cities, and with what came to be known as the "Green Revolution" in rural areas, the migration of people from rural to urban areas became inevitable. In industrial production, based on the logic of a capitalist market economy, an overarching value is "control through domination." This was indeed prioritized with industrialization, corresponding to the marginalization and denigration of women and social characteristics associated with women [27]. One example is the exploitation of female "low-skilled" workers, the high supply of whom drives down the labor price. On the other hand, the increase in women's labor market participation, away from the home, results in the double burden of providing labor at home, such as child and elder care and household management, and at work.
>
> This double burden decisively characterizes the predicament of women across social and economic boundaries, as does the threat to masculinity that occurs when women start earning their own incomes. This then leads to multiple levels of violence, from intimate partner violence (IPV) at home, and sexual harassment on the streets [28, 29]. In particular, gender role transforming institutional interventions such as microfinance was first thought to be empowering until it became clear that there are unintended consequences in terms of IPV when women's status rise poses a threat to men's status [30]. Additionally, women's increased presence in the public domain, amidst patriarchal structures that are discriminatory and violent towards women, mean

that they are more likely to experience violence in all kinds of locations: at their workplaces, at factories, and on public transportation, to name a few.

Other drivers of GBV in Bangladesh include individual-level factors such as young age, low levels of education, income, and household wealth, as well as exposure to violence, justification of violence, and substance use [31–36]. Community-level factors include traditional gender norms and social norms that are based on religious ideology that are often focused on controlling the bodies of women [37]. Commodification and sexualization of women's bodies, amidst religion-sanctioned notions of what it means to be "pure and good" women create further avenues for their oppression.

The Potential for Leveraging New, Under-Utilized, or Unexpected Disciplines/Approaches

Historically, legal and political institutions did not recognize nor aim to ameliorate gender-based violence (GBV) through public policy. Rather than view such violence as a problem affecting society, it was categorized as "women's" problem. When various feminist factions mobilized officially to end institutionalized discrimination around the globe, such factions often faced internal and external undermining by dominant societal and policy discourses premised upon liberal individualism [38, 39]. Assisted by religious moralists, victims of GBV, sexual or otherwise, tend to face accusation, denial, shaming, and official suppression rather than institutional empowerment [40].

According to Tong, female "oppression is not the result of individuals' intentional actions, but is the product of the political, social, and economic structures within which individuals live." [41] Scholars suggest that improvements in status of women, de-normalizing violence, and poverty reduction are just a few of the ways GBV might be reduced [42]. One way to address GBV is to encourage those working in both public and private spheres to consider, adopt, and support the use of feminist praxis as a mode of daily social interaction. This means understanding that systems, policy, and practice fundamentally are not value neutral, but rather are sanctioned by power dynamics. Adopting such a view is argued to help buttress gender-equitable norms and behaviors at the individual and localized levels toward larger scale transformative change [43].

Within the field of psychology and social work, feminism-oriented practitioners have provided individual-level interventions or "direct services" for both women and others who experience violence. These services fall under three broad categories: crisis interventions, counseling by professionals, and medical and legal advocacy, the latter being challenging given that most individuals who experience violence do not seek medical services or report their experience of violence to law enforcement. Direct practitioners who provide services to women experiencing violence have been advocating for what they call "comprehensive services." This entails going "beyond meeting immediate crisis needs to supporting more in-depth healing, empowerment, and integration of pre-assault and post-assault identity... informed by... neurobiological, social, and emotional impacts of trauma." [44] This can be done by providing trauma-informed services and advocating for trauma-informed policy, that includes elements of safety, choice, collaboration, empowerment, transparency and accountability, and intersectionality [45].

In terms of addressing GBV through policy, feminist critical policy analysis (FCPA) could prove an insightful theoretical tool for advocates and practitioners working on the ground. Feminist policy analysis incorporates a distinct focus on how sex and gender are addressed within policy processes, toward the end of revealing value neutral, and taken-for granted values, when operationalized silence, undermine, and give the false appearance of addressing equity concerns. In reviewing feminist policy studies, Hawkesworth, for example, demonstrated how the state maintains androcentric, impartial interpretations of equality to the detriment of females [46].

Finally, in the field of education, FCPA has been utilized to address various forms of gender disparity, discrimination, and violence. Concerned with gender inequity in policy and practices across the elementary, secondary, and higher education pipeline, Marshall discussed FCPA as a tool for unpacking how policy creates and maintains male-normed systems—and when gender is a factor in such systems it often is one of contest and dispute [47, 48]. Shaw analyzed the gendered nature of welfare reform to understand roadblocks to postsecondary education experienced by 22 women [49]. Ackerman examined how gendered perceptions of child care workers shape low wages, quality, and worker access to policy agendas [50]. Using a feminist lens, Salinas and Reidel looked at how educational values in Texas favor business-power elites and undermine the Texas public education system as a whole [51]. Lemke used FCPA to interrogate human trafficking dynamics and whether a landmark Texas education policy on a specific form of gender violence fulfilled its promises, and based on that answer, what that meant for K-12 public education and trafficking policy more broadly [14, 52]. In light of the continued backlash against feminist research, leadership, and political advances, these studies and multiple other feminist analyses are keen to highlight the significance that a gendered lens brings to understanding GBV [47, 53–56].

Conclusion

Bangladesh represents a specific geographic, sociopolitical, and economic context. Still, a feminist framing of GBV permits one, regardless of geographic location, to be attentive to political discourses, power, and how false consciousness legitimates the status quo. Such feminist, and as previously discussed, interdisciplinary framing, is instructive in their ability to highlight the ways that policy, practice, and various disciplines can limit and/or reproduce power toward the end of reforming society.

Given the deep penetration of mobile phones in Bangladesh, other practical ways by which GBV might be addressed includes mobile phone-based social messaging. Several mobile phone-based applications such as Maya Apa (which means elder sister) act as a messaging helpline for sexual and reproductive health problems for women. Through this app they also are provided help and resources for GBV. Such platforms are well-placed to be a useful resource for all genders in the future.

Another way to increase awareness about gender norms and GBV involves including men as part of empowerment programs such as microfinance organizations given they are deeply involved in the communities within which they work [57]. This does the following: (1) it directly becomes an avenue for sharing content and (2) indirectly includes men into microfinance programming and reduces the threat that they may experience when their partners access microfinance. This would be in line with previous research that shows that women often have to involve their partners in their micro businesses given that they do not receive any or adequate training in skills needed to become an entrepreneur, or, that they become microfinance entrepreneurs at the behest of their husbands' ambitions to start a business [57, 58].

Finally, connecting women, men, and other genders with education programs aimed at changing mindsets about gender roles and associated status also could occur. Such educational programming or a messaging campaign could focus on specific content that builds solidarity among women and other genders, while concomitantly transforming male dominated cultures of violence.

Appendix

Table 6.1 Glossary of terms

Key terms	Definition
Androcentric	Either consciously or not, centered on males or masculine interests thereby marginalizing females or feminine interests
Cis-gendered	When gender expression and identity are aligned with gender assigned at birth. It is argued the more they are aligned the more social privilege one receives
Commercialized exploitation	The act of being discriminatory, utilizing fraud, inducing fear, being coercive, or causing harm to benefit from another monetarily. The term is most associated with labor and sexual exploitation wherein exploiters aim to exact maximum monetary or nonmonetary gain from another through work or sexual acts
Culture of fear	Refers to a perceived or actual presence of social and political discourse designed to instill fear of individuals, groups, actions, behaviors, and/or events toward some expected outcome and which may affect human interaction
Culture of impunity	Refers to the presence of discourse, behaviors, and/or actions that together create conditions that exempt individuals and/or groups from punishment for harmful actions, and in this case gender or sexual violence

(continued)

Table 6.1 (continued)

Key terms	Definition
Gender	The social construction of biologic sex, which is demonstrated through fluid expressions, identities, and roles. Individuals can accept or reject such constructions and as a result can face gender stereotyping, discrimination, and violence
Hegemonic	Dominant, ruling, or authority figure or entity in a given cultural, economic, geographic, institutional, political, and/or social context
Institutionalize	To establish a norm, idea, set of beliefs, values, and/or activity within a given institution or organization
Intersex	Umbrella term describing a person with sexual/reproductive anatomy that does not match female and male medical definitions
Male-female binary	A socially constructed and norm-driven system having only male and female gender. This does not account for a diversity of gender expression/identity and thus can be viewed as oppressive by those outside it
Neocolonial	Following twentieth century colonial independence, the practice of using oppressive economic, political, social, and cultural forces to indirectly or directly dominate another nation or people
Neoliberal	Global political, economic, and cultural processes that work to privatize the public sector, restrict or use government regulation to the benefit of capital, and maintain a highly fixed understandings of identity including a heterosexual understanding of sex and gender
Normalize	To make normal or to conform to a set of dominant group beliefs, values, or ideas, and in this case regarding gender violence
Patriarchy	Though it looks different dependent on historic location, a system wherein men hold primary political, economic, educative, legal, and military power, and they utilize this privilege to assert moral and cultural authority to the disadvantage of women
Transgender	Umbrella term describing a person whose gender expression/identity are not aligned with birth assigned gender

References

1. Women, UN. Progress of the World's women 2015–2016. New York: United Nation Publication; 2015. p. 69. Chapter 2.
2. WHO. Violence against women. Geneva: UN World Health Organization; 2016.
3. Murshid NS. Men's report of domestic violence perpetration in Bangladesh: correlates from a nationally representative survey. J Interpers Violence. 2017;32(2):290–307.
4. Campbell J. Health consequences of intimate partner violence. Lancet. 2002;359:1331–6. https://doi.org/10.1016/s0140-6736(02)08336-8.
5. Campbell JC, Lewandowski LA. Mental and physical health effects of intimate partner violence on women and children. Psychiatr Clin N Am. 1997;20(2):353–74.
6. Smith PH, Homish GG, Leonard KE, Cornelius JR. Intimate partner violence and specific substance use disorders: findings from the national epidemiologic survey on alcohol and related conditions. Psychol Addict Behav. 2012;26(2):236.
7. St. Vil NM, Carter T, Johnson S. Betrayal trauma and barriers to forming new intimate relationships among survivors of intimate partner violence. J Interpers Violence. 2018. https://doi.org/10.1177/0886260518779596.

8. Schuler SR, Lenzi R, Yount KM. Justification of intimate partner violence in rural Bangladesh: what survey questions fail to capture. Stud Fam Plan. 2011;42(1):21–8. https://doi.org/10.1111/j.1728-4465.2011.00261.x.
9. Murshid KAS, Murshid NS. Adolescent exposure to and attitudes toward violence: empirical evidence from Bangladesh. Child Youth Serv Rev. 2019;98:85–95.
10. Murshid NS, Critelli FM. Empowerment and intimate partner violence in Pakistan: results from a nationally representative survey. J Interpers Violence. 2017. https://doi.org/10.1177/0886260517690873.
11. Murshid NS, Murshid N. Intergenerational transmission of marital violence: results from a nationally representative sample of men. J Interpers Violence. 2018;33(2):211–27.
12. Bandura A. Social learning theory of aggression. J Commun. 1978;28(3):12.
13. Davis AY. Freedom is a constant struggle: Ferguson, Palestine, and the foundations of a movement. Chicago: Haymarket Books; 2016. p. 101.
14. Lemke M. Educators as the "frontline" of human-trafficking prevention: an analysis of state-level educational policy. Leadersh Policy Sch. 2018;18:1–21.
15. Kelly L. The continuum of sexual violence. In: Women, violence and social control. New York: Springer; 1987. p. 46–60.
16. Hooks B. Choosing the margin as a space of radical openness. Framework J Cine Media. 1989;36:15–23.
17. Hussain A. Towards a political economy of racism and colonialism: re-reading Frantz Fanon's the wretched of the earth. In: Young J, Braziel JE, editors. Race and the foundations of Knowledges: cultural Amneisia in the Acadmy. Illinois: University of Illinois Press; 2006. p. 127.
18. Žižek S. Violence: six sideways reflections. London: Profile books; 2008.
19. Giles-Corti B, Kerr J, Pratt M. Contributing to helping to achieve the UN sustainable development goals: truly shifting from niche to norm. Amsterdam: Elsevier; 2017.
20. McClintock A. Imperial leather: race, gender, and sexuality in the colonial contest. New York: Routledge; 2013.
21. Katz J. The macho paradox: why some men hurt women and and how all men can help, vol. Book, Whole. Naperville: Sourcebooks, Inc.; 2006.
22. Price JM. Structural violence: hidden brutality in the lives of women. Albany: SUNY Press; 2012.
23. Zizek S. A plea for ethical violence. Bible Crit Theory. 2004;1(1):02. https://doi.org/10.2104/bc040002.
24. Crenshaw KW. Close encounters of three kinds: on teaching dominance feminism and intersectionality. Tulsa Law Rev. 2010;46(1):151.
25. Finn G. Limited edition: voices of women, voices of feminism//review. Resour Feminist Res. 1994;23(3):62.
26. Mohanty CT, Russo A, Torres L, Societies American Council of Learned. Third world women and the politics of feminism. Bloomington: Indiana University Press; 1991. vol. Book, Whole.
27. Menzies H. The politics of science: keeping women in their place. In: Finn G, editor. Voices of women, voices of feminism. Halifax: Fernwood Publishing; 1993.
28. Murshid NS, Akincigil A, Zippay A. Microfinance participation and domestic violence in Bangladesh: results from a nationally representative survey. J Interpers Violence. 2016;31(9):1579–96. https://doi.org/10.1177/0886260515569065.
29. Murshid NS, Zippay A. Microfinance participation and marital violence in Bangladesh: a qualitative inquiry. Violence Against Women. 2016;23(14):1752–70. https://doi.org/10.1177/1077801216665480.
30. Murshid NS, Akincigil A, Zippay A. Microfinance participation and domestic violence in Bangladesh: results from a nationally representative survey. J Interpers Violence. 2015;31(9):1579–96. https://doi.org/10.1177/0886260515569065.

31. VanderEnde K, Amin S, Naved RT. Community-level correlates of physical violence against unmarried female adolescents in Bangladesh. BMC Public Health. 2014;14(1):1027. https://doi.org/10.1186/1471-2458-14-1027.

32. Naved RT, Persson LA. Dowry and spousal physical violence against women in Bangladesh. J Fam Issues. 2010;31(6):830–56. https://doi.org/10.1177/0192513x09357554.

33. Silverman JG, Decker MR, Gupta J, Kapur N, Raj A, Naved RT. Maternal experiences of intimate partner violence and child morbidity in Bangladesh: evidence from a national bangladeshi sample. Arch Pediatr Adolesc Med. 2009;163:700–5. https://doi.org/10.1001/archpediatrics.2009.115.

34. Sambisa W, Angeles G, Lance PM, Naved RT, Curtis SL. Physical and sexual abuse of wives in urban Bangladesh: husbands' reports. Stud Fam Plan. 2010;41:165–78. https://doi.org/10.1111/j.1728-4465.2010.00241.x.

35. Murshid NS, Murshid N. Intergenerational transmission of marital violence results from a nationally representative sample of men. J Interpers Violence. 2015;33(2):211–27. https://doi.org/10.1177/0886260515604413.

36. Murshid NS. Men's report of domestic violence perpetration in Bangladesh correlates from a nationally representative survey. J Interpers Violence. 2015;32(2):290–307. https://doi.org/10.1177/0886260515585544.

37. Naved RT, Samuels F, Le Masson V, Talukder A, Gupta T, Yount KM. Understanding intimate partner violence in rural Bangladesh. London: Overseas Development Institute; 2017.

38. Lemke M. Feminist organizations. In: Naples N, Hoogland RC, Wickramasinghe M, Wong A, editors. The Wiley-Blackwell encyclopedia of gender and sexuality studies. Hoboken: Wiley; 2016. p. 1–3.

39. Faludi S. Backlash: the undeclared war against American women, vol. Book, Whole. New York: Crown; 1991.

40. Pisani E. The wisom of whores: bureaucrats, brothers, and the business of AIDS. New York: WW Norton & Company; 2008.

41. Tong R. Feminine and feminist ethics. Social Philosophy Today. 1995;10:94.

42. Jewkes R. Intimate partner violence: causes and prevention. Lancet. 2002;359(9315):1423–9.

43. Kabeer N. Paid work, women's empowerment and gender justice: critical pathways of social change. Brighton: Pathways of Women's Empowerment RPC; 2008.

44. Campbell R, Townsend SM. Defining the scope of sexual violence against women. In: Renzetti CM, Edleson JL, Bergen RK, editors. Sourcebook on violence against women. Los Angeles: Sage Publisher; 2011. p. 359.

45. Bowen EA, Murshid NS. Trauma-informed social policy: a conceptual framework for policy analysis and advocacy. Am J Public Health. 2016;106(2):223–9. https://doi.org/10.2105/AJPH.2015.302970.

46. Hawkesworth M. Policy studies within a feminist frame. Policy Sci. 1994;27(2/3):97–118. https://doi.org/10.1007/BF00999883.

47. Marshall C, Young M. Policy inroads undermining women in education. Int J Leadersh Educ. 2013;16(2):205–19. https://doi.org/10.1080/13603124.2012.754056.

48. Marshall C. Feminist critical policy and anaysis I: a perspective from primary and secondary schooling. London: The Falmer Press; 1997.

49. Shaw KM. Using feminist critical policy analysis in the realm of higher education: the case of welfare reform as gendered educational policy. J High Educ. 2004;75(1):56–79.

50. Ackerman DJ. The costs of being a child care teacher: revisiting the problem of low wages. Educ Policy. 2006;20(1):85–112.

51. Salinas CS, Reidel M. Agenda: examining who gets what, when, and how. Anthropol Educ Q. 2007;38:1.

52. Lemke M. The politics of 'giving student victims a voice': a feminist analysis of state trafficking policy implementation. Am J Sex Educ. 2019;14(1):1–36. https://doi.org/10.1080/15546128.2018.1524805.

53. Blackmore S. The meme machine. Oxford: Oxford Paperbacks; 2000.

54. Young MD, Skrla L. Reconsidering feminist research in educational leadership. Albany: State University of New York Press; 2003. vol. Book, Whole.
55. Brown M. Eighteenth-century literary history: an MLQ reader. Durham: Duke University Press; 1999. vol. Book, Whole.
56. Shapiro I, Adams RM. Integrity and conscience/edited by Ian Shapiro and Robert Adams. New York: New York University Press; 1998.
57. Murshid NS, Zippay A. Microfinance participation and marital violence in Bangladesh: a qualitative inquiry. Violence Against Women. 2017;23(14):1752–70.
58. Karim L. Analyzing women's empowerment: microfinance and garment labor in Bangladesh. Fletcher F World Aff. 2014;38:153.

Chapter 7
Gender, Disability, and Access to Health Care in Indonesia: Perspectives from Global Disability Studies

Michael Rembis and Hanita P. Djaya

Introduction

Disability is a global health crisis. About 15% of the world's population or approximately one billion people are living with a disability—usually defined as a motor, sensory, speech, learning, developmental, or intellectual impairment or chronic illness. The World Health Organization (WHO) and United Nations (UN) estimate that between 110 and 190 million people live with what they define as "severe disabilities" [1]. If we add mental illnesses and autism to the list of disabilities, the number rises. A 2011 WHO report states that "Across the globe 450 million people suffer from a mental or behavioral disorder. The estimate is that one in five persons will suffer from a mental illness in a given year" [2]. The WHO predicts that "depression will be the number one global burden of disease by 2030, surpassing heart disease and cancer, and [it is] anticipated to be the number two burden by 2020" [2].

Disability or mental illness can affect anyone, anytime. We are all likely to encounter disability at some point during our lives, whether it is through our families, our jobs, our recreational activities, or through our own bodies. Approximately 85% of disability is acquired throughout our lifetime. This means that only 15% of disabled people are born with their disabilities [3]. Disabilities are usually caused by unsanitary, unhealthy, overcrowded living conditions, malnutrition, war, environmental hazards, disease, poverty, accidents, and childbirth, among other social factors.

M. Rembis (✉) · H. P. Djaya
University at Buffalo, State University of New York (SUNY), Buffalo, NY, USA
e-mail: marembis@buffalo.edu

© Springer Nature Switzerland AG 2020   97
K. H. Smith, P. K. Ram (eds.), *Transforming Global Health*,
https://doi.org/10.1007/978-3-030-32112-3_7

People in the developing world and women are more likely to experience disability; 80% of people with disabilities live in developing countries. Both the UN and the WHO have found that there is a higher representation of women among the world's disabled population [1]. The disparity in occurrence of disability between women and men is the product of a complex mix of factors. However, it is clear that one of the root causes of a higher rate of disability among girls and women is the general devaluing of girls and women in society.

In this chapter, we will use a close reading of secondary literature and a collection of open-ended interviews of disabled Indonesian women conducted in part by co-author Hanita Djaya to analyze the social, economic, and structural barriers women and girls with disabilities face when attempting to access healthcare in Indonesia. We will use an interdisciplinary global disability studies approach that engages with the ways in which social class or caste, gender, ethnicity, language, and religion affect the lived experiences of girls and women in South and East Asia, and more specifically in Indonesia. It is our contention that a careful qualitative analysis of the lived experiences of disabled people is an important way to build what Rapp and Ginsburg have called the "social fund of knowledge" concerning disability and disabled people [4].

A Global Disability Studies Approach [1]

Before we begin our discussion of disabled women and girls in Indonesia, we will provide a sketch of what it means to take a global disability studies approach to considering disabled people's access to healthcare. Disability studies is an interdisciplinary academic endeavor that analyzes disability and the lived experiences of people with disabilities from the perspective of the humanities, social sciences, and arts. Disability studies has a tenuous, often critical relationship with the medical and applied fields (such as rehabilitation science or other health related fields).

One of the basic tenets held by disability studies researchers is that "disability" extends beyond individual bodies. Disability is not merely something that an individual has (I have cerebral palsy), or is (I am a paraplegic). Disability is socially created by built environments, cultures, and social norms that devalue, stigmatize, segregate, or discriminate against people who are considered "abnormal." Disability studies scholars have labeled their focus on the interactions between disabled people and the world around them the *social model* of disability, which they define in opposition to an older, well-entrenched *medical model* of disability that sees it in more negative and individualizing terms. The medical model defines disability as a deficit or defect that must be fixed, cured, or eliminated in order for an individual to function "normally" in society. The social model of disability does not view disability in the same way.

The Social Model of Disability

The social (also called the socio-political) model of disability originated among disability rights activists in the United Kingdom in the early 1970s. Its primary purpose was to separate the conceptualization of disability from that of impairment; to say that disability was something that was socially created, while impairment was merely a biological fact with no cultural values attached to it.

Under the social model, what became disabling for people was not their inability to walk, see, or hear (for example), but rather the inaccessibility of a physical, social, and cultural environment that remained hostile to their presence in it. As the British Union for the Physically Impaired Against Segregation (UPIAS) explained, disability is "a form of [socially created] disadvantage which is imposed on top of one's impairment, that is, the disadvantage or restriction of activity caused by a contemporary social organization that takes little or no account of people with physical impairments" [5]. Put simply, the social model of disability makes a critical distinction between impairment (body) and disability (society). It roots disabled people's limitations in societal barriers that disable them, not in any individual embodied deficit. Disability studies scholars refer to this system of exclusion as "ableism." They argue that ableism and ableist attitudes are present in all societies that are built by and for nondisabled people [6, 7].

The social model of disability redefines disability as something created in the social world, and not through biology (or genes or neurochemistry). This new way of thinking about disability has enabled scholars (and activists and artists) to move disabled people away from their historical place in society as individuals in need of medical, rehabilitation, welfare, and other services and interventions to that of an oppressed social minority in need of recognition of its civil and human rights. By discarding the notion that disability is negative and rooted in the individual, and by thinking critically about the power of various social arrangements to disable, social model theorists have been able to develop a powerful understanding of what it means to live differently in the world. Part of the success of the social model derives from its ability to expand the definition of disability to include a broad range of impairments, illnesses, and conditions, and to show that disability will touch everyone at some point in their life. The tremendous diversity among the world's disabled population and the broad range of experiences we all have with disability have been a source of empowerment for disability rights activists and academics alike.

Global Disability Studies

One critique of the social model that has emerged within disability studies comes from scholars interested in global disability studies. Disability studies scholars working on non-Western topics and those working outside of the Western English-speaking world (alternatively referred to as *the global North*) are finding that disability studies theories that are dominant in the West or global North, including the social model of disability, are often ineffective, or in some cases only partially effective, in helping to explain the lived experiences of disabled people in other parts of the world (referred to as *the global South*). Global disability studies theorists make strong arguments for avoiding the uncritical exportation of global North disability studies theories to the global South, and, instead, argue for situating analyses of the lived experiences of disabled people in their own local cultural and historical contexts, social relations, and governing structures, as well as larger international political and economic systems. Rather than dismiss global North disability studies, global South scholars encourage collaborative and constructive dialogue between North and South, which they argue will build stronger disability studies analyses and more powerful disability politics in both parts of the world [8]. It is with this view of global disability studies and its critique of the British and North American social model of disability in mind that we turn to our discussion of disabled women and girls in Indonesia.

Disability and Gender in Indonesia

Indonesia is a South East Asian island nation with a population of about 260 million (in 2016) people. It is ethnically and linguistically diverse. More than half, almost 60%, of Indonesians live on the island of Java, the most populous island in the world and the home of Jakarta, Indonesia's capital. The World Bank considers Indonesia a "lower-middle income" country with a Gross Domestic Product (GDP) of about $932 billion US dollars annually. By comparison, Canada, India, and the United Kingdom have GDPs that range from 1.5 to $2.6 trillion US dollars annually. About 10% or one out of every 10 Indonesians live in poverty. Until recently and relative to other countries, Indonesia spent very little on publicly funded healthcare services, only about 3.1% of its 2013 GDP.[1] The global average for publicly funded healthcare services is about 10% of GDP. Until 2014, most Indonesians relied primarily on a privatized "out-of-pocket" system to fund healthcare, where individuals paid directly for the healthcare they received [1, p. 13]. Although both systems were regulated with the intent of providing adequate and accessible healthcare to all Indonesians, the reality was that great disparities existed among different segments of the population, with poor, rural, and disabled people suffering the most severe

---

[1] Poor Indonesians used a mix of regional and national public health insurance schemes. Although coverage varied and it was means tested, most beneficiaries received a comprehensive package that included inpatient and outpatient care, as well as generic prescriptions, with no premium payments or copayments upon visiting the doctor. 76.4 million people used Indonesia's national public health insurance program known as *Jamkesmas* in 2009. For more information on health insurance see: Adioetomo, Sri Moertiningsih, Daniel Mont and Irwanto. 2014. *Persons with Disabilities in Indonesia: Empirical Facts and Implications for Social Protection Policies.* Jakarta: Demographic Institute, Faculty of Economics, University of Indonesia. pp.107–117.

inequities. In 2014, the Indonesian government implemented a national healthcare program called the Social Security Administrator for Health (BPJS), providing "broad-ranging" coverage to all of its citizens, including those experiencing mental health problems and those living on the autism spectrum [1, p. 143], [9, pp. 3–19]. Even within this system, however, disabled people continue to face significant challenges and constraints.[2]

Although Indonesia is showing signs of improvement with respect to disability and disabled people, negative attitudes continue to persist and they remain gendered, especially within poor and rural communities. In general, disability and disabled people are something to be pitied in Indonesia. Disability is considered "shameful" and "dirty" [9]. Many parents believe that "if my child is handicapped, he/she cannot live a normal life" [9]. If a married woman becomes disabled, her husband can legally divorce her [9, p. 3.76]. Although the language is changing—the official word for disability is "disabilitas"—words like "cacat" (incomplete, disfigured, defect) and "Orang gila" (crazy person) can still be heard in public [9, p. 3.3]. Disabled women and girls are statistically more likely to experience violence, abuse, discrimination, and isolation in Indonesia. Studies have shown that some disabled people have internalized these negative attitudes and report feelings of low self-confidence and low self-esteem [9]. There is, however, a growing group of disabled people and their allies, who will be discussed in more detail below, who are fighting to change these attitudes. This social and economic context is critical for understanding disability and gender in Indonesia.

The Indonesian Ministry of Health estimates that approximately 11% of the population, or 27 million people are living with "at least one moderate to severe disability" [1, p. 12]. It is likely that the number of disabled people living in Indonesia is higher than the official count would indicate. Most reports only consider "vision, hearing, mobility or climbing stairs, concentration, cognition, and self-care" when tabulating disability [1, p. 13]. Behavioral disorders such as those related to autism spectrum disorders and mental illnesses go relatively unrecognized in most studies. Social and cultural stigmas, which will be discussed in more detail below, also mean that many disabled people and their families will hide disability from public view. That most of the Indonesian population is rural and 10% live in poverty, also means that in most cases disability will go unmarked and untreated. Additionally, some studies indicate that disabled women (and presumably girls) outnumber disabled men and boys by about 3:1. Because women and girls, specifically disabled women and girls, are not valued as highly as men in Indonesian society, the likelihood is

---

[2]The Indonesian government has passed a number of laws that directly address access to transportation, education, employment, and healthcare. These laws, however, remain unfulfilled due to a lack of funding and political interest. Disabled people in Indonesia, like disabled people in most countries do not possess significant social or cultural capital that would enable their demands to be heard by politicians and government leaders. For an extensive list of laws see: Japan International Cooperation Agency (JICA). 2015. *Data Collection Survey on Disability and Development in Indonesia Final Report.* (KRI International Corporation, Tekizaitekisho, LLC.) p. 3.10–23.

that their numbers are also underrepresented in official counts. One study has found that only about 10% of disabled children are registered and enrolled in school and that there is "a high possibility that persons with disabilities are not registered as citizens at birth, as mandated by law, and are thus not eligible to receive education, medical, employment, care, and welfare services" [9]. This means that they also go uncounted in national studies.

Recognizing that at least 11% of its population is living with some form of disability, the Indonesian government has implemented measures to ameliorate the conditions under which most of its disabled people live. On 30 March 2007, it signed the UN Convention on the Rights of Persons with Disabilities (CRPD). In 2009, the government issued Undang–Undang No. 36, which one researcher describes as legislation that requires the government "to look after the health of elderly and disabled people by providing an accessible health service, and other facilities, to enable this sector of the population to live independently and productively" [1, p. 12]. The Indonesian government ratified the CRPD in November 2011 [9, p. 1.1]. Officially recognizing the UN CRPD and passing legislation that specifically addresses the needs of disabled people is an important first step. However, as long as public-sector healthcare remains underfunded and Indonesians continue to rely even in part on a "pay-as-you-go" system, the needs and desires of disabled people, especially disabled women and girls will not be met by these largely symbolic efforts toward inclusion. Indonesia is a decentralized island nation, where culture and language, as well as poverty and underdeveloped infrastructure outside of major urban areas prohibit efficient transportation and communication and limit access to healthcare for most Indonesian people, especially disabled women and girls.

Health Related Disparities of Girls and Women with Disabilities in Indonesia

Studies confirm that women with disabilities living in Indonesia have negative experiences when receiving healthcare services, especially if those women live in rural areas [1]. The unequal treatment of disabled women, and in some cases their exclusion from healthcare services are experienced in several important and interrelated areas: (1) inadequate healthcare systems and services, which can include ignorance and inexperience among professionals and staff; (2) stereotyping, stigma, and discrimination concerning the need for access to healthcare, especially reproductive and sexual healthcare; (3) a lack of physical access, which can include inadequate transportation, inaccessible buildings, and inaccessible medical equipment; (4) general attitudinal barriers among the community, which leads to increased segregation and isolation for women with disabilities; (5) a lack of recognition of disabled women's desires to be considered "normal"; and (6) difficulties creating a positive self-image [1, pp. 112–113].

Although conditions appear to be improving in the years since Indonesia signed the UN CRPD in 2007, disabled women and girls continue to face significant challenges in accessing adequate, affordable, safe, and effective healthcare. Disabled girls and women are more likely to experience violence, sexual assault, and isolation, even within their own families. They often have to travel great distances to access healthcare. Public transportation is unreliable and often inaccessible. In some locations and under some provisions, disabled women are grouped with the elderly in terms of healthcare provision, despite their vastly different needs and concerns. Disabled women of childbearing age—especially those who have already had a child—are more likely than other women to face pressure to undergo sterilization procedures to prevent pregnancy. In general, negative attitudes toward disabled women and girls continue to persist both within healthcare settings and in the community.

Many disabled Indonesian women, like disabled people in other parts of the world, rely on a range of disabled people's organizations (DPOs), non-profit organizations (NPOs), and non-governmental organizations (NGOs) that focus specifically on disability related issues, and in some cases are led by disabled people themselves, to advocate for their rights and provide them with important material support and a sense of community. Disability focused DPOs, NPOs, and NGOs are more active in Indonesia's urban areas. Djaya's interviews confirm that disabled women and girls living in rural areas have a difficult time utilizing these organizations, which are critical in the lives of disabled people who wish to improve their access to education, healthcare, and employment.

While it is not as robust as it is in other countries, like neighboring Japan, for example, Indonesia has a strong and growing disability rights movement that is led at the national level by prominent disabled people's organizations or DPOs. One of the oldest and most active rights-based organizations is the Indonesian Disabled Persons Association (PPDI), which was established in 1989 and is a member of Disabled Peoples International (DPI). The PPDI works closely with relevant governmental ministries to create policy and enact legal changes designed to ensure the rights of all Indonesians living with disabilities. In addition to the PPDI, disabled Indonesians and their allies have also formed the Movement of the Welfare of Deaf Indonesia and the Indonesia Blind Union. The most prominent organization that focuses specifically on issues related to women with disabilities is the Indonesian Association of Women with Disabilities (HWDI).[3]

The HWDI, formerly known as the Indonesian Association of Handicapped Women was established in September 1997 as a socially oriented, rights-based organization for women living with various types of disabilities, including physical, sensory, and mental disabilities, as well as the parents of children with disabilities. The national organization is based in Jakarta. According to one of Djaya's informants, Maria, the HWDI was established to answer to global demands for the

---

[3] For a more extensive list of major DPOs in Indonesia see: *Data Collection Survey...* 2015, p. 3.82.

empowerment of women (and girls) with disabilities. Part of the mission of the HWDI is to ensure that the issues of women and girls with disabilities, and gender equality, are not forgotten in the movement for the rights of people with disabilities and also in other women's organizations in Indonesia. While the HWDI does not focus specifically on healthcare related issues, it does focus on promoting physical accessibility and advocating for legislative measures designed to ensure the rights of women and girls with disabilities, which as we have argued has a direct impact on healthcare and health related issues. The HWDI has a keen interest in promoting the reproductive and sexual health of women and in preventing women and girls from becoming the victims of violence and sexual abuse through legislation and other activist means.

The fourth largest country in the world, behind China, India, and the USA, Indonesia has made significant gains in securing the rights of people with disabilities. But significant work remains to be done in the area of changing the general population's attitudes toward disability and the equal acceptance of disabled people in education, employment, and in the provision of healthcare. Negative attitudes toward disability can be seen clearly in oral interviews with disabled women living in Indonesia (Fig. 7.1).

Notes from the Field

In this final section of the essay, we will move from the macro-level, where we provided a large-scale assessment of disability and gender in Indonesia to a closer look at the lives of specific women with disabilities living in various parts of Indonesia. The interviews were conducted by Djaya in 2017 and 2018. It will become evident upon reading these brief excerpts that women's lived experiences confirm many of the arguments presented in this essay, which are further supported by larger-scale demographic and policy studies conducted primarily by NGOs. Because the women interviewed for this essay were self-selected and in most cases are active in the disabled community, the reader will find that some interview subjects present their disabilities and their experiences in a positive way that may not be representative of the larger disabled population living in Indonesia. Some of the women interviewed for this study benefit from a class status that enables them to attend school, usually in mainstream settings, and also attend college or university or receive advanced training in particular employment sectors. Some of these women possess the social and cultural capital as well as the technological skills and access to resources and technology that enable them to succeed. These are benefits that are not readily available to many Indonesian people, especially disabled women and girls and those

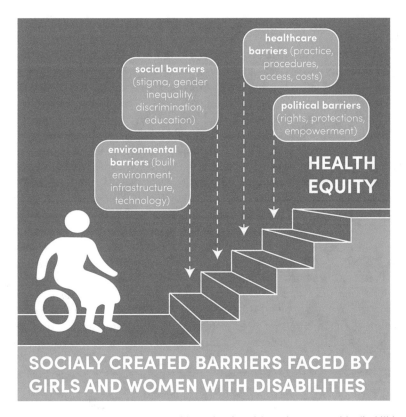

Fig. 7.1 Socially created barriers to health equity for girls and women with disabilities in Indonesia. (Source: Nicole Little and Michael Rembis)

people living in rural areas. A study that explores the experiences of a broader range of disabled women and girls, and especially focuses on their access to and utilization of healthcare, is strongly needed in the field of global disability studies. Despite their limitations, however, the following oral interviews contain much-needed insight into the experiences of disabled women in Indonesia.

Anggi (21 Y.O)

Anggi has been deaf since age 3, when a high fever led to her hospitalization and eventually the loss of her hearing. When she was old enough to go to school, her parents applied for her to attend kindergarten, but she was rejected do to her disability. Her parents enrolled her in a "special school" where she was in a class with other deaf children. She remained in that school until she was old enough to attend high school, at which time she entered a mainstream school, where her teachers gave her junior high textbooks instead of high school textbooks because they assumed she could not learn. Anggi remembers that many of her teachers were "nice," but they were unable to communicate with her because of language barriers. They did not know how to sign and there were not any interpreters for deaf students.

Despite these disadvantages, Anggi was able to succeed. At the time of the interview, she was studying as a freshman in the Makassar Public University, majoring in special education. Anggi claims to experience "no discrimination" in her university education, yet she is forced to attend class without a sign language interpreter, because none is available. The local deaf people's organization relies upon informal networks of support like friends and family to provide interpreting for people. In class, Anggi counts on her friends to interpret for her, and on her own ability to learn from slide presentations during class. She also relies on her friends to help her complete assignments for class. Anggi reports that she has never considered asking for accommodations and is "reluctant to ask for such things." Like her high school teachers who gave her junior high textbooks, her university lecturer gave her fewer questions on her final exam, which can be interpreted as a sign of the negative attitudes of the general population toward disabled people.

Overall, Anggi reports a "positive attitude" toward the medical system in Indonesia. She states that she has never encountered blatant discrimination or insults from health professionals. However, although she is 21 years old, she has never been to the doctor on her own. Her mother has always accompanied her to doctor's appointments to interpret for her. Anggi says that she is "not brave enough" to go to the doctor by herself because she is afraid that miscommunication will occur and result in her receiving the "wrong medicine." As a young healthy woman, Anggi has few health concerns.

Maria (55 Y.O)

Maria was born and raised in a remote area in East Nusa Tenggara. She became disabled in 1974 when a strong wind caused the bookshelves in her school to fall over and crush her leg. She was taken to the hospital, where they amputated her leg. She remained in the hospital for 3 months. "Feeling embarrassed" about her disability, the young Maria did not return to school for another 3 months.

Maria remembers resuming her daily activities "without so many changes." She used a "Canadian crutch" to move around. She stated that the "biggest barrier" in her life was the attitude of her friends, who mocked and insulted her for having only one leg. She recalls that the road that she had to travel to get to school was "harsh"—hilly and full of rocks—but she was able to manage using her crutches.

When she was in seventh grade, Maria's family moved to Atambua, the capital city of a regency in East Nusa Tenggara Province, where she attended Catholic school. There were only four disabled students in the entire school. She remembers that she "didn't receive any discrimination and insult from other students." However, the school's headmaster "warn the students in the beginning of the school that whoever insult these four disabled students will be expelled."

Maria moved to Makassar, South Sulawesi, where she met her husband, who also has a disability. She was married in 1984. When Maria became pregnant, health professionals asked: "Why do you want to have kids?" And "other people were looking [at] her like she is really weird."

Maria attended university in 1994 for one and a half years but quit when she became pregnant with her second child. She adds that the university building was "quite inaccessible." Her classes were held on the ninth floor and sometimes the elevator was not working. "But," Maria states, "the professors were nice."

At the time of the interview, Maria was the head of the Indonesian Association of Women with Disabilities, South Sulawesi. Her husband was the head of the Indonesian Association of Disabled People, South Sulawesi.

Maria prefers receiving healthcare in a private clinic, where she feels she receives the best possible treatment.

Narti (Between 40 and 50 Y.O)

Narti's age is unclear. She claims to be 40 years old, but "clearly remembered that she finished her high school in 1985." She was raised in a rural part of South Sulawesi Province, which is now an independent province, in a place called Mandar.

After finishing high school, Narti moved to Makassar to pursue her degree in Education. She became sick and quit after one semester, returning to her home. In 1991, skin sores started to appear on Narti's back. Raised in a poor, rural environment, with no mother—she died during childbirth—Narti's father and other kin—she has 12 siblings from different mothers—initially attempted to heal the sores using "traditional methods." After 2 years, the skin condition progressed to a point that it impaired her mobility. This was when "she knew that she was having leprosy and started to take medicine." She "was better, she could walk again, and her skin sores started to heal although it did leave scars in her extremities." She continued taking medication for another 2 years.

Narti was declared "clean" of the bacterium in 1995 or 1996, but the effects of leprosy continued to affect her life. In 2003, she was hospitalized for a year for a "leg problem." She remembers that most of the time the "healthcare providers were very nice and understanding." However, she could tell that some of them "was feeling disgusted by her skin sores." Narti recalls that although "they did not tell her right in the face, she could tell from the way they hold her extremities, their expression, and the way they cleaned her skin sores." She said that the most disgusted healthcare providers were usually the "newbies" or interns.

At the time of the interview, Narti was married to a man who also had leprosy, her second husband. Although she refused to talk about her first marriage, it was evident that Narti's first husband left her because of her disability.

Narti stated—with tears in her eyes—that "many people were disgusted by her condition, including her own family members." She met her husband in the hospital in 2003 and they got married in 2004. They moved to Makassar in 2006, where they live in a "leprosy community."

Narti describes herself as a "homemaker." Her husband earns his income through begging, which is a common way of living for people with leprosy and other "disfigurements."

Eka (26 Y.O)

Born in 1991, Eka began wearing glasses when she was a young girl. During her second year of high school, she started to feel pain behind her right eye. Physicians diagnosed her with glaucoma. She began taking medications, but the condition continued to progress. Eka completely lost her sight in 2011, when she was 20 years old.

Initially she was sad, but when she met other blind people and "started learning how to operate computer" she started to "feel better." She decided to quit the university she had been attending, even though she states that because her vision loss was gradual, she was able to become accustomed to it and there was "no adaptation whatsoever."

Through her new friendships in the blind community, Eka was able to obtain a recommendation from a prominent blind organization, Kartunet, which enabled her to attend university in Semarang, the capital city Central Java, where she resumed her studies in English literature. Having access to computer technologies, and knowing how to use them, greatly enhanced her educational experience at university.

Originally from the Brebes Regency of Central Java, Eka had to move to attend school, which meant she would no longer be living with her parents. She was able to live on her own in a kos-kosan, which is much like the dorm rooms in which university students in the USA live.

Eka states that she is able to access healthcare services and that the services are "good." However, she relates "bad experiences" with nurses and doctors who were "rude" to her and inconsiderate of her inability to see. Eka recalls that one time when she visited a "famous eye clinic" in Bandung for a regular checkup, the nurse seemed as though she "refused to believe" that Eka was completely blind, and she proceeded with tests that Eka could not perform. When Eka complained to the physician about the nurse's behavior, he seemed indifferent and suggested she file a complaint with the clinic.

Sometimes healthcare professionals' dismissive and negligent behavior had serious health consequences for Eka. She states that sometimes when she saw healthcare professionals for "simple conditions (like flu, for example), the doctors… did not give her the 'right' medication. She always warned the doctors that she had glaucoma and please do not give her medications that would be contradictive with her condition, but every time she took the prescribed medication, she suffered adverse effect (her eye pressure increases, which leads to headache and stomachache)."

Before 2014, Eka did not have any kind of health insurance and all her health treatments had been paid for by her parents. Eka currently lives in Bangka Belitung, South Sumatra, where she is a freelance translator and "website writer."

Conclusion

In this brief study we have noted that a significant minority of the world's population is disabled and that most disabled people live in developing countries. We have outlined the current situation in Indonesia with respect to disability and disabled people. We have shown that significant progress has been made in certain areas. Indonesia has ratified the UN CRPD and passed legislation meant to ensure the rights of people with disabilities. A number of important disabled people's organizations, non-profit organizations, and non-governmental organizations have been created in Indonesia. Significant work remains, however, in the area of changing the predominantly negative culture of disability in Indonesia. We have shown these negative attitudes in the oral interviews provided by disabled women living in Indonesia, especially as they relate to healthcare and to a lesser extent education.

While professionals such as occupational therapists and social workers have worked to empower individual disabled people and change the culture of their own work environments, and designers and architects have worked to build a more accessible built environment, we contend that global disability studies provides new avenues of intervention and new theories that will further empower disabled people, especially women and girls, in places like Indonesia. We argue that a powerful way to create more positive attitudes is through the consideration of a diverse range of life stories that include the lived experiences of disabled people. Thus, we have concluded our essay with four brief vignettes that highlight the obstacles faced by disabled women living in Indonesia, specifically in the areas of healthcare and education, and to a lesser extent housing and employment.

Questions moving forward might include: How do we build more effective communication networks among various disability communities in Indonesia? How do we bridge the significant gaps between rural and urban settings, and between middle- or upper-class Indonesians and their poor counterparts? How do we mitigate the difficulties posed by ethnic and linguistic diversity, as well as differing levels of religious conviction? What might a comparative study of disabled women and disabled men look like in Indonesia? What role can researchers play in advocating for the equal treatment of disabled people in Indonesia? And finally, how can new information that includes the thoughts and experiences of disabled people be conveyed to healthcare professionals and educators in a way that they will find useful in their daily practice?

References

1. de LaRoche D, Florentina. A qualitative investigation of the healthcare challenges of women with disabilities in Yogyakarta (Indonesia): Implication for health policy. Thesis submitted in partial fulfillment of the requirements for the degree of Master of Health Sciences, School of Health Sciences. Canterbury: University of Canterbury; 2017. p. 9.
2. Titchkosky T, Aubrecht K. WHO's MIND, whose future? Mental health projects as colonial logics. J Soc Identities. 2015;21(1):72.

3. Charlton JI. Nothing about us without us: disability oppression and empowerment. Berkeley: University of California Press; 2000.
4. Rapp R, Ginsburg F. Enabling disability: rewriting kinship, reimagining citizenship. Public Culture. 2001;13(3):533–55. (Duke University Press).
5. Tremain S. On the government of disability: Foucault, power and the subject of impairment. In: Davis LJ, editor. The disability studies reader. 2nd ed. New York: Routledge; 2006. p. 187.
6. Goodley D. Dis/ability studies: theorising disablism and ableism. Abingdon: Routledge; 2014.
7. Goodley D. Disability studies: an interdisciplinary introduction. Los Angeles: SAGE; 2011.
8. Mehrotra N. Disability, gender and state policy: exploring margins. Jaipur: Rawat Publications; 2013.
9. Japan International Cooperation Agency (JICA). Data collection survey on disability and development in Indonesia final report. Chiyoda-ku: KRI International Corporation, Tekizaitekisho, LLC; 2015. p. 3.4.

Chapter 8
Low-Tech Innovations to Prevent Neonatal Mortality: Perspectives from Public Health, Neonatology, and Biomedical Engineering

Pavani Kalluri Ram, Sara K. Berkelhamer, and Anirban Dutta

Anaya's Big Save

Anaya did not notice the sounds from the streets, or the playful voices of children outside the health center as she focused intensely on the mother-to-be in front of her. "I can see the head!" she helped explain, as she encouraged the woman in labor to take deep breaths and push. Anaya felt queasy in her excitement, running through what she needed to watch for and do. She had been studying for this for over a year, quietly standing behind her mentor, midwife Mama Safiya, as she confidently assisted at deliveries after her many years as a birth attendant. But this time it was just Anaya. And in just minutes the baby would be there.

With the mother's final push, the baby delivered with its umbilical cord wrapped around its neck. Anaya quickly and carefully unwrapped the cord and set the baby down on the blanket that she had carefully laid out along with her clean and organized equipment. She looked at the beautiful baby with dark hair and round cheeks and for a moment froze. The pale infant lay on the blanket silent, arms and legs limp at its sides. Anaya quickly began to dry the baby, aggressively rubbing its back in an effort to stimulate breathing as she had watched Mama Safiya do for many other babies who were not crying at birth. Her heart raced as she waited for a sound from the quiet infant but none came. Anaya looked down at her neatly arranged equipment and fumbled as she grabbed the funny looking bag-mask ventilator that had been given to her with the Helping Babies Breathe training that she took part in, under Mama Safiya's mentoring. Her hands shook as she placed the mask over the baby's nose and mouth, but she somehow found herself counting out loud as she

P. K. Ram (✉)
United States Agency for International Development, Washington, DC, USA
e-mail: pkram@buffalo.edu

S. K. Berkelhamer · A. Dutta
University at Buffalo, State University of New York (SUNY), Buffalo, NY, USA

© Springer Nature Switzerland AG 2020
K. H. Smith, P. K. Ram (eds.), *Transforming Global Health*,
https://doi.org/10.1007/978-3-030-32112-3_8

repeatedly squeezed the bag to deliver breaths—"Breathe, two, three, breathe, two, three, breathe two three…."

In amazement, Anaya watched as the baby's pale color improved, its small arms and legs began to move, and finally, as if to voice discontent, a muffled cry could be heard under the mask. Anaya stopped and stared as the baby gasped, took several deep breaths, and finally cried out with vigor. She felt her own tears well up as she turned and proudly announced "it's a baby girl. And she is beautiful."

Epidemiology of Newborn Survival

Mothers, fathers, and family and community members eagerly await the birth of a newborn baby the world over. Nearly 260 babies were born every minute in 2017, approximately 135 million births in total for the year. Although often a time of great joy for families, the first month of life is the most vulnerable of a child's life, with an estimated 2.6 million babies dying before reaching the 28th day of life in 2016, approximately 2% of all newborn babies. Within the first month of life, the first week and especially the first day are the most dangerous times [1].

The world has experienced a dramatic decline in global neonatal mortality (death prior to 28 days of life) in recent decades, falling 51% from about 37 deaths per 1000 live births in 1990 to 19 deaths per 1000 live births in 2016 [2]. Even though newborns everywhere are more likely to survive today than they were a decade or two ago, there are substantial global disparities in the risk of neonatal mortality. Whereas only 1 in 333 newborns die in high-income countries, the risk is much higher in sub-Saharan Africa, where 1 in 36 newborns die annually. Unlike the global neonatal mortality rate of about 2% or 20 deaths per 1000 live births, newborns in some countries face far greater hurdles in growing to older infancy. Nearly 80% of newborn deaths occur in either South Asia or sub-Saharan Africa, with 50% of all newborn deaths occurring in the following five countries: the Democratic Republic of Congo, Ethiopia, India, Pakistan, and Nigeria. Smaller countries that are especially low-resourced experience dramatically higher neonatal mortality rates than the global average: in South Asia, Afghanistan and its neighbor Pakistan (40 and 46 deaths per 1000 live births); and in sub-Saharan Africa, Chad and its neighbor, the Central African Republic (35 and 42 deaths per 1000 live births, respectively).

The major causes of neonatal mortality are well characterized, including complications associated with prematurity or low birth weight, intrapartum related events (or asphyxia), and infection. Collectively these causes account for approximately 80% of all deaths, with the residual attributed to congenital malformations (9%) and other less common causes. Notably, evidence-based interventions to address many of these conditions exist and efforts to strengthen the quality of maternal and newborn care remain a global agenda as highlighted by the *Every Newborn Action Plan* (ENAP). ENAP has been endorsed by the ministers of health of the member nations of the United Nations at the World Health Assembly and establishes ambitious global targets to reduce neonatal mortality to below 12 deaths per 1000 live births [3].

Getting a Head Start Before and During Pregnancy

Protecting a newborn from dying demands implementing interventions across the continuum of care, beginning even before conception, through pregnancy, the day of birth, and through the remainder of the vulnerable newborn period. In the *pre-conception period*, complications, as well as adverse social, economic, and educational outcomes, by delaying the first pregnancy until they are more fully physically developed [4]. Increasing folate consumption among women of childbearing age before and during the early weeks of pregnancy can reduce the risk of congenital malformations, specifically neural tube defects such as spina bifida [5].

During pregnancy, numerous interventions have been shown to reduce the risk of adverse birth outcomes. Identifying pregnancy early and estimating the gestational age are central to high-quality antenatal care. Life-threatening neonatal tetanus can be prevented by administering tetanus toxoid injections to the pregnant woman. Maternal infections, such as syphilis, must be identified and managed early and appropriately. Major global health gains have been made in reducing the risk of newborn and child mortality through prevention of mother to child transmission (PMTCT) of HIV, now a core antenatal intervention irrespective of national HIV prevalence. Promoting smoking cessation and improving maternal nutrition during pregnancy can reduce the risk of low birth weight, or weight at birth less than 2500 g, a major underlying cause of neonatal mortality [6]. Routine screening among pregnant women can lead to identification and management of potentially life-threatening pregnancy conditions, such as preeclampsia, through basic measurements such as blood pressure and protein in the urine. Throughout the antenatal period, the health provider must provide comprehensive counseling for the pregnant woman to encourage adherence to antenatal care, facility delivery, birth planning with allocation of funds, and decision-making regarding transportation to the facility for the delivery. The mother must be able to recognize danger signs in herself and her soon-to-be-born baby, so that she may seek care early for potentially life-threatening conditions. Given the complex set of interventions that must be administered in the antenatal period, the World Health Organization increased the minimum number of antenatal care visits to eight from the previously recommended four [7]. Unfortunately, coverage of antenatal care remains low, with 86% of pregnant women having at least one antenatal visit and 62% of women having at least four antenatal visits in 2010–2016 [8]. In the regions where maternal and neonatal mortality are highest, four antenatal care visits are accessed by only 52% of pregnant women in sub-Saharan Africa and 46% in South Asia.

Low-Technology, High-Impact Interventions

The majority of newborn deaths are preventable. Preventing these deaths on the day of birth or during the newborn period does not call for scaling up advanced and costly technology or clinical care services, such as those delivered in a neonatal intensive care unit. Rather, a number of low-technology solutions applied by skilled health workers, preferably in a clean and equipped health facility, have been shown to yield high impacts for prevention of neonatal mortality. Here, we discuss four such solutions: breastfeeding, thermal support, delayed cord clamping, and care of the umbilical cord stump.

The most ancient of interventions to prevent newborn and child mortality is *breastfeeding*. Extensive scientific research and decades of experience in low- and middle-income countries have confirmed that newborns and young infants who are only fed breast milk through 6 months of life, a practice referred to as "exclusive breastfeeding," are at significantly reduced risk of mortality than those who are partially breastfed or who only receive breast milk substitutes, such as infant formula. By conferring immune protection against infections and improving nutritional status, breastfeeding is estimated to reduce respiratory infection risk by one-third and diarrhea risk by one-half; reductions in severe infections that lead to hospitalization are even greater [9]. Moreover, breastfed babies benefit from improved brain development and, later in life, protection from overweight and obesity. Breastfeeding even within the first hour of life confers impressive health benefits [10]. Newborn babies that breastfeed within the first hours and days of life receive colostrum, a thin liquid replete with immunologic and other bioactive substances. Delaying breastfeeding beyond the first hour can lead to increasing the risk of neonatal mortality by 33%; delaying beyond the first day of life can increase the risk of death even further [11]. The health benefits extend well beyond the early hours and days of life, with significant reductions in infections and respiratory symptoms and promote bonding between the mother and baby. Unfortunately, despite robust evidence supporting its health benefits, there are notable gaps in early initiation of breastfeeding in many of the countries in which newborns are at high risk of dying. Globally, only about 40% of newborn babies were offered early initiation of breastfeeding in 2017. Among the regions from which data are available, the prevalence of early initiation of breastfeeding ranges from 35% in the Middle East and North Africa to 65% in Eastern and Southern Africa. Culturally grounded traditions may lead to discarding the nutrient rich colostrum in favor of pre-lacteal feeds such as water, cow's milk, or other liquids. Health workers too may inadvertently play a part in delaying breastfeeding. Commonly, birth attendants separate the newborn from the mother immediately after birth for various reasons, including a perceived need to clean the baby before reuniting him/her with mother and newborn; unfortunately, such unnecessary separation results in delaying breastfeeding.

Washing the baby immediately after birth to remove body fluids is a common practice in many cultures. However, this practice can be very harmful to newborns who do not regulate their body temperatures as well as older humans and, thus, heat loss can occur quickly. Hypothermia, or low body temperature, increases the risk for neonatal mortality, although the specific physiologic mechanisms for this increased risk are not fully understood. *Thermal care* of the newborn includes the suite of clinical practices that protect a newborn baby's body temperature from dropping too low. Drying, immediate placement of the unwrapped newborn on the skin of the mother, and delaying bathing until after the first 24 h following delivery are important elements of thermal care [12]. Increasingly, mothers in low- and middle-income countries are reporting that their newborns were dried and placed skin-to-skin immediately after delivery. But, in several countries with high neonatal mortality rates, one-third or more mothers continue to report inappropriate thermal care practices, indicating the need to continue to improve the quality of care delivered at birth [13]. Indeed, thermal care is especially important for the millions of small babies who are born weighing 2500 g or less at birth. Small babies are at increased risk of mortality because they are prone to infections and more susceptible to hypothermia, compared to larger newborn peers. For babies weighing less than 2000 g at birth, an expanded approach to thermal support called *kangaroo mother care* is especially powerful, placing the newborn in skin-to-skin care for an extended duration along with support to the mother for exclusive breastfeeding [14] (Fig. 8.1). Some mothers need support for expressing breast milk and feeding it to the newborn using a special type of cup. Kangaroo care is not relegated to mothers alone, and can be successfully delivered by fathers or other family members as well. Robust data suggest that kangaroo care promotes growth and adherence to exclusive breastfeeding, both of which protect the small newborn from infection and death. As easy as it sounds, adhering to the strict regimen of continuous and prolonged skin-to-skin contact in kangaroo care can be challenging for mothers; it can be physically uncomfortable and demands persistent encouragement and support from health workers and other family members. Relief from family members, who step in to provide care while the mother attends to her own needs for personal hygiene and

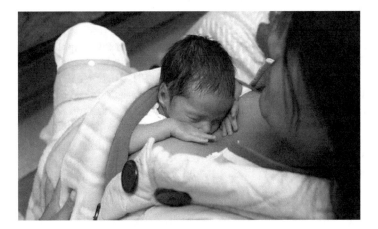

Fig. 8.1 Stages of care and examples of low-technology, high-impact interventions. (Source: concept: Pavani Ram; graphic design: Nicole C. Little)

rest, is crucial, as is continued vigilance on the part of the health providers overseeing kangaroo care.

Appropriate care of the umbilical cord can significantly impact the well-being of newborns. Whereas the fetus receives its nutrition largely via the umbilical cord that connects to the placenta, newly born babies rely on exogenous sources of nutrition, preferably from breast milk. Thus, the umbilical cord serves little purpose after delivery and is cut from the placenta soon after birth. Separation of the umbilical cord from the placenta is typically done by placing two clamps a few centimeters apart on the cord (or tying the cord using string) and cutting between the clamps using a clean blade. There is ample data suggesting that *delaying the clamping of the cord* until at least one minute after birth improves the newborn's nutritional status, especially for body stores of iron, an essential mineral that is important for blood oxygen transport [15]. This important benefit staves off iron deficiency anemia, which can predispose children to developmental delays. Further, premature infants serve to benefit from delayed cord clamping with more stable circulation, less need for transfusions and fewer complications of prematurity. Importantly, delayed cord clamping also benefits the mother, by reducing her risk of postpartum hemorrhage, the leading cause of maternal mortality [3, 16].

Infections, including those resulting from pathogenic microorganisms acquired with unhygienic *care of the umbilical cord* stump, account for approximately one-quarter of newborn deaths. The residual umbilical stump is an important portal of entry for pathogens before it withers and leaves behind the belly button. Around the world, traditional practices typically intended to promote healing result in the application of various types of substances to the umbilical stump, including oils, herbs or plants, minerals, and animal dung [17]. Contamination from the hands or from other topical applications can result in localized infections of the umbilical stump or the skin surrounding it, or evolve into systemic infectious process such as life-threatening sepsis. For many years, in an effort to prevent contamination with the application of unclean substances or contact with contaminated hands, the World Health Organization recommended dry cord care, meaning that no substances should be applied after the cord is cut. An antiseptic used routinely to clean the skin in healthcare facilities in high-income countries, chlorhexidine applied to the umbilical cord stump of newborns in rural Nepal was shown to reduce umbilical infections by 75% and neonatal mortality by 24% [18]. Together with data from the successful trial in Nepal, additional studies in Bangladesh and Pakistan showing mortality reduction benefits led the World Health Organization to issue guidance recommending that chlorhexidine should be applied to the umbilical stump for babies born outside health facilities in countries with high neonatal mortality (defined as greater than 30 deaths per 1000 live births or 3% of babies dying in the first 28 days after delivery) [19, 20]. Additional data suggesting a benefit for babies born in health facilities is also compelling, suggesting that the benefits may transcend location of birth. However, as with any new product, as straightforward or "simple" a technology as it may be, the pathway to scaling up the intervention so that it is available for and used by all the populations that would benefit is arduous. The Chlorhexidine Working Group (CWG), a collaborative partnership involving a range of institutional actors including government agencies, academia, international organizations, non-governmental organizations, and for-profit pharmaceutical companies, was formally established in 2012 after informally coalescing over approximately a decade [21]. With donor agency funding support, the CWG functioned to advocate for adoption of chlorhexidine into newborn health programs in high neonatal mortality countries and to provide technical assistance to expand the availability, quality, and implementation of chlorhexidine in these countries. By 2017, with extensive support from governments, CWG members, and donor agencies, 27 countries were reportedly advancing chlorhexidine programs to prevent neonatal mortality. However, the pathway to scale, meaning availability and utilization of chlorhexidine for all the newborns at risk in high neonatal mortality settings is long and arduous. Extensive efforts and funding support will need to be invested over many years to ensure that health workers, mothers, and family members attending the birth have access to and make appropriate use of chlorhexidine for umbilical cord care.

Quality of Care

Globally, concerted efforts have been invested to mobilize pregnant women to deliver at health facilities instead of at home in order to reduce or mitigate the harms that can arise when women deliver without access to the skilled care, equipment, and medicines available in a health facility setting. However, despite such efforts and resultant increases in facility delivery, there have not been commensurate decreases in neonatal mortality [22]. A key concern has been whether health facilities have been able to provide the quality of care necessary to prevent deaths among mothers and newborns.

Intrapartum events, meaning events that occur during labor and delivery, account for nearly one-quarter of neonatal mortality [3]. Many of these deaths occur because of perinatal asphyxia, or oxygen deprivation at the time of birth. A highly skilled birth attendant with the necessary equipment, such as Anaya or Mama Safiya in the example at the beginning of this chapter, can effectively resuscitate the asphyxiated newborn, thereby reducing the baby's risk of mortality significantly. But, if the health worker lacks the necessary skills or the equipment, the newborn faces a high risk of death or considerable brain damage from insufficient oxygen circulation. The program Helping Babies Breathe (HBB) grew out of increasing global demand for a program that would feasibly improve the skills of health providers during the critical moments at the time of birth to manage asphyxiated newborns [23]. HBB's developers constructed an intervention package that innovatively took into account low-resource contexts, in which there may only be a single healthcare worker caring for both the woman and her baby and limited available infrastructure in a frontline health facility. In addition to accessing the basic equipment necessary for resuscitation, a reusable suction device and simplified bag for ventilating the newborn, HBB training strengthens the skill level of healthcare workers to assess babies immediately

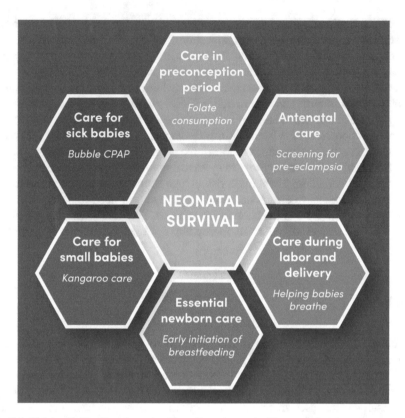

Fig. 8.2 Helping Babies Breathe action plan. (Image reused with permission from the American Society of Pediatrics. Helping Babies Breathe, 2nd Edition, 2018, Global Health: Science and Practice)

upon birth and to respond efficiently and effectively to ensure that every baby is either breathing spontaneously or receiving ventilation support within The Golden Minute after birth (Fig. 8.2). Deployment of the HBB training involves not only passive learning in the classroom but also active learning through participation in simulations of perinatal asphyxia to reinforce assessment, response, and resuscitation skills. An extensive research agenda has accompanied the rollout of HBB and has documented significant reductions in neonatal mortality. For example, neonatal deaths in the first 24 h after birth were reduced in Tanzania by approximately 47% following HBB training [24]. HBB has since become a cornerstone of training packages for strengthening the quality of care at the time of birth in low-resource settings (Fig. 8.3).

In 2018, the Lancet Global Health Commission on High Quality Health Systems in the Sustainable Development Goals Era wrote "the human right to health is meaningless without good quality care because health systems cannot improve health without it [25]." How can the quality of health care delivery be discerned? Key features of high-quality care include that it is safe, effective, timely, efficient,

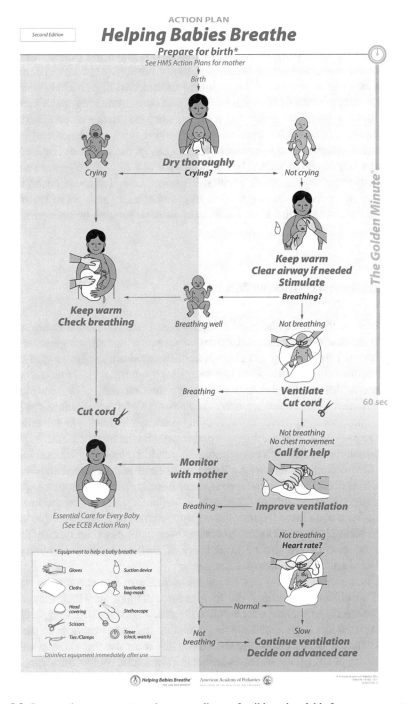

Fig. 8.3 Interventions to prevent newborn mortality are feasible and scalable from pre-conception to the end of the newborn period

equitable, and people-centered [22]. In 2016, the World Health Organization issued Standards for Improving Quality of Maternal and Newborn Care in Health Facilities. The underlying framework includes three standards about the provision of care, including the evidence-based clinical care, robust health-related data systems, and linkages between communities, lower-level facilities, and higher-level facilities to provide a strong continuum of care. Three additional standards make clear that women and newborns are entitled to a high-quality experience of care, with clear communication from health workers, respect and dignity, and emotional support from health workers and companions of choice. Lastly, the standards emphasize the import of a skilled and motivated health workforce as well as the physical resources and infrastructure of the health facility. This landmark document enshrines the dimensions of quality of care for women and newborns, irrespective of whether the facility is operating at the most basic level in communities or delivering complex care.

Multidisciplinary Innovation

In the earlier sections, we discussed evidence-based interventions to reduce neonatal mortality that did not demand extensive scientific or technological advancements. Surely, breastfeeding and placement of the newborn in skin-to-skin contact with the mother may be considered low-technology solutions. However, some newborns need higher-level technology to improve their chances of survival, growth, and development. Thus, there is a need for innovation of novel solutions applicable to the settings in which babies are still dying preventable deaths, or suffering lifelong consequences of complications of pregnancy and the time of birth.

Equipment that is affordable is essential to providing high-quality care for newborns in low-income countries. As Richards-Kortum has written, band-aid solutions, such as donations of surplus equipment from high-income countries, do not solve the gaps in availability of equipment in health facilities in low-income countries [26]. Such equipment requires accessory materials or spare parts that are difficult to replenish when stocks run low; moreover, equipment can quickly fall into disrepair because of the limited availability of technicians in the country with appropriate training in maintenance and repair.

There is, hence, a need for new technologies that take into account not only the technical specifications but also the low-resource settings in which they will be used. Considerations include cost but also the challenges to distributing technologies and related to a large and disparate population that may benefit from them; the size and capacity of the healthcare workforce that must use them; the availability and reliability of electricity, water, and other infrastructure; sociocultural motivators and barriers to adoption and use; and demonstration of compelling enough benefit to motivate change from current practice [27].

With the intent to improve the health and well-being of pregnant women and newborns, the Saving Lives at Birth (SLAB) Grand Challenge program has been funded by a multilateral cooperation between the Bill and Melinda Gates Foundation, the US Agency for International Development, Grand Challenges Canada, and the governments of the United Kingdom, Norway, and the Republic of Korea (South Korea) [28]. Between 2011 and 2018, SLAB has sparked 3500 ideas from diverse disciplines. An excellent example of the impact of SLAB in fostering multidisciplinary innovation is a low-cost bubble continuous positive airway pressure (CPAP) device to provide respiratory support for babies suffering from respiratory distress. Bubble CPAP has been shown to prevent neonatal mortality from respiratory failure in high-income countries but an individual device can cost thousands of dollars. Key members of the SLAB-funded bCPAP team included bioengineers from Rice University in the USA and pediatricians from Queen Elizabeth Central Hospital in Blantyre, Malawi. Together, this multidisciplinary team developed the Pumani bCPAP, a bCPAP device that improved survival among newborns with respiratory failure by 27% *and* reportedly cut the cost of the bCPAP device by 97% [29]. This remarkable innovation rendered an otherwise infeasible technology accessible to hospitals caring for newborns in one of the world's poorest countries [30].

Technology alone will not address the complex challenges faced by newborns in resource-limited health systems. Rather, healthcare service delivery must be strengthened throughout low-income countries to address the motivation and skill level of healthcare workers. Novel, effective, and scalable approaches are needed to motivate improved preventive and promotive health behaviors and early careseeking for pregnant women, new mothers, and newborns. Innovation is essential to developing feasible solutions to address the technological, systemic, and individual and organizational behavioral barriers to newborn survival and well-being. Such innovation demands the energy and creativity of diverse disciplinary perspectives.

An Innovation to Protect Infants at Risk of Brain Injury
Perinatal asphyxia can be caused by one or more complications, including injury to the umbilical cord or placenta, failure to breathe at birth, and prolonged difficult labor. An extended duration of oxygen deprivation can lead to hypoxic-ischemic encephalopathy (HIE), a dangerous neurological condition that requires immediate medical intervention.

Initial Resuscitation and Stabilization involves stimulation, resuscitation, and management of appropriate oxygen delivery. In resource-limited environments, Helping Babies Breathe (HBB), described in the chapter, provides the initial steps of neonatal resuscitation with assisted breaths provided by a bag and mask within one minute of birth when an infant is not spontaneously breathing. Several clinical trials indicate that resuscitation of full-term infants with perinatal asphyxia without supplemental oxygen is as effective as resuscitation with supplemental oxygen. HBB neonatal resuscitation training has been shown to reduce neonatal mortality by up to 47% and fresh stillbirths by 24%. Routine non-invasive measurement of blood oxygen concentrations, via pulse oximetry, is essential to ensure adequate oxygen perfusion to the brain and other tissues in the body.

Following standard clinical protocols, additional supportive measures to care for babies at risk for HIE include careful fluid management, avoidance of hypoglycemia and hyperglycemia, treatment of seizures, and avoidance of hyperthermia (high body temperature). Indeed, there is evidence that actively lowering the baby's body temperature (therapeutic hypothermia) can reduce the potential harmful effects of HIE.

Therapeutic hypothermia involves mild cooling (3–4 °C below baseline temperature) of the body applied within the first few hours after birth in the early stages of potential brain injury. An effective way to initiate early hypothermia therapy may include use of widely available icepacks to cool an infant while awaiting transfer to a tertiary neonatal intensive care unit. In resource-limited environments, a low-cost device powered by a microprocessor and two AAA batteries called Cooling Cure, a clay pot containing a burlap basket lined with plastic, sand, instant ice-pack powder, and temperature sensors [31]. Therapeutic hypothermia has been shown to improve the neurodevelopmental outcomes of neonates with moderate-to-severe HIE, as reported in a 2013 Cochrane review that included 11 randomized controlled trials and 1505 infants [32].

Early detection of the severity of HIE is challenging at the point-of-care in low-resource settings. The severity of HIE and the potential for benefit from therapeutic hypothermia can be determined by the clinical presentation, and can be graded using a clinical classification scale called Sarnat staging. The clinician assesses an infant's alertness, muscle tone, positioning, pupillary response, respiratory effort, and the presence of seizures to determine a Sarnat grade of mild, moderate, or severe. However, Sarnat staging is somewhat subjective, and the score can change over time with clinical progression. Also, seizures are difficult to detect clinically but are associated with a poor prognosis. Electroencephalogram (EEG) measures the electrical activity of the brain from the scalp non-invasively and assists in the evaluation of HIE. However, conventional EEG requires at least 60 min of careful handling by highly trained personnel and, in low-income countries, this diagnostic procedure is not available outside of tertiary care referral hospitals in urban areas. Therefore, a tool for continuous monitoring of brain activity that is feasibly applied in lower-level facilities or by healthcare workers with less training is needed. In the early stages of development is an innovative, portable, diagnostic approach that combines a simplified EEG approach with near infrared spectroscopy (NIRS) for detection of brain tissue viability. This technology has the potential to identify those in need of the treatment, not only in the referral hospitals that serve the very wealthy in resource-limited countries, but also in the lower-level facilities that serve most babies. This is a tangible example of technology advancing equity in global health.

References

1. Lawn JE, Cousens S, Bhutta ZA, Darmstadt GL, Martines J, Paul V, Knippenberg R, Fogstadt H, Shetty P, Horton R. Why are 4 million newborn babies dying each year? Lancet. 2004;364(9432):399–401. https://doi.org/10.1016/S0140-6736(04)16783-4.
2. Hug L, Sharrow D, Zhong K, You D, UN Inter-agency Group for Child Mortality Estimation. Levels & trends in child mortality: report 2017. New York: United Nations Children's Fund; 2017.
3. Chou D, Daelmans B, Jolivet RR, Kinney M, Say L, Plan Every Newborn Action, and groups Ending Preventable Maternal Mortality working. Ending preventable maternal and newborn mortality and stillbirths. BMJ. 2015;351:h4255. https://doi.org/10.1136/bmj.h4255.
4. Obare F, Kabiru C, Chandra-Mouli V. Reducing early and unintended pregnancies among adolescents. In: Family planning evidence brief. Geneva: Department of Reproductive Health and Research, World Health Organization; 2017.
5. Pitkin RM. Folate and neural tube defects. Am J Clin Nutr. 2007;85(1):285S–8S. https://doi.org/10.1093/ajcn/85.1.285S.
6. Grieger J, Clifton V. A review of the impact of dietary intakes in human pregnancy on infant birthweight. Nutrients. 2015;7(1):153.
7. HRP. WHO recommendations on antenatal care for a positive pregnancy experience. Geneva: World Health Organization; 2016.
8. Databases, UNICEF Global. Antenatal care. New York City: UNICEF; 2018.
9. Victora CG, Bahl R, Barros AJD, França GVA, Horton S, Krasevec J, Murch S, Sankar MJ, Walker N, Rollins NC. Breastfeeding in the 21st century: epidemiology, mechanisms, and lifelong effect. Lancet. 2016;387(10017):475–90. https://doi.org/10.1016/S0140-6736(15)01024-7.
10. Khan J, Vesel L, Bahl R, Martines JC. Timing of breastfeeding initiation and exclusivity of breastfeeding during the first month of life: effects on neonatal mortality and morbidity—a systematic review and meta-analysis. Matern Child Health J. 2015;19(3):468–79. https://doi.org/10.1007/s10995-014-1526-8.
11. Smith ER, Hurt L, Chowdhury R, Sinha B, Fawzi W, Edmond KM, Group on behalf of the Neovita Study. Delayed breastfeeding initiation and infant survival: a systematic review and meta-analysis. PLoS One. 2017;12(7):e0180722. https://doi.org/10.1371/journal.pone.0180722.
12. Library, WHO Reproductive Health. WHO recommendation on bathing and other immediate postnatal care of the newborn. Geneva: The WHO Reproductive Health Library; 2012.
13. Mallick L, Yourkavitch J, Allen C. Thermal care and umbilical cord care practices and their associations with newborn mortality. In: International Health and Development, editor. DHS Analytical Studies No. 68. Rockville: United States Agency for International Development; 2018.
14. Organization, World Health. WHO recommendations on interventions to improve preterm birth outcomes. Geneva: Department of Reproductive Health and Research; Department of Maternal, Newborn, Child & Adolescent Health; 2015.
15. Organization, World Health. Guideline: delayed umbilical cord clamping for improved maternal and infant health and nutrition outcomes. Geneva: WHO; 2014.
16. Practice, American College of Obstetricians and Gynecologists' Committee on Obstetric. Delayed umbilical cord clamping after birth. In: Committee opinion no. 684. Washington, DC: American College of Obstetricians and Gynecologists; 2017.
17. Coffey PS, Brown SC. Umbilical cord-care practices in low- and middle-income countries: a systematic review. BMC Pregnancy Childbirth. 2017;17(1):68. https://doi.org/10.1186/s12884-017-1250-7.
18. Mullany LC, Darmstadt GL, Khatry SK, Katz J, LeClerq SC, Shrestha S, Adhikari R, Tielsch JM. Topical applications of chlorhexidine to the umbilical cord for prevention of omphalitis

and neonatal mortality in southern Nepal: a community-based, cluster-randomised trial. Lancet. 2006;367(9514):910–8. https://doi.org/10.1016/S0140-6736(06)68381-5.

19. Mullany LC, El Arifeen S, Winch PJ, Shah R, Mannan I, Rahman SM, Rahman MR, et al. Impact of 4.0% chlorhexidine cleansing of the umbilical cord on mortality and omphalitis among newborns of Sylhet, Bangladesh: design of a community-based cluster randomized trial. BMC Pediatr. 2009;9:67. https://doi.org/10.1186/1471-2431-9-67.

20. Soofi S, Cousens S, Imdad A, Bhutto N, Ali N, Bhutta ZA. Topical application of chlorhexidine to neonatal umbilical cords for prevention of omphalitis and neonatal mortality in a rural district of Pakistan: a community-based, cluster-randomised trial. Lancet. 2012;379(9820):1029–36. https://doi.org/10.1016/S0140-6736(11)61877-1.

21. Coffey PS, Hodgins S, Bishop A. Effective collaboration for scaling up health technologies: a case study of the chlorhexidine for umbilical cord care experience. Glob Health Sci Pract. 2018;6(1):178–91. https://doi.org/10.9745/ghsp-d-17-00380.

22. Tunçalp Ö, Were WM, MacLennan C, Oladapo OT, Gülmezoglu AM, Bahl R, Daelmans B, et al. Quality of care for pregnant women and newborns—the WHO vision. BJOG. 2015;122(8):1045–9. https://doi.org/10.1111/1471-0528.13451.

23. Niermeyer S. From the Neonatal Resuscitation Program to Helping Babies Breathe: Global impact of educational programs in neonatal resuscitation. Semin Fetal Neonatal Med. 2015;20(5):300–8. https://doi.org/10.1016/j.siny.2015.06.005.

24. Msemo G, Massawe A, Mmbando D, Rusibamayila N, Manji K, Kidanto HL, Mwizamuholya D, Ringia P, Ersdal HL, Perlman J. Newborn mortality and fresh stillbirth rates in Tanzania after helping babies breathe training. Pediatrics. 2013;131(2):e353–60. https://doi.org/10.1542/peds.2012-1795.

25. Kruk ME, Gage AD, Arsenault C, Jordan K, Leslie HH, Roder-DeWan S, Adeyi O, et al. High-quality health systems in the Sustainable Development Goals era: time for a revolution. Lancet Glob Health. 2018;6(11):e1196–252. https://doi.org/10.1016/S2214-109X(18)30386-3.

26. Richards-Kortum R. Tools to reduce newborn deaths in Africa. Health Aff. 2017;36(11):2019–22. https://doi.org/10.1377/hlthaff.2017.1141.

27. Howitt P, Darzi A, Yang G-Z, Ashrafian H, Atun R, Barlow J, Blakemore A, et al. Technologies for global health. Lancet. 2012;380(9840):507–35. https://doi.org/10.1016/S0140-6736(12)61127-1.

28. Birth, Saving Lives at. 2019. Innovators. https://savinglivesatbirth.net/innovatorshome.

29. Kawaza K, Machen HE, Brown J, Mwanza Z, Iniguez S, Gest A, O'Brian Smith E, Oden M, Richards-Kortum RR, Molyneux E. Efficacy of a low-cost bubble CPAP system in treatment of respiratory distress in a neonatal ward in Malawi. PLoS One. 2014;9(1):e86327. https://doi.org/10.1371/journal.pone.0086327.

30. Birth, Saving Lives at. 2011. Low-cost respiratory support, reducing early neonatal death in rural Malawi. https://www.youtube.com/watch?v=2tVjo7ejR8M.

31. Sneiderman P. Low-cost 'cooling cure' could avert brain damage in oxygen-starved babies. Baltimore: Office of Communications: Johns Hopkins University; 2013.

32. Jacobs SE, Berg M, Hunt R, Tarnow-Mordi WO, Inder TE, Davis PG. Cooling for newborns with hypoxic ischaemic encephalopathy. Cochrane Database Syst Rev. 2013;1 https://doi.org/10.1002/14651858.CD003311.pub3.

Chapter 9
The Global Crisis of Antimicrobial Resistance: Perspectives from Medicine, Geography, Food Science, and Chemistry

Shamim Islam, Jared Aldstadt, David G. White, and Diana Aga

Introduction

Antimicrobials are perhaps the greatest public health gain ever of humankind. As 2000 approached, many claimed penicillin to be the greatest discovery of the millennium; commercially available only from the early 1940s, the drug had already saved an estimated 200 million lives [1].

Currently, however, the emergence and spread of multidrug-resistant pathogens threatens to reverse these benefits, and experts believe we are in the dawn of a post-antibiotic era. An estimated two million antibiotic resistant (AR) infections occur annually in the USA, accounting for 23 thousand deaths, $20 billion in direct costs, and an additional $35 billion in lost productivity [2]. The global projections are alarming: AR infections may cause ten million deaths/year and lead all causes of mortality by 2050 [3]. Similar to other pressing global issues, the burden of AR is inequitably distributed, with 90% of AR infections occurring in Asia and Africa. Various forces—natural and anthropogenic—contribute to AR, and successful containment will demand coordinated efforts of diverse constituents, from environmental, behavioral, and medical scientists, to public health educators, and leaders in agricultural and international policy [4].

S. Islam (✉) · J. Aldstadt · D. Aga
University at Buffalo, State University of New York (SUNY), Buffalo, NY, USA
e-mail: sislam@upa.chob.edu

D. G. White
University of Tennessee, Knoxville, TN, USA

© Springer Nature Switzerland AG 2020
K. H. Smith, P. K. Ram (eds.), *Transforming Global Health*,
https://doi.org/10.1007/978-3-030-32112-3_9

Lessons in Antibiotic Resistance from the Earliest Antibiotics and *Staphylococcus aureus*

Antibiotics, which are active against bacterial pathogens, are the most important and widely used class of antimicrobials. The sulfonamides and penicillin were the first antibiotics available to the general public. The serendipitous discovery of penicillin in Alexander Fleming's laboratory in 1928 is now common lore, but the drug's tremendous potential was not truly realized until World War II, when it was found to cure wound infections in soldiers. Penicillin soon after proved effective against a number of other major infectious diseases of the time, such as syphilis and rheumatic fever.

However, penicillin's original miracle drug property, against *Staphylococcus aureus*, was largely lost within 30 years of the antibiotic's introduction. *S. aureus*, a nearly ubiquitous germ which lives on people's skin, causes a wide spectrum of infections outside of those on the battlefield, including boils, pneumonia, and endocarditis. Penicillin-resistant (PCN-R) *S. aureus* was first described in 1942, and with the drug's widespread use, followed a common pattern for antibiotic resistance: initially proliferating in hospitals and then spreading to and within the community [5]. By the late 1960s, greater than 80% of all *S. aureus* were PCN-R. As a response to PCN-R *S. aureus,* methicillin, a semisynthetic penicillin, was developed in 1960; within 2 years, methicillin-resistant *S. aureus* (MRSA) was recognized, and currently, MRSA is among the most aggressive and challenging-to-treat pathogens. The introduction of each new, "more powerful" antibiotic—ciprofloxacin, linezolid, daptomycin—has replicated this sequence of prompt emergence and expansion of drug-resistant *S. aureus* (Fig. 9.1).

The sulfonamides were actually put into clinical use before penicillin, first in 1935. They are most frequently found in drug combinations, primarily with trimethoprim as TMP-SMZ. Sulfonamides intrinsically have broad antimicrobial properties, against common bacterial pathogens, malaria (a leading global parasitic disease), and also opportunistic infections seen in AIDS patients. The antibiotic's varied therapeutic applications, along with low cost and worldwide availability, have made TMP-SMZ a heavily utilized drug in several low- and middle-income country (LMIC) contexts, including empiric treatment for diarrhea, and routine prophylaxis for HIV patients. But ongoing, generalized use of TMP-SMZ has predictably come with the double edged sword of increasing antibiotic resistance; the drug is now largely obsolete for bacterial gastroenteritis and lung infections. In addition, many of the greater than 20 known TMP-SMZ resistance mechanisms are easily shared between bacteria, and can co-migrate with genes encoding resistance to other antibiotics on mobile genetic elements [6].

Antibiotic Resistance: Biological Roots and Basic Mechanisms

Bacteria may resist antibiotic activity by several mechanisms, including altering the antibiotic's binding sites, expelling the drugs with cellular pumps, and degrading the drugs by enzymes. Resistance mechanisms can develop naturally through

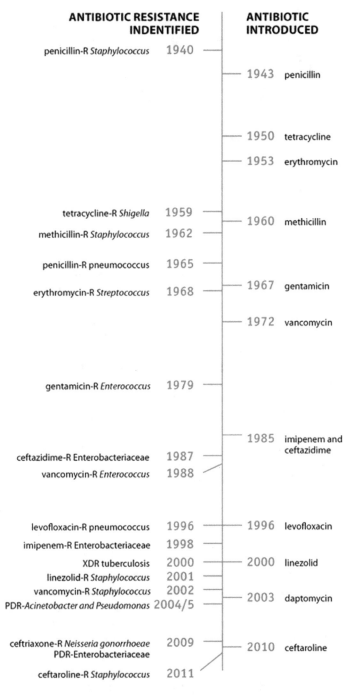

Fig. 9.1 An antibiotic resistance timeline. (Source: Centers for Disease Control and Prevention, Office of Infectious Diseases, Antibiotic Resistance Threats in the United States, 2013, page 28. https://www.cdc.gov/drugresistance/pdf/ar-threats-2013-508.pdf)

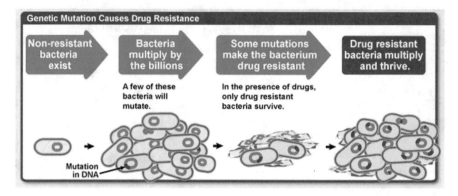

Fig. 9.2 Antibiotic resistance development by genetic mutation, and expansion after antibiotic exposure. (Source: National Institute of Allergy and Infectious Diseases, NIH, 2010)

genetic mutations, which inevitably occur during microorganisms' frequent rounds of replication (Fig. 9.2). As such, resistance may be present within bacteria independent of any antibiotic exposure, and genetically derived AR mechanisms have been detected in archaic environmental samples (DNA from 30,000-year-old permafrost). With antibiotic exposure—or "pressure" in a Darwinian evolutionary framework—populations of resistant-bacteria are specifically selected for and can expand.

Antibiotic resistant genes (ARGs) can be passed not only to an organism's own progeny, but also between microorganisms, which is known as horizontal gene transfer. An important mechanism of horizontal transfer is via plasmids, rings of DNA on which ARGs can fluidly be exchanged between bacteria of the same and differing species (as shown in Fig. 9.3). Multiple ARGs may collect on the same plasmid and co-migrate, rendering a recipient bacteria multidrug resistant. Most of the specific bacterial resistance issues focused upon in the following sections are plasmid associated and have the capacity for rapid dissemination [4].

Contemporary Antibiotic Resistance

Current leading AR issues echo the historical features described previously, but are more complex and dire. Of numerous potential examples, recent rises in resistant-*Neisseria gonorrhea* are already having a substantial impact on disease treatment and control, not only in low- and middle-, but also high-income countries. Gonorrhea, a sexually transmitted disease (STD) affecting ~80 million persons/year globally, may silently be transmitted between individuals who initially have few symptoms, but can also cause significant morbidity, such as pelvic inflammatory disease and infertility [7]. Gonorrhea has demonstrated progressive resistance to basic, oral antibiotics, to the extent that international guidelines now suggest empiric treatment with two different agents. The regimen includes an intramuscular antibiotic (ceftriaxone),

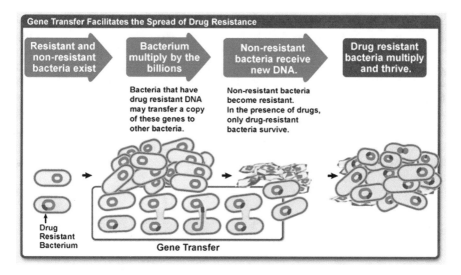

Fig. 9.3 Antibiotic resistance spread via plasmid-mediated gene transfer between bacteria. (Source: National Institute of Allergy and Infectious Diseases, NIH, 2010)

which to administer in the resource-limited, public health settings in which STD patients typically present, is cost-prohibitive and operationally challenging.

Capturing the greatest attention from the medical community currently are newer resistance mechanisms against several broad-spectrum antibiotics; each are known by acronyms, which reflect the antibiotic classes involved. ESBLs, or extended-spectrum beta-lactamases, are a large family of enzymes that confer resistance to nearly all penicillins and cephalosporins. CREs are carbapenem-resistant *Enterobacteriaceae*. Carbapenems are considered the most powerful antibiotics, and are highly active against numerous, life-threatening pathogens. CRE infections have been associated with various healthcare settings, including nursing homes and community hospitals, and have carried mortality rates of 50%. Of CREs, NDM-1 (New Delhi metallo-beta-lactamase 1) is the most infamous; first described in a Swedish national who acquired a urinary tract infection in India in 2008, NDM-1 soon after was identified in several other countries, mainly in patients with foreign healthcare exposure [8].

Colistin, a dated antibiotic with significant side effects, has recently been revived as a drug of last resort, particularly for CRE infections which have no other thera-peutic options. In 2015, the first mobilized colistin-resistance gene, *mcr*-1, was first described in a retail-meat producing pig in China. The emergence of the "*mcr*'s" (which by 2019 is up to *mcr*-9) is suspected to be related to the widespread use of colistin in animal agriculture; the resistance genes have now been reported in live-stock and food samples throughout Asia and other parts of world [9]. Most worri-some colistin-resistance genes can be found in bacterial isolates causing human infection, which portends the potential for pathogens that are resistant to all avail-able antibiotics (i.e., pan-resistant).

It should be noted the resistance issues focused upon here are primarily related to just one category of microorganisms—gram-negative bacteria—while pressing antimicrobial resistance issues exist across the span of globally important pathogens. For example, though various antibiotics are active against MRSA, these are often more expensive, more toxic, and less effective than the simpler drugs which treat susceptible *S. aureus*. Tuberculosis (TB) is a leading global killer that requires prolonged, combination therapy, and drug-resistant TB is endemic to several regions of the world with limited healthcare infrastructure. Individual treatment and global disease control efforts for malaria and HIV, parasitic and viral diseases, respectively, have been complicated by evolving drug resistance throughout the past few decades.

Keys Drivers in the Expansion of Antibiotic Resistance

The drivers of AR are diverse, complex, and at times interconnected. Some are directly related to human healthcare, but many are independent. Antibiotic misuse in formal medical settings is an important component, but international commerce and travel may be greater drivers. AR pathogens can be spread within hospitals and clinics, but public food and water sources are higher potential vehicles. The selection and propagation of AR in the environment, especially related to agriculture and waste management, are critical but incompletely defined areas. Clinicians and the general public's perception on the role, risks, and benefits of antibiotics, deeply influences antibiotic seeking and utilization behavior, which contributes to AR issues.

Resistance Related to Antibiotic Use in Human Healthcare and Medical Facilities

Antibiotics are among the most frequently prescribed and utilized drugs. The majority of hospitalized patients in the USA will receive at least one antibiotic during their inpatient course; for some—children, those in intensive care units (ICUs)—the rate can be as high as 90% [10]. Nevertheless, antibiotics are often administrated in scenarios for which there is no ultimate evidence of a bacterial infection, and an estimated 20–50% of all antibiotic use is considered to be inappropriate or unnecessary [11].

Hospital data have clearly demonstrated that antibiotic resistance patterns are directly associated with antibiotic utilization. In general terms, AR infections are consistently greater where and when antibiotics are more intensely used. AR pathogens are more prevalent in hospital-acquired infections (HAIs) than those found in the community, and within a hospital, the locations with the highest antibiotic utilization have the highest AR infection rates (e.g., ICUs). Temporal findings are similarly indicative: when ceftriaxone-resistance in *Enterobacteriaceae* infections increased from 10 to 27% at a French hospital, a review revealed that ceftriaxone utilization had also tripled during the same 4 years [12]. In individual patients, and at a population level, use of the quinolone antibiotics directly increases the risk for subsequent carriage and infection with resistant gram-negative bacteria, and also MRSA [13].

While making direct correlations is difficult, outpatient antibiotic utilization likely has a larger impact on AR than in-hospital use. In high-income countries (HIC), up to 90% of antibiotic use is in the ambulatory setting [14]. In the USA, roughly 20% of all pediatric and 10% of adult clinic visits result in antibiotic prescriptions. Numerous studies have demonstrated antibiotics are often prescribed for illnesses for which they are not indicated, such as influenza, and other non-bacterial respiratory conditions. A recent US pediatric analysis estimated ten million such inappropriate antibiotic courses are given per year [15]. Also, even for diagnoses which mandate antibiotics, young children frequently receive broader-spectrum agents than medical society endorsed guidelines would recommend, potentially due to convenience.

Outpatient antibiotic use in LMIC is likely even more critical to global AR expansion. The world's largest consumer of antibiotics for human health are India and China; in 2010, in billions of units, their antibiotic consumption was 12.9 and 10.0, respectively, compared to 6.8 in the USA [16]. While formal data are limited, antibiotic use in LMICs is further skewed towards the outpatient sector than in HIC, as healthcare delivery is yet more community, rather than hospital, based. In addition to upper respiratory presentations ("colds"), outpatient antibiotics are frequently inappropriately used for simple diarrhea in LMICs.

Several factors may particularly facilitate AR in LMIC. Regulations and oversight of drugs are often limited: physicians can receive substantial rewards for their prescribing practices from drug companies and pharmacies, and nearly all antibiotics can be obtained over-the-counter without a prescription. Very broad-spectrum agents, which are not approved elsewhere, may come in oral form and be used relatively unrestricted by the general population. For example, faropenem, an oral antibiotic structurally similar to the carbapenems, had a 150% increase in consumption in India in the 5 years after being introduced [17]. Also, the availability (and intake) of inferior quality drugs, and also fixed-dose antibiotic combinations which have not been adequately evaluated, can promote AR pathogen selection. Lastly, suboptimal dosing and failure to complete outpatient antibiotics courses can allow AR organisms to persist, an issue which is relevant also in HIC.

As antibiotic use drives resistance, it follows that healthcare settings where antibiotics are most intensely used are centers for AR pathogen presence and dissemination. Several important AR pathogens, such as MRSA, vancomycin-resistant *Enterococci* (VRE), and *Acinetobacter*, are notorious for persisting in hospitals, on linens, medical equipment, and in water sources. Basic infection prevention (IP) measures by medical staff—especially consistent hand hygiene—along with attentive environmental cleaning, can be highly effective in preventing the spread of AR between patients and hospital wards; however, even in high-resource settings, adherence to such practices can be suboptimal and hospital-acquired infections are significant.

Low-resource inpatient settings, where overcrowding of patients, inadequate staffing, and limited clean water and IP knowledge are common, create a theoretical perfect storm for AR dissemination. While data is limited, based on an analysis of estimates from South Africa, hospital-acquired infections (HAI) rates could occur in 15% of hospitalized patients in LMICs, and HAIs may be among the most important causes of mortality [18].

Finally, while AR transmission in outpatient settings lacks extensive study, it is well documented to occur. For example, MRSA acquisition has been linked to dialysis and cystic fibrosis clinics [19]. Again, characteristics typical of outpatient facilities in developing countries—a high volume and density of patients, many with contagious symptoms and prolonged waiting times—are particularly conducive to AR contamination and propagation.

Antibiotics in Agriculture and Environmental Contributors to Resistance

The volume of antibiotics used in agriculture well exceeds that related to human healthcare. 70 to 80% of the total volume of antibiotics sold in the USA is used for food animal production; in LMIC, the disproportion towards agriculture is even more pronounced. Antibiotics have long been used as a growth promoter, which traditionally has underlain the bulk of its administration to animals (rather than for disease treatment). Notably, the growing demand for protein food by economically ascending middle classes in LMICs, especially in Asia, is fueling a rapid increase in antimicrobial use in global agriculture. By 2030, an astonishing one-third of all antibiotics produced in the world are projected to be consumed by livestock in China [20].

Agricultural antibiotic use may expand global AR by various mechanisms, including the promotion of AR reservoirs in animals and the environment, and via direct transmission to humans. Robust studies from the USA and Europe demonstrate that livestock given feed with antibiotics are prone to asymptomatically harbor AR bacteria (ARB), in their intestines and feces [21]. These ARB may then enter the food-chain and be acquired by humans, resulting in either immediate illness, or again latent carriage. Individuals who carry, or are "colonized" with ARB, may pass them to other persons or spaces, and also have the risk of later being personally infected by the organisms.

Foodborne outbreaks with drug-resistant enteric pathogens, such as *Salmonella*, *Campylobacter*, and *E. coli*, are not uncommon. In a striking example, urinary tract infections occurring in six states between 1999 and 2000, caused by a distinct clone of multidrug-resistant *E. coli* known as CgA, was suspected to be related to a common food source [22]. Sophisticated molecular analysis found that the specific CgA *E. coli* detected in the human infections was highly similar (94% identical by pulse gel electrophoresis) to the *E. coli* found in a single cow sampled in 1998. In a university community where the MDR-*E. coli* infections had occurred, 40% of sampled women were found to be intestinally colonized with the specific CgA *E. coli* at some point during the 6 months of testing (Figs. 9.4 and 9.5).

Outside of the direct connection to humans via food, agricultural practices have a profound impact on AR in our natural environment. Food animals' ARB-containing fecal waste may be used as land-fertilizer, and otherwise reach bodies of water, from which they can further disperse. In one US study, concentrations of fecal ARB were up to 30 times higher in surface and groundwater downstream vs. upstream of a swine production facility [21]. Transmissible ARGs themselves may be exchanged, disseminate, and persist in environmental compartments.

Also, drug metabolites and residue from agricultural antibiotics may select for AR bacterial populations in the natural environment. Non-antibiotic chemicals, including nitrogen fertilizers and heavy metals, may similarly influence ARG and microorganism levels [9]. These last impacts are of particular concern in the rapidly growing global aquaculture sector, in which antibiotics and chemicals are frequently administered. Fish are known to be poor metabolizers of anti-infective agents; as a result, most antibiotics utilized in aquaculture are likely to remain in the environment [23]. In addition, fisheries may use open caged systems, in which farmed water fluidly flows into surrounding natural water bodies, and contamination can potentially be far reaching (Fig. 9.6).

Fig. 9.4 Poultry farm. (Source: Chicken Profile, Laura Hadden, 2002, CC BY 2.0. Image: www.flickr.com/photos/raucousrage/47614557. License: https://creativecommons.org/licenses/by/2.0/)

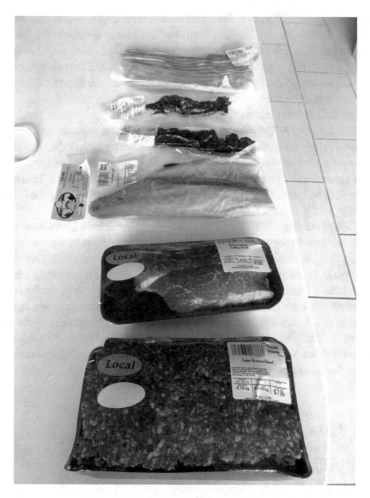

Fig. 9.5 Retail meats. (Source: Alan Levine, 2014, CC BY 2.0. Image: www.flickr.com/photos/
cogdog/15544154067. License: https://creativecommons.org/licenses/by/2.0/)

Independent of agriculture, water and waste are important environmental sources
of antibiotic resistance, especially in LMIC. NDM-1 (the plasmid-mediated
carbapenem-resistance mechanism) has been found in several environmental and
tap water sources in South Asia [24, 25]. Of great concern, sewage treatment centers
in much of the world lack the technology—and also regulatory oversight—to ensure
ARGs and ARB are effectively removed from municipal water supplies [26].
Investigations of areas near such wastewater treatment plants (WWTPs), and also
surrounding major hospitals in the developing world, frequently detect multidrug
resistant organisms and ARGs at high concentrations. Effluents from pharmaceutical and other industries in LMICs may also impact AR in the environment.

Fig. 9.6 Fisheries in Laguna Lake, the Philippines. The 90,000-hectare Laguna Lake, the largest in-land lake in the Philippines, has 13,000 fishpens and fishcage operations. These fisheries are artificial and stationary water enclosures for the culture of fish and other aquatic animal species, made up of bamboo poles, wood, screen, and other construction materials, are arranged to prevent the escape of fish. In addition to being used by the aquaculture industry, the lake also serves as source of domestic water, irrigation water, and food for the growing duck industry. It is also a recreational area and tourist spot

Perspectives from the Field: Environmental Chemistry
Conventional treatment technologies used in municipal and hospital WWTPs are not designed to remove organic contaminants such as antibiotics, which allows their release into surface waters [27]. Similarly, disinfection with chlorination or ultraviolet radiation is ineffective in eliminating antibiotic residues in effluents.

Influent and effluent WWTP samples from several nations were recently analyzed. Antimicrobial levels were generally higher in samples coming from lower income countries, such as Hong Kong and the Philippines, relative to the samples from Sweden, Switzerland, and the USA. Also, the differences in profiles reflected drug consumption patterns; e.g., influent and effluent samples had strikingly high levels of quinolone antibiotics in Hong Kong, and reached concentrations over 95,000 ng/L. [28] Antibiotic concentrations from Asian LMIC were consistently very high. In China, multiple antibiotics were around 900 ng/L, and 1066 ng/L was found in a sewage-impacted urban canal in the Philippines, compared to <200 ng/L in US and Swedish samples.

Humans as Critical Reservoirs and Vehicles in Antimicrobial Resistance

As alluded to, animals and humans may be asymptomatic carriers of ARBs. The human gut is densely populated with diverse microorganisms: the colon typically has >100 bacterial species at 10^{12} bacteria/gram of content [29]. Organisms which may harbor AR mechanisms, and also be pathogenic, such as *E. coli* and other *Enterobacteriaceae*, are normal members of intestinal microbiota. The human GI tract has been described as "an open system, which every day encounters a myriad of bacterial acquisitions originating from the environment (e.g. from food, water, soil, and other humans and animals)" [30]. As such, the human gut is an important locus for ARG exchange and endurance, and also ARB selection—particularly in the face of direct antibiotic exposure [29, 30].

Humans can acquire ARB with new interpersonal and environmental contact. In one study, a quarter of French families became colonized with the same AR enteric bacteria as their recently adopted children [31]. In particular, international travel has proven to be a significant risk factor for carriage and infection with ESBL-producing bacteria. Stool investigations of Scandinavian travelers have found that greater than 20% acquired ESBL-producing bacteria during trips abroad; rates were the highest for those visiting Asia, and up to 80% for those who took antibiotics for diarrhea during their journey [32]. Perhaps most concerning, ESBL intestinal colonization rates are already strikingly high in many parts of the world: in South East Asia, community rates of 50–80% are typically found, and the densely populated region is estimated to have over 1.1 billion ESBL carriers [33]. In many senses, the increasing interconnectivity of the world—particularly by food and people—potentiates rapid and widespread AR dissemination (Fig. 9.7).

Combatting Antibiotic Resistance: Antibiotic Stewardship

Increasing conscientiousness and limiting antibiotic use are fundamental to contain antibiotic resistance. In human medicine this falls under the umbrella of antibiotic stewardship, which may be broadly characterized as "optimizing antimicrobial selection, dosing, route, and duration of therapy to maximize clinical cure, while limiting unintended consequences, such as the emergence of resistance, adverse drug events, and cost." [34] Stewardship programs employ various strategies, including guidelines on appropriate antibiotic use, need for prior approval on specific broad-spectrum agents, and regular monitoring of antibiotic use with direct feedback to clinicians.

Antibiotic stewardship efforts have proven to be effective, but ongoing experience has revealed important challenges [13]. In one large medical system, an 80% decline of cephalosporin use through active institutional restriction led to a 44% decrease in ESBL-producing *Klebsiella*. A recent analysis of children's hospitals with formal antibiotic stewardship programs (ASPs) found they had a 6% monthly decrease in overall antibiotic use [35]. However, the study also demonstrated that in the absence of such ASPs, use of new, broad-spectrum agents tends to steadily increase, as much as 15% per year. In a striking outpatient example, when community pediatricians were given a 1-h educational session, followed by a year of monitoring and direct feedback on their individual prescription patterns, broad-spectrum antibiotic utilization was cut in half; however, within 18 months of study completion, the rates of inappropriate prescribing rebounded to above the pre-intervention levels [36].

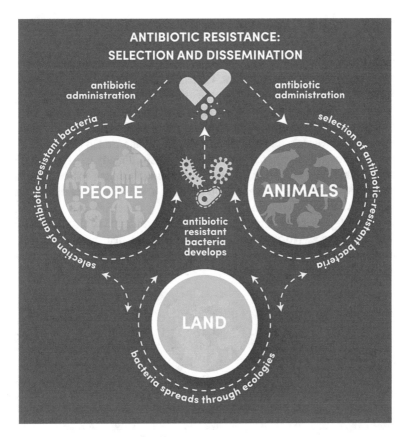

Fig. 9.7 Interrelations between humans, land, and animals in the development of antibiotic resistance. (Source: Shamim Islam and Nicole C. Little)

Healthcare providers clearly have a strong inclination to use antibiotics, and to achieve meaningful reduction, oversight needs to be comprehensive and sustained. Patient's awareness on antibiotics roles and potential risks, and their expectations on receiving them in clinic visits, also strongly influence antibiotic utilization. Providers are more likely to prescribe antibiotics when they believe their patients are expecting them, even when this has not been communicated, or the perception is incorrect [37]. Providers and patients' both are often unaware of the rate of adverse effects of antibiotics, which in everyday clinical scenarios, may exceed the potential benefit. For example, for upper respiratory presentations, antibiotics are estimated to prevent a serious complication in only 1 in 4000 cases; in contrast, the risk of presenting to the ED for a drug reaction is 1/1000, and for experiencing antibiotic-associated diarrhea, up to 25% [38]. Greater attention to understanding and modifying health-seeking behaviors, along with improved education to the general public on antibiotics, are all critical to impacting AR.

Surveillance as a Cornerstone of AR Containment Activities

Systematic monitoring and international political commitment are perhaps the most essential pillars for global AR control. Collecting standardized data and examining trends from surveillance systems is an important foundation for understanding and communicating the magnitude of, and responding to, AR issues. In 2015 the World Health Organization launched a Global Action Plan (GAP) to standardize AR surveillance, which received formal political endorsement by the United Nations in 2016.

Robust national and regional surveillance systems already exist in Europe and other areas, and the newly formed Global Antimicrobial Resistance Surveillance System (GLASS) partners with and supports these to capture clinical and epidemiological data. However, the commitment and capacity to perform AR surveillance in LMIC varies considerably. Current global AR surveillance is also challenged by defining what exactly to measure and track—indices that are meaningful from human, animal, and planetary health perspectives, and that also capture the potential impact of AR at health, economic, and societal levels [39].

Perspectives from the Field: Food and Veterinary Science

The development of integrated surveillance systems for tracking antimicrobial resistance among foodborne bacteria is a critical component of AR control. One such program is the US National Antimicrobial Resistance Monitoring System (NARMS), a collaborative network of the Food and Drug Administration (FDA), CDC, Department of Agriculture, local public health departments, and several universities. NARMS monitors changes in susceptibility/resistance of select zoonotic bacterial pathogens and commensal organisms, recovered from food producing animals, retail meats and humans, to antimicrobials of human and veterinary importance.

One important example of how active surveillance influenced regulatory policy was the identification of emerging resistance to third generation cephalosporins among *E. coli* and *Salmonella* in Canada and the USA. Between 2003 and 2008, the Canadian Surveillance Program (CIPARS) identified a strong correlation between ceftiofur-resistant *Salmonella* Heidelberg isolated from retail chicken and the incidence of ceftiofur-resistant *S.* Heidelberg human infections [40]. In Québec, ceftiofur resistance changes in chicken *S.* Heidelberg and *E. coli* isolates appeared related to changing levels of ceftiofur use in chicken hatcheries, from highest to lowest levels before and after a voluntary withdrawal, to increasing levels after reintroduction of ceftiofur use. These events provided strong evidence that ceftiofur use in chicken hatcheries selected for extended-spectrum cephalosporin resistance in foodborne bacteria recovered from chicken and ill humans. Contemporaneously, NARMS observed a similar increase in ceftiofur resistance among certain strains of *Salmonella* spp. recovered from poultry, retail chicken, and humans, which led the FDA (in 2012) to prohibit extralabel cephalosporin use in various livestock, including the practice of in ovo (egg embryo) chick injections.

There are numerous additional areas critical to combatting AR, including better understanding the environmental dimensions of AR, the development of enhanced, accessible diagnostics for ARB and ARGs, and also improving mitigation technologies. Such and all AR efforts are best approached through a One Health framework of "collaborat[ion] of multiple disciplines—working locally, nationally, and globally—to attain optimal health for people, animals, and our environment [41].

References

1. New World Encyclopedia. Alexander Fleming. 2013. http://www.newworldencyclopedia.org/entry/Alexander_Fleming. Accessed 5 Mar 2018.
2. Centers for Disease Control and Prevention. 2017. https://www.cdc.gov/drugresistance/about.html. Accessed 5 Mar 2018.
3. O'Neill J. Review on antimicrobial resistance antimicrobial resistance: tackling a crisis for the health and wealth of nations. 2014. https://amr-review.org/sites/default/files/AMR%20Review%20Paper%20-%20Tackling%20a%20crisis%20for%20the%20health%20and%20wealth%20of%20nations_1.pdf. Accessed 5 Mar 2018.
4. Holmes AH, Moore LSP, Sundsfjord A, et al. Understanding the mechanisms and drivers of antimicrobial resistance. Lancet. 2016;387:176–87.
5. Lowy FD. Antimicrobial resistance: the example of *Staphylococcus aureus*. J Clin Investig. 2003;111:1265–73.
6. Huovinen P. Resistance to trimethoprim-sulfamethoxazole. Clin Infect Dis. 2001;32:1608–14.
7. Wi T, Lahra MM, Ndowa F, et al. Antimicrobial resistance in Neisseria gonorrhoeae: global surveillance and a call for international collaborative action. PLoS Med. 2017;14:e1002344.
8. Mochon AB, Garner OB, Hindler JA, et al. New Delhi Metallo-β-lactamase (NDM-1)-producing Klebsiella pneumoniae: case report and laboratory detection strategies. J Clin Microbiol. 2011;49:1667–70. https://doi.org/10.1128/JCM.00183-11.
9. Sarah Z. Resistance to the antibiotic of last-resort is silently spreading. The Atlantic; 2017. https://www.theatlantic.com/health/archive/2017/01/colistin-resistance-spread/512705/. Accessed 5 Mar 2018.
10. Gerber JS, Newland JG, Coffin SE, et al. Variability in antibiotic use in Children's hospitals. Pediatrics. 2010;126:1067–73.
11. Centers for Disease Control and Prevention. 2017. https://www.cdc.gov/antibiotic-use/healthcare/implementation/core-elements.html. Accessed 5 Mar 2018.
12. Conus P, Francioli P. Relationship between ceftriaxone use and resistance of Enterobacter species. J Clin Pharm Ther. 1992;17:303–5.
13. Patterson DL. "Collateral damage" from cephalosporin or quinolone antibiotic therapy. Clin Infect Dis. 2004;38(Suppl 4):S341–5.
14. Sanchez GV, Fleming-Dutra KE, Roberts RM, Hicks LA. Core elements of outpatient antibiotic stewardship. Morb Mortal Wkly Rep Recomm Rep. 2016;65:1–12.
15. Hersh AL, Shapiro DJ, Pavia AT, Shah SS. Antibiotic prescribing in ambulatory pediatrics in the United States. Pediatrics. 2011;12:1053–61.
16. Laxminarayan R, Chaudhury RR. Antibiotic resistance in India: drivers and opportunities for action. PLoS Med. 2016;13:e1001974.
17. Gandra S, Klein EY, Pant S, et al. Faropenem consumption is increasing in India. Clin Infect Dis. 2016;62(8):1050–2.
18. Okeke IN, Laxminarayan R, Bhutta ZA, et al. Antimicrobial resistance in developing countries part: recent trends and current status. Lancet. 2005;5:481–93.

19. Matlow AG, Morris SK. Control of antibiotic-resistant bacteria in the office and clinic. CMAJ Can Med Assoc J. 2009;180:1021–4.
20. van Boeckel TP, Brower C, Gilbert M, et al. Global trends in antimicrobial use in food animals. Proc Natl Acad Sci. 2015;112:5649–54.
21. Landers TF, Cohen B, Wittum TE, Larson EL. A review of antibiotic use in food animals: perspective, policy, and potential. Public Health Rep. 2012;127:4–22.
22. Ramchandani M, Manges AR, DebRoy C, et al. Possible animal origin of human-associated, multidrug-resistant, uropathogenic Escherichia coli. Clin Infect Dis. 2005;40:251–7.
23. Romero J, Feijoó CG, Navarette P. Antibiotics in aquaculture – use, abuse and alternatives. In: Carvalho E, editor. Health and environment in aquaculture. London: InTech; 2012. p. 159–99.
24. Walsh TR, Weeks J, Livermore DM, Toleman MA. Dissemination of NDM-1 positive bacteria in the New Delhi environment and its implications for human health: an environmental point prevalence study. Lancet Infect Dis. 2011;11:355–62.
25. Shah TA, Rabaab Z. Screening of environment water for the presence of blaNDM-1 gene containing microorganisms. J Coll Physicians Surg Pak. 2014;9:695–7.
26. Berendonk TU, Manaia CM, Merlin C, et al. Tackling antibiotic resistance: the environmental framework. Nat Rev Microbiol. 2015;13:310–7.
27. Batt AL, Bruce IB, Aga DS. Evaluating the vulnerability of surface waters to antibiotic contamination from varying wastewater treatment plant discharges. Environ Pollut. 2006;142:295–302.
28. Shimizu A, Takada H, Koike T, et al. Ubiquitous occurrence of sulfonamides in tropical Asian waters. Sci Total Environ. 2013;452:108–15.
29. Donskey CJ. The role of the intestinal tract as a reservoir and source for transmission of nosocomial pathogens. Clin Infect Dis. 2004;39:219–26.
30. Penders J, Stobberingh EE, Savelkoul PHM, Wolffs PFG. The human microbiome as a reservoir of antimicrobial resistance. Front Microbiol. 2013;4:87. https://doi.org/10.3389/fmicb.2013.00087.
31. Tande D, Boisrame-Gastrin S, Munck MR, et al. Intrafamilial transmission of extended-spectrum-B-lactamase-producing Escherichia coli and Salmonella enterica Babelsberg among the families of internationally adopted children. J Antimicrob Chemother. 2010;65:859–65.
32. Kantele A, Laaveri R, Mero S, et al. Antimicrobials increase travelers' risk of colonization by extended-Spectrum Beta-lactamase-producing Enterobacteriaceae. Clin Infect Dis. 2015;60:847–6.
33. Karanika S, Karantanos T, Avanitis M, et al. Fecal colonization with extended-Spectrum Beta lactamase-producing Enterobacteriaceae and risk factors among healthy individuals: a systemic review and meta-analysis. Clin Infect Dis. 2016;63:310–8.
34. Dellit TH, Owens RC, McGowan JE Jr, et al. Infectious Diseases Society of America and the Society for Healthcare Epidemiology of America guidelines for developing an institutional program to enhance antimicrobial stewardship. Clin Infect Dis. 2007;44:159–77.
35. Hersh AL, De Lurgio SA, Thurm C, et al. Antimicrobial stewardship programs in freestanding children's hospitals. Pediatrics. 2015;135:33–9.
36. Gerber JS, Prasad PA, Fiks AG, et al. Durability of benefits of an outpatient antimicrobial stewardship intervention after discontinuation of audit and feedback. JAMA. 2014;312:2569–70.
37. Mangione-Smith R, McGlynn EA, Elliott MN, et al. Parent expectations for antibiotics, physician-parent communication, and Satisfaction. Arch Pediatr Adolesc Med. 2001;155:800–6.
38. Linder JA. Editorial commentary: antibiotics for treatment of acute respiratory tract infections: decreasing benefit, increasing risk, and the irrelevance of antimicrobial resistance. Clin Infect Dis. 2008;47:744–6.
39. Wernli D, Jørgensen PS, Harbarth S, et al. Antimicrobial resistance: the complex challenge of measurement to inform policy and the public. PLoS Med. 2017;14:e1002378. https://doi.org/10.1371/journal.pmed.1002378.

40. Dutil L, Irwin RJ, Finley R, et al. Ceftiofur resistance in *Salmonella* enterica Serovar Heidelberg from chicken meat and humans, Canada. Emerg Infect Dis. 2010;16:48–54.
41. Robinson TP, Bu DP, Mas-Carrique J, et al. Antibiotic resistance is the quintessential one health issue. Trans R Soc Trop Med Hyg. 2016;110:377–80.

Chapter 10
The Importance and Challenge of Oral Health in LMICs: Perspectives from Dental Medicine, Dental Librarianship, and Dental Administration

Joseph E. Gambacorta, Elizabeth Stellrecht, and James M. Harris

Introduction

Everyone will be impacted by oral disease at some point in life. Dental caries (tooth decay) is the most widespread chronic disease worldwide, impacting nearly 100% of adults and 60–90% of school children. In addition, about 30% of people aged 65–74 have no natural teeth, and severe periodontal disease impacts 15–20% of middle-aged (35–44 years) adults globally [1]. These staggering statistics demonstrate evidence of significant oral health neglect across the general population.

The *FDI World Dental Federation* defines oral health as "multi-faceted and includes the ability to speak, smile, smell, taste, touch, chew, swallow, and convey a range of emotions through facial expressions with confidence and without pain, discomfort, and disease of the craniofacial complex" [2]. The definition further states that oral health:

- Is a fundamental component of health and physical and mental well-being. It exists along a continuum influenced by the values and attitudes of individuals and communities.
- Reflects the physiological, social and psychological attributes that are essential to the quality of life.
- Is influenced by the individual's changing experiences, perceptions, expectations and ability to adapt to circumstances. [2]

Oral health is defined by the *World Health Organization* (WHO) as "a state of being free from mouth and facial pain, oral and throat cancer, oral infections and sores, periodontal disease, tooth decay, tooth loss and disorders that limit an

---

The objective is to provide the reader with a basic understanding of the relationship between oral health and systemic health, define social determinants of oral health, and the impact dental public health initiatives can have on overall health and lifestyle choices.

J. E. Gambacorta (✉) · E. Stellrecht · J. M. Harris
University at Buffalo, State University of New York (SUNY), Buffalo, NY, USA
e-mail: jeg9@buffalo.edu

© Springer Nature Switzerland AG 2020
K. H. Smith, P. K. Ram (eds.), *Transforming Global Health*,
https://doi.org/10.1007/978-3-030-32112-3_10

individual's capacity in biting, chewing, smiling, speaking, and psychosocial well-being." [1]

These two widely recognized definitions are complemented by the acknowledgement that oral health is an essential component of overall health and is essential to the individual's quality of life [3, 4].

History of Dentistry

Many individuals do not give much thought to how their teeth and oral health affect daily life. Have you ever thought of how your daily life would be impacted if you needed to replace a missing front tooth, or if you had an infected tooth and were unable to find a dentist to provide treatment? Would you feel comfortable smiling or speaking in public with a missing tooth? Imagine eating, working, or going to school if you were constantly in pain.

Dentistry and oral health care have existed for thousands of years in some way, shape, or form across all cultures, but were not more formally recognized until the sixteenth and seventeenth centuries when barber-surgeons included dental procedures such as tooth extractions in their wide range of medical services. Barber-surgeons were a guild of barbers who provided shaves and haircuts as well as surgeons who mostly learned their trade while practicing in the military [5]. While the individual practices were typically kept separate from one another, neither profession had a claim to dental practices, so both would often act as what we know as dentists. These barber-surgeons, many of which were questionable in their skills as medical practitioners as there were no enforced professional standards as this time, often earned nicknames such as "tooth breaker" and "snatcher of teeth."

Dental medicine became an organized profession in the nineteenth century with both the first dental college opening in 1840 as well as the formation of the first national organization of dentists in the world [5]. This formal organization has led to what we recognize as dental medicine today: professionally educated practitioners practicing dental medicine in a clean, sterile environment. Despite the importance and formal organization of dentistry, the profession has evolved independently from medicine. Globally, this disconnect has contributed to health care policy lacking an oral care component [6].

Research has proven that correlations exist between oral and systemic health. These discoveries have propelled dental professionals to assume a more active role in the provision of primary medical services by monitoring health risks, expanding preventive services, and screening for specific conditions such as diabetes and cardiovascular disease [7–9]. This shift in patient care has reinforced the importance of early intervention in diagnosing disease, and how integrating oral care into primary medical care can impact the proliferation of the burden of chronic and oral disease on already strained health systems in *low- and middle-income countries.*

Collaborative practice is the key to resolving what challenges currently exist in oral health care. It is important to focus energies to create educational programs to limit the impact of disease on strained health care systems [10].

The Oral Systemic Connection and the Burden of Disease

The mouth is a central element in interpersonal communication, a major sensory organ, the portal through which virtually all sustenance is ingested and, very importantly, a "window to the body" through which many diseases can be diagnosed [10]. Associations between oral conditions and general health include periodontal (gum) disease with preterm and low-birth-weight babies, cardiovascular disease, diabetes, and cancers of the gastrointestinal tract and pancreas. Oral infections are linked to pneumonia, stomach ulcers and infections of the heart, brain, and other organs (see Fig. 10.1) [2]. "It is likely that systemic exposure to oral bacteria impacts upon the initiation and progression of cardiovascular disease through triggering of inflammatory processes" [8]. It has been reported inflammation by bacteria can accelerate atherosclerosis, and that oral bacteria has been found in human atherosclerotic plaques [11].

Most oral diseases share common risk factors with non-communicable diseases (NCDs), including poor diets (especially those high in sugar, saturated fat, and salt), tobacco, and alcohol use. A NCD is a medical condition that is not caused by an infectious agent. Typically, it takes a long time to develop, with no apparent signs or symptoms until the disease has reached an advanced stage. The WHO reports that NCDs kill 40 million people annually, equivalent to 70% of all deaths worldwide. Cardiovascular disease is the most prevalent NCD, accounting for 17.7 million deaths annually. Other examples of highly prevalent NCDs are cancers, chronic respiratory diseases (i.e., chronic obstructive pulmonary disease (COPD) and asthma), and diabetes. Of the 15 million individuals who die from an NCD between the ages of 30 and 69 years, 80% (12 million) reside in low- and middle-income countries [1].

According to a 2015 report from the FDI World Dental Federation, "Dental caries is the most common childhood disease as well as affects people throughout their lifetime, and is the common NCD worldwide" [2]. In 2010, it was estimated that three billion people had untreated dental decay while in comparison, one billion suffered from chronic migraine headaches, 549 million suffered from diabetes, and 150 million suffered from asthma (see Fig. 10.2) [2].

Besides dental caries, there are several other common oral conditions which negatively affect the world's population on a daily basis. For instance, the primary cause of tooth loss in adults is *periodontal disease* that affects 5–20% of populations globally; *oral cancers* are the eighth most common cancer; 50% of *HIV-positive* patients have symptoms that manifest themselves as oral lesions and ulcers that cause pain and represent a source of opportunistic infections. *Noma*, a

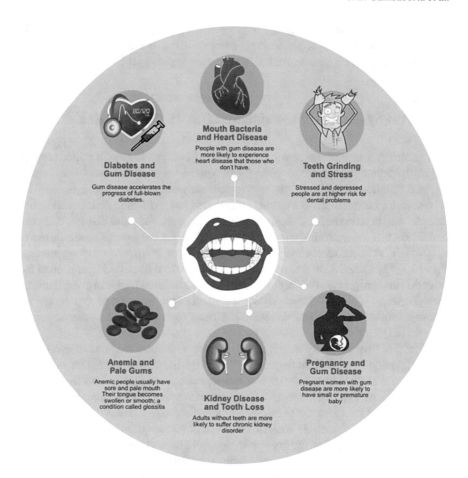

Fig. 10.1 The mouth and the body are connected. (Source: leonardforsyth284, 2018, CC BY 2.0. Image adapted from: https://www.flickr.com/photos/160681424@N07/26312519447. License: https://creativecommons.org/licenses/by/2.0/)

"disfiguring gangrene that rapidly spreads and destroys facial soft tissue and bone" is primarily associated with poor hygiene, malnutrition, and compromised immunity. *Oral cleft lips and palates* are a common congenital problem that occurs most commonly in Asian countries. Finally, *tooth trauma*, commonly associated with unsafe environments and sports activities, is also common in automobile accidents and violent environments; these conditions often require specialty treatment and rehabilitation, which in many cases is either unaffordable or unavailable [12].

In 2011, the United Nations (UN) General Assembly adopted the Political Declaration of the High-level Meeting of the General Assembly on the Prevention and Control of Non-communicable Diseases (NCDs), which recognized oral diseases as significant public health problems for the first time in a UN resolution [13]. For oral diseases to be included in this declaration, all 193 UN member states

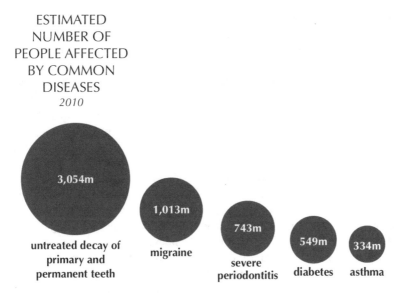

ESTIMATED
NUMBER OF
PEOPLE AFFECTED
BY COMMON
DISEASES
2010

3,054m

untreated decay of
primary and
permanent teeth

1,013m

migraine

743m

severe
periodontitis

549m

diabetes

334m

asthma

Fig. 10.2 Estimated number of people affected by common diseases 2010. (Source: The Challenge of Oral Disease—A call for global action. The Oral Health Atlas. 2nd ed. Geneva: FDI World Dental Federation; 2015 p.16)

needed to come to a consensus acceptable to all member state governments. In article 19 of the Declaration, member states recognize that "renal, oral, and eye diseases pose a major health burden for many countries and that these diseases share common risk factors and can benefit from common responses to non-communicable diseases" [14]. This recognition of the importance of treating oral disease in conjunction with NCDs within an official UN resolution will motivate the global health care community, educators, and governments to allocate resources to improve the prevention and treatment of all NCDs.

Social Determinants of General and Oral Health

Oral health is impacted by a wide range of social determinants, defined by WHO as "the circumstances in which people are born, grow up, live, work and age" [2]. These factors are often influenced by wider socioeconomic and political conditions over which populations typically have little or no control over and largely determine the choices people make and the behaviors they adopt.

All major NCDs, including oral diseases, share the same social determinants and common risk factors that include poverty, sugar, tobacco, alcohol, and poor diet. Worldwide consumption of sugar has tripled over the past 50 years, and this increase is expected to continue rising in low- and middle-income countries. Sugar consumption changes the mix of normal bacteria in the mouth to bacteria that converts sugars to acid. The acid demineralizes a tooth's protective covering, enamel, which leads to dental caries. Repeated exposure to sugar throughout the day, in conjunction with little or no oral care, increases the frequency of acid attacks and the risk of developing tooth decay. In 2011, WHO reported that average sugar consumption per individual per day is 109 grams globally and 166 grams in the USA [2]. There are two types of sugars in our diets: naturally occurring, which are found naturally in foods such as fruit (fructose) and milk (lactose), and added, which include any sugars or caloric sweeteners (high fructose corn syrup) added to foods during processing or preparation. The American Heart Association recommends limiting the amount of added sugars consumed to no more than 100 calories (6 teaspoons or 24 grams) per day for women and no more than 150 calories (9 teaspoons or 36 grams) per day for men [15]. Several countries have implemented initiatives to educate people on the importance of reducing sugar intake, limiting the content of sugar in food and drinks, and enhancing regulations to ensure proper labelling of food that clearly states the amount of sugar products contain.

WHO reports that tobacco use is the most common cause of preventable death globally. In fact, as reported in the 2015 Oral Health Atlas, "cigarettes kill half of all lifetime users, and in the 20th century, tobacco use caused 100 million deaths. This number is expected to rise to 1 billion in the 21st century if smoking patterns remain unchanged. Moreover, exposure to secondhand smoke accounts for approximately 600,000 deaths each year." [2] It has been proven that smoking is linked to oral cancer, periodontal disease, and congenital defects in newborns. Smoking impacts the quality of life in several ways, as it causes bad breath, premature loss of teeth, slowed wound healing, suppression of immune response to oral infections, and promotion of gum disease in diabetics. It is estimated that over half of the cases of gum disease in the USA are caused by smoking [16], and that 90% of cancers of the oral cavity are caused by tobacco use [3].

The consumption of alcohol can be a large detriment to overall health. The excessive use of alcohol often results in substantial health, social, and economic liabilities on infrastructures in society. Alcohol by itself or in conjunction with tobacco can be a major risk factor for cancers of the mouth, larynx, pharynx, and esophagus, and is linked to other oral diseases such as periodontal disease [2]. Alcohol use increases the risk of facial and dental injuries through falls, traffic accidents, or interpersonal violence. Some alcoholic drinks are high in sugar which can contribute to increased risk for dental caries.

Oral diseases pose public health issues due to their prevalence, expense of treatment, and impact on individuals and society. Since NCDs and oral diseases are progressive, patients are often undiagnosed until symptoms such as pain or swelling present. The sudden onset of symptoms forces patients to seek urgent care, which in many cases is not readily available. Health and dental care systems in most countries, including the USA, are not sufficient to cope with the demands on oral health care. While 45% of the 194 WHO member states report to have less than one physician per 1000 population, over 93% report to have less than one dentistry personnel per 1000 population [17]. These shortages represent the necessity to shift basic primary oral care to non-oral health care personnel. Providing additional oral health training for doctors, nurses, clinical officers, teachers, and *community health workers* can help meet some of the unmet need for dental services (see Fig. 10.3) [18].

For low- and middle-income countries, the ability to meet the current need for treatment is more than most health care budgets can accommodate, which often makes prevention the most viable option to address the increasing burden of oral disease [19]. Governments and health care systems need to focus their efforts in designing programming that involves conducting *risk assessments*, providing preventive services (i.e., fluoride treatments and *pit and fissure sealants*) along with educating and training the target population. Providing education and training will empower communities to better understand how lifestyle choices can impact health, the importance of incorporating prevention into daily life, and how to manage disease. Interventions that can be introduced into communities include incorporating education on brushing after meals with a fluoride toothpaste into schools, reducing snacks that contain refined sugars, proper hand hygiene, and training local clinicians on the importance of infection control protocols in preventing *cross contamination* of patients.

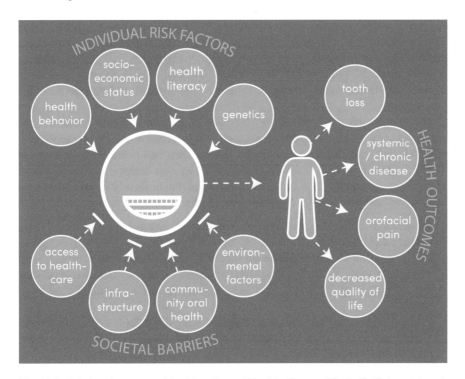

Fig. 10.3 Relations between oral health and overall health. (Source: Nicole C. Little and Joseph E. Gambacorta)

The correlation between oral disease and NCDs is not only limited to systemic factors. Social, economic, and environmental factors impact disease prevalence as well, such as poverty, unemployment, access to providers, and education. Oral health care providers, public health workers, and government policy creators need to coordinate efforts together to diminish the burden of oral disease worldwide [10].

Capacity Building

Many low- and middle-income countries are changing. Development, population migration, enhanced availability of resources like electricity and the internet, combined with a shift in the availability and consumption of processed foods rich in carbohydrates, fats, and sugars are helping drive the transformation. Progress has brought many advances, but unfortunately education on how to use and manage new-found materialism has not occurred [20].

According to Wagner Marcenes, director of research at Barts Health NHS Trust in London, "if left untreated, dental disease can cause severe pain, infection and negatively impact the quality of life, children's growth, school attendance and performance, and can lead to poor productivity at work and absenteeism in adults." [19] There are biological, individual, family, and community level influences that impact the development of disease [2]. Policy makers need to understand that causation is multifactorial and that care delivery systems must be designed not only to address disease, but to introduce strategies to assist communities to better provide for themselves.

Despite progress and advances, the state of the world's oral health is still neglected. Among global health care issues, oral health is often a low priority as it is not viewed with the same urgency by governments and national health systems even though almost all individuals are impacted by oral health care issues in their lifetime [2]. Introducing oral health care initiatives will positively impact communities they are directed toward. Improvements in quality of life include:

- Treatment of active disease: Dentists have the capacity to relieve pain caused by untreated disease that will positively impact those fortunate enough to receive care. Providing quality clinical care including fillings, extractions, replacement of missing teeth, fluoride treatments, and management of infectious processes can alleviate pain, suffering, restore confidence, and enhance patient's ability to chew.
- Addressing risk factors: Programs demonstrating the links between diet, social habits, and health can be impactful in decreasing the burden of oral and other chronic diseases [19]. Training sessions must focus on the importance of the following: (1). reducing sugar intake to prevent tooth decay and diabetes; (2). consuming more fruits and vegetables to protect against some cancers; (3). eliminating tobacco and alcohol use to reduce the risk of developing respiratory conditions, heart problems, periodontal disease, and tooth loss; and (4). oral

hygiene instruction focusing on the importance of prevention by developing consistent brushing habits, seeking care before problems present, and using toothpaste with therapeutic levels of fluoride or introducing a fluoride supplement.

- Promotion of infection control standards: Evidence suggests that infection control protocols followed by healthcare workers in developing countries are less than optimal; for example, the reuse of needles, syringes, and other instruments between patients is commonplace [21]. The recent Ebola outbreak in West Africa and worldwide panic that ensued was linked to healthcare workers not following proper infection control guidelines. Working to improve infection control standards and providing hands on training along with educational materials will decrease treatment complications, assist practitioners in enhancing credibility within the community, and help protect the public against the spread of life-threatening infections such as HIV and hepatitis.
- Establishing school based education and screening programs: Established school based health programs offer an ideal platform to introduce oral health promotion through education as well as daily tooth brushing with a fluoride toothpaste [22]. These sustainable interventions are designed not only to educate school children and their families on the importance of oral care, but to promote good habits, conduct risk assessments, provide preventive services, screen for early stage disease, refer for treatment when appropriate, and to help reinforce the relevance of oral health to overall health.
- Training non-dentists: Oral healthcare is best delivered by a team led by a trained dentist, but availability and affordability are major barriers in accessing care. "Differences in disease burden, inequalities in access to care, and the unequal distribution of dentists within nations present major challenges to global healthcare systems" [2]. Collaborating with local practitioners on the delivery of care during an outreach event provides the perfect setting for volunteers to educate non-dentists on how to recognize and treat dental disease. Examples of programs that have been established include training teachers on how to apply pit and fissure sealants to their students while in school, and expanding the scope of practice of existing public healthcare workers to include simple extractions, treatment of infections, and the restoration of teeth. The gap between burden of disease and availability of care can be addressed by having a mix of *mid-level providers*, and others to assist trained dentists in the delivery of care [2].

Organizing a Dental Outreach Intervention

There are no data published on how many dental professionals or dental teams serve on short term-mission trips annually [23]. Several volunteer opportunities exist locally, domestically, and internationally where practitioners can partner with organizations such as church groups, dental schools, *non-governmental organizations* (NGOs), and professional organizations to lend their talents to alleviating the pain and suffering attributed to untreated dental disease.

When soliciting volunteers, it is important to remember that the success of any outreach is not solely dependent upon recruiting trained healthcare workers specializing in direct patient care. You do not have to be a physician, nurse, or dentist to participate in an outreach intervention. Non-medical volunteers provide essential functionality including but not limited to translation, crowd control, transportation, food preparation, equipment repair, sterilization, and custodial services.

Several approaches exist when coordinating volunteers to participate in outreach missions. No matter what method is utilized, every organizing committee must first establish the goal of the intervention and determine if capacity exists to achieve it. When evaluating capacity, certain essential factors must be considered before anyone makes travel plans such as:

- Identifying local partners: As the previously mentioned data indicate, there is no shortage of need for oral health care across the world. Selecting a destination with established local contacts is essential to meeting established goals and assuring the well-being of volunteers. Local partners act as liaisons with governmental agencies, secure clinic sites, coordinate translation services, and recruit patients, as well as provide security, transportation, food, lodging, and if necessary, emergency support of patients and volunteers.
- Scope of practice: Before determining what treatment options the team plans on offering, a *health needs assessment* of the target population must be conducted. Having a comprehensive understanding of the needs and customs of a community is essential to building trust with the host community, empowering volunteers, guiding educational programming, assuring acceptance of proposed treatment, and enhancing outcomes. Another benefit to establishing the scope of practice before departure is that it will assist organizers in determining what materials, instruments, and equipment are needed to deliver care.
- Selecting a team: Everyone has strengths and weaknesses. Only by first establishing the scope of practice as well as associated logistical and clinic workflows will organizers be able to select team members. A comprehensive review of how the individual strengths of practitioners and support staff align with mission goals, clinic site capacity, population needs, and logistics of the mission will help maximize the impact of the outreach as well as the cohesiveness of the group.
- Training Volunteers: Once the selection process is complete, the focus will shift toward building the team and preparing for departure. Establishing programming to educate volunteers will assist in coordinating travel plans, how to meet governmental requirements for tourism and clinical practice, understanding roles in established workflows, developing educational messaging, and creating *cultural competency*. The goal is to build a team that works well together and that is prepared, confident, focused, cohesive, and adaptive in order to best meet the needs of the community they will be serving.

Concern has been expressed as to the real benefits of volunteer actions. Due to the temporary nature of outreach missions, it is possible that there might be more harm than benefit to the host community [24]. This is based on the fact that dental outreach missions have traditionally been organized as short term clinically focused interventions which function primarily to relieve pain by extracting teeth. In addition, preventive services, such as cleanings and fluoride treatments, and direct restorative procedures (fillings) comprise the scope of care that can also be delivered to most populations during an outreach event. These actions address immediate needs, but on their own, these treatments provide little or no long-term benefits and do nothing to address the global epidemic of dental caries (tooth decay) [1].

WHO allocates resources to implement oral health education and preventive programs to assist populations on becoming more self-reliant and to decrease the burden of disease. For example, WHO has established programs to integrate oral health into school health nutrition programs, and has sponsored initiatives to introduce fluoride into milk and the water supply as a measure to prevent dental caries [25]. Unfortunately, these programs, designed to create new cultural norms, take time, if meaningful change happens at all. These factors, in conjunction with a lack of training programs, adequate supplies, and proper facilities, make the need for short term missions a reality for several communities.

Since the time spent and services typically offered are limited, the number of individuals who actually receive care is low. Without providing long-term solutions, this raises several ethical issues, as single visit outreach missions will not provide a sustainable methodology to address oral disease. Mission teams must collaborate with local agencies to provide education, training, and resources for patients to receive care once the volunteers have departed [26]. Because of this, the following things should be kept in mind:

- Providing dental care is vastly different than offering primary medical care. Dentistry is predominately a procedurally based discipline, where in a majority of cases patients undergo minor surgical procedures at each encounter. In any setting with limited resources, the provision of dental care requires additional extensive planning, materials, and equipment not required by primary care providers [23]. Delivery of this specialized care mandates not only competent practitioners but numerous trained and untrained support staff. Clinic infrastructure must have the physical space to support the entire team and the delivery of care by providing adequate shelter, reliable electricity, water, waste disposal, safety, and if necessary, equipment repair.
- Radiographs (X-rays) are instrumental in assisting with the diagnosis of disease. The likelihood that any technology is available on site especially in rural settings is remote. Portable equipment is available, but is expensive, difficult to travel with, and requires a methodology to view captured images. If necessary equipment is unavailable, disease can remain undiagnosed, which will negatively impact patients after the volunteers leave.
- Organizers must assure that all volunteers wear appropriate *personnel protection equipment* (PPE) and follow *universal precautions* to help prevent accidents in the clinic. *Percutaneous exposures* or other accidents typically occur when volunteers are in a hurry, tired, or do not follow established policies and procedures. The best protection for volunteers to avoid exposure or injury is to thoroughly wash hands, use gloves, masks, protective eyewear, and gowns, take regular

breaks, remain hydrated, minimize blood splattering during procedures, and exercise extreme caution when handling sharps (blades and needles) or potentially infectious materials [27].

Globally, healthcare professionals and NGOs have the capabilities of making a substantial impact on the burden of oral and systemic disease. First, a cultural shift from short, service-orientated missions to capacity building of health care systems and communities through the implementation of education and training programs must take place in order to maximize the outcomes of outreach interventions. It is recommended that anyone interested in organizing or participating in a dental outreach should first learn about the laws of the country, speak with local dentists to determine their needs and challenges, and aim for sustainable results through education and prevention [28]. Volunteering can be an incredibly rewarding experience and it will be even more rewarding if it brings lasting change to the host community [20].

Key Terms

1. *community health workers (CHW)*—are frontline public health workers who have a close understanding of the community they serve.
2. *cross contamination*—the process by which bacteria or other microorganisms are unintentionally transferred from one person to another, with harmful effect.
3. *cultural competency*—ability to interact effectively and be respectful and responsive to the beliefs and practices of diverse population groups.
4. *FDI World Dental Federation*—the principal representative body for over one million dentists worldwide, developing health policy and continuing education programs, speaking as a unified voice for dentistry in international advocacy, and supporting member associations in oral health promotion activities worldwide.
5. *health needs assessment*—a systematic method of identifying unmet health and healthcare needs of a population and making changes to meet these unmet needs.
6. *low- and middle-income countries*—economies with a gross national income (GNI) per capita between $1026 and $4035 annually.
7. *mid-level providers*—healthcare providers who are trained to provide care in a discipline beyond their established scope of practice or level of expertise.
8. *non-governmental organization (NGO)*—any non-profit, voluntary citizens' group which is organized on a local, national, or international level.
9. *percutaneous exposure*—the penetration of the skin by a needle or other sharp object, which has been in contact with blood, tissue, or other body fluids before the *exposure*.
10. *personnel protection equipment (PPE)*—refers to protective clothing, helmets, goggles, or other garments or equipment designed to protect the wearer's body from injury or infection.

11. *pit and fissure sealant*—material that is introduced into the *pits and fissures* of caries susceptible teeth, thus forming micromechanically bonded protective layer cutting access of caries producing bacteria from their source of nutrients.
12. *risk assessments*—a condition targeted evaluation instrument designed to access contributing factors to disease. A common example in dentistry is the caries risk assessment.
13. *universal precautions*—is an approach to infection control to treat all human blood and certain human body fluids as if they were known to be infectious for HIV, HBV, and other blood borne pathogens.
14. *World Health Organization (WHO)*—a specialized agency of the United Nations established in 1948 that directs and coordinates international public health initiatives and policy.

Voices from the Field: Evolution of an Outreach
Senegal is a country in West Africa with a population of over 16 million people. In Senegal, oral health and dentistry are negatively impacted by the same factors as most other developing nations: poverty, malnutrition, child mortality, and the lack of national policy or funding dedicated to oral health. The Ministry of Health reported in 2014 that there is one dentist per 27,000 inhabitants, and 74% of practicing dentists are working in the capital region of Dakar [29]. This dearth of practitioners creates a huge disparity in access to care across the remainder of the country.

In 2014 the University at Buffalo School of Dental Medicine (UB-SDM) began coordinating service learning experiences with faculty, staff, and students from the Ecole Dentaire Internationale (ECD) of the El Hagji Ibrahima Niasse University, and a public health dental clinic in Dakar, Senegal. Through the years, strengthening partnerships, improving workflows, enhancing treatment options offered, and increasing the number of faculty, staff, and student volunteers have positively impacted the total number of patients who have received treatment. Delivering care in conjunction with local providers has also benefited patients in assuring that reliable follow-up care is available after volunteers depart. Although positive steps have been made, the scope of these interventions has remained short term and primarily clinically focused.

The clinic design has evolved from a small five-station clinic arranged in a conference center to utilizing multiple treatment areas at the ECD and/or the public health clinic. Preventive, restorative, and exodontia services are delivered free of charge to both pediatric and adult patients in need. All patients are assigned a tracking number, screened with the assistance of a translator, and referred to one of three clinics (pediatric, restorative, and oral surgery) for care. Student providers rotate through each clinic and provide care under the supervision of discipline specific faculty. In-progress and completed treatment is documented on the screening form, and retained by representatives

continued

from the public health clinic to assist in providing follow-up care. Security, assisting, and sterilization services are provided by staff and local volunteers.

Location, infrastructure concerns, access to care, and the necessity to transform the current short term, clinically focused model to one that supports sustainable training programs, embraces care delivered by an interdisciplinary team, and brings care to underserved areas outside of Dakar has driven organizers to revise the goals of the next intervention. In preparation for departure, faculty and student volunteers will attend educational sessions offered by the University at Buffalo Community for Global Health Equity. This new partnership will expose team members to cultural competency training, help in developing educational programs focused on the needs of the host community, and assist in relationship building with NGOs, community clinics, and governmental agencies in Senegal to deliver direct patient care alongside local providers. Faculty from the University at Buffalo Schools of Nursing and Medicine have also been invited to participate for the first time this year. The goal is to determine how a multidisciplinary team can develop a care delivery model that not only complements the existing dental outreach but enhances the healthcare already available to the host community.

As previously stated oral health is an essential component of overall health and a fundamental human right. The dental profession, governments, NGOs, and other healthcare disciplines have an obligation to coordinate their efforts to address persisting oral health cost and access to care inequalities in many regions, empower populations to better meet their own needs by implementing educational and training programs, and the needs of a dynamic, growing, and aging global population. The primary objective of any intervention should be to collaboratively address challenges to engender meaningful and measurable change that preserves the integrity, and positively advances the target population.

References

1. World Health Organization. Oral health. 2012. http://www.who.int/oral_health/publications/factsheet/en/. Accessed 5 Feb 2018
2. FDI World Dental Federation. The challenge of Oral disease – a call for global action. Geneva; 2015.
3. Petersen PE, Bourgeois D, Ogawa H, Estupinan-Day S, Ndiaye C. The global burden of oral diseases and risks to oral health. Bull World Health Organ. 2005;83(9):661–9.
4. World Health Organization. The objectives of the WHO Global Oral Health Programme (ORH). 2018. http://www.who.int/oral_health/objectives/en/. Accessed 11 Sept 2018.
5. Ring ME. Dentistry: an illustrated history. New York, St. Louis: Abrams, Mosby; 1985.
6. Simon L. Overcoming historical separation between oral and general health care: interprofessional collaboration for promoting health equity. AMA J Ethics. 2016;18(9):941–9. https://doi.org/10.1001/journalofethics.2016.18.9.pfor1-1609.

7. Kim J, Amar S. Periodontal disease and systemic conditions: a bidirectional relationship. Odontology. 2006;94(1):10–21. https://doi.org/10.1007/s10266-006-0060-6.
8. Rautemaa R, Lauhio A, Cullinan MP, Seymour GJ. Oral infections and systemic disease - an emerging problem in medicine. Clin Microbiol Infect. 2007;13(11):1041–7. https://doi.org/10.1111/j.1469-0691.2007.01802.x.
9. Zarco MF, Vess TJ, Ginsburg GS. The oral microbiome in health and disease and the potential impact on personalized dental medicine. Oral Dis. 2012;18(2):109–20. https://doi.org/10.1111/j.1601-0825.2011.01851.x.
10. Beaglehole R, Benzian H, Crail J, Mackay J. Oral health and general health. In: The oral health atlas. Brighton: FDI World Dental Federation; 2009. p. 18–9.
11. Kholy KE, Genco RJ, Van Dyke TE. Oral infections and cardiovascular disease. Trends Endocrinol Metab. 2015;26(6):315–21. https://doi.org/10.1016/j.tem.2015.03.001.
12. FDI World Dental Federation. Oral health worldwide. FDI World Dental Federation: Geneva; 201.
13. Benzian H, Bergman M, Cohen LK, Hobdell M, Mackay J. The UN high-level meeting on prevention and control of non-communicable diseases and its significance for oral health worldwide. J Public Health Dent. 2012;72(2):91–3.
14. United Nations General Assembly. Political declaration of the high-level meeting of the general assembly on the prevention and control of non-communicable diseases. New York: United Nations; 2011.
15. American Heart Association. Added sugars. 2018. https://www.heart.org/en/healthy-living/healthy-eating/eat-smart/sugar/added-sugars. Accessed 3 Jan 2019.
16. Tomar SL, Asma S. Smoking-attributable periodontitis in the United States: findings from NHANES III. J Periodontol. 2000;71(5):743–51. https://doi.org/10.1902/jop.2000.71.5.743.
17. World Health Organization. Health workforce. Available online: https://doi.org/10.2471/BLT.15.020915 Accessed 4 Oct 2018.
18. WHO Regional Office for Africa. Promoting oral health in Africa: prevention and control of oral diseases and noma as part of essential noncommunicable disease interventions. Geneva: World Health Organization; 2016.
19. Treerutkuarkul A, Gruber K. Prevention is better than treatment. Bull World Health Organ. 2015;93:594–5. http://www.who.int/gho/health_workforce/en/.
20. Tepe JH, Tepe LJ. A model for mission dentistry in a developing country. Front Public Health. 2017;5:119. https://doi.org/10.3389/fpubh.2017.00119.
21. Simonsen L, Kane A, Lioyd J, Zaffran M, Kane M. Unsafe injections in the developing world and transmission of bloodborne pathogens: a review. Bull World Health Organ. 1999;77:789–800.
22. Benzian H. Revitalizing school health programs worldwide. Compend Contin Educ Dent. 2010;31(8):580–2.
23. O'Callaghan MG. Implementation of an international short-term dental mission. Gen Dent. 2012;60(4):348–52; quiz 353–344.
24. Illich I. To hell with good intentions. In: Speech to the conference on InterAmerican student projects. Mexico: CIASP; 1968.
25. Petersen PE, Ogawa H. Prevention of dental caries through the use of fluoride--the WHO approach. Community Dent Health. 2016;33(2):66–8.
26. Holmgren C, Benzian H. Dental volunteering – a time for reflection and a time for change. Br Dent J. 2011;210(11):513–6. https://doi.org/10.1038/sj.bdj.2011.426.
27. American Dental Association. International dental volunteer guide. Chicago: American Dental Association; 2014.
28. American Dental Association. Principles of ethics & code of professional conduct. Chicago; 2018.
29. Diop M, Kanouté A, Diouf M, Ndiaye AD, Lo CMM, Faye D, Cissé D. Behavior of the access to oral health care in Senegal. Edorium J Public Health. 2017;4:58–68.

Chapter 11
The Hidden Risks of E-Waste: Perspectives from Environmental Engineering, Epidemiology, Environmental Health, and Human–Computer Interaction

Nirupam Aich, Katarzyna Kordas, Syed Ishtiaque Ahmed, and Tara Sabo-Attwood

Brief Overview

Electronic waste (e-waste) is a rising global environmental and health inequity issue. Rapid and excessive manufacture and use of electronics is causing global e-waste buildup. While there is an opportunity to recover important and/or expensive resources (e.g., recovery of plastic, copper, gold, and platinum) via recycling, these discarded electronics contain many hazardous contaminants including heavy metals (e.g., lead, chromium, copper, mercury, nickel, zinc) and organic compounds (e.g., halogenated flame retardants). Each of these chemicals has been linked with adverse health effects, i.e., respiratory diseases, impairment of central nervous systems, carcinogenesis, and others. Because proper and safe e-waste recycling is expensive, informal recycling abounds, and illegal flows of e-waste (~60–90% of globally produced e-waste) occur from high- to low- and middle-income countries (LMICs). The informal repair and recycling of electronic devices in LMICs often occur without implementing proper protective measures for the workers or their environment. Commonly, e-waste repair/recycling workers are from poor and marginalized populations and in many cases, represent highly susceptible groups, such as pregnant women and children. These individuals are resource limited and cannot afford or demand protective safety measures and adequate health-care. Therefore, the risks of environmental pollution and health hazards associated with exposure to toxic elements and compounds used in electronic components are high. In this

N. Aich (✉) · K. Kordas
University at Buffalo, State University of New York (SUNY), Buffalo, NY, USA
e-mail: nirupama@buffalo.edu

S. I. Ahmed
University of Toronto, Toronto, ON, Canada

T. Sabo-Attwood
University of Florida, Gainesville, FL, USA

© Springer Nature Switzerland AG 2020
K. H. Smith, P. K. Ram (eds.), *Transforming Global Health*,
https://doi.org/10.1007/978-3-030-32112-3_11

chapter, we discuss how the technological advancement of the electronics field has given rise to a worldwide problem of rapid growth and inadequate management of e-waste; how the current practices of informal e-waste repair and recycling in LMICs result in environmental pollution and adverse health effects for marginalized population; and how the challenges associated with e-waste management, environmental issues, and health may be addressed.

Progress in Nanotechnology, Growing E-Waste Burden, and Inadequate E-Waste Management

Continuous scientific and technological advancements help us to lead comfortable lives, but come at a high price by directly or indirectly creating burdens on the environment (i.e., pollution, climate change) and human populations (i.e., health, economic). For example, outdoor and indoor air pollution created by industries and transportation sectors is the cause of 1 in 9 deaths worldwide according to the World Health Organization (WHO) [1]. But air pollution does not affect all people equally; there is evidence that poor communities bear a disproportionate health burden in many parts of the world [2]. Rapid technological advancement contributes to other equally important but not well known or not so easily-detectable problems.

The field of modern-day electronics has been undergoing rapid change due, in part, to recent advancement in nanotechnology which offers the ability to manipulate materials at the nanometer level, or more precisely, at a length scale of one billionth of a meter. To put it into perspective, 80,000 nanometers can fit within the width of a single human hair. The invention and advancement of electron microscopy and other sophisticated instrumentation during the early 1980s allowed us to visualize, discover, and work at the nanoscale and with nanomaterials, paving the way for the shrinking of electronics from room-size computers to smartphones. Such rapid growth of nanotechnology has contributed to the creation of a large volume of diverse electronic devices that quickly become obsolete, thus contributing to the accumulation of electronic waste.

By definition, e-waste or waste electrical and electronic equipment (WEEE) refers to any device that operates with electrical power [3]. E-wastes are classified into six categories: (1) Temperature exchange equipment: refrigerators, freezers, air conditioners, heat pumps; (2) Screens and monitors: televisions, monitors, laptops, notebooks, and tablets; (3) Lamps: fluorescent, high intensity discharge, and LED; (4) Large equipment: washers/dryers, dish-washer, electric stoves, printers/copiers, solar panels; (5) Small equipment: vacuum cleaners, microwaves, toasters, electric kettles, electric shavers, digital scales, calculator, radio sets, video cameras, electronic toys, small medical devices; (6) Small IT and telecommunication equipment: cell phones, GPS, pocket calculators, personal computers [4]. Currently, e-waste is the fastest growing waste stream in the world. The amount of e-waste generated in 2016 was 44.7 million metric tons (MMT), an 8% increase from 2014, and is projected to reach 52.2 MMT by 2021 [5]. To provide a perspective, the 44.7 MMT e-waste is equivalent to 4500 Eiffel Towers and equals 5.8 kg per capita globally. While this mass of e-waste provides a source for harvesting valuable materials, lack of proper management and recycling has created a growing environmental and health hazard on a global scale.

Electronics contain different parts, made from metals and plastics, including protective casing and structural support, displays/screens, memory devices, lighting devices, battery/energy storage, wires and cables, etc. These parts provide a great source from which we can recover precious metals (e.g., gold, silver, and other rare earth elements) that have economic value. With such potential, in 2016, e-waste was valued at an estimated 70 billion dollars [5]. This "urban mining of e-waste" can reduce the cost of raw material production up to 90% compared to the "conventional mining of natural resources." However, these same electronics that provide valuable resources contain a range of toxic contaminants like heavy metals (e.g., lead, mercury, copper, cadmium) and various persistent organic pollutants (POPs), e.g., polychlorinated biphenyls (PCBs) and brominated flame retardants, in many cases, at concentrations that pose significant health concerns.

Both the excellent promise for resource recovery and the concern about potential toxicity from e-waste derived contaminants have led to the creation of formal and high-tech e-waste recycling facilities in many parts of the world. The formal e-waste repair/recycling facilities are typically large scale industries that often use either state-of-the-art machineries/tools or highly skilled/trained laborers with proper personal protective equipment (e.g., facemasks, hard hats, gloves, protective suit) to perform e-waste breakdown and extraction of resources. Unfortunately, formal e-waste recycling processes are expensive and contribute to recycling of only ~20% of e-waste, primarily in high-income countries [5]. In parallel, e-waste is illegally exported to countries in Africa and Asia, including China, Ghana, Nigeria, Thailand, Philippines, Vietnam, India, Pakistan, and Bangladesh. E-waste recycling in LMICs is primarily informal, characterized by manual labor in discrete small shop-setting (both indoor and roadside open-to-outdoor types) without proper personal protective equipment for dismantling, separation, burning, melting, and chemical degradation of components.

Such informal e-waste recycling—largely performed in open spaces by bare hands with simple tools such as screwdrivers and pliers, and without adequate personal protection—poses a significant health hazard to workers via direct exposure to toxic chemicals, and to people living close to e-waste sites through contaminated air, water, and soil. Rapid growth of e-waste globally, lack of proper management and regulation strategies in LMICs, economic incentives to perform low-cost informal recycling, and the absence of solutions to protect workers have resulted in e-waste derived environmental and health hazards, as discussed below.

Informal E-Waste Recycling and Environmental Pollution

Figure 11.1d–g shows the steps of informal recycling documented during a survey in Bangladesh. These practices are similar to those observed in other LMICs. The informal e-waste recycling process starts with physical dismantling of different materials including plastics, metal parts, and glass using hammers, chisels, and screwdrivers [6]. Printed circuit boards obtained from computers/laptops/phones are often heated on coal-fired grills in open spaces to separate the metal-containing components (e.g., iron, copper) from the plastic parts manually using hand-held tools. The epoxy, fiberglass, aluminum, and/or woven papers that make the substrates are burned in the process which can release metals and POPs into the surrounding environment. Precious metals are often recovered by dissolving the electronic parts in strong acids which in itself is a highly hazardous process, and the resulting waste acids that contain toxic heavy metals are often discarded into open water bodies and soil. Manual stripping and burning of plastics are performed to recover copper, a practice which often releases organic contaminants including polychlorinated dibenzo-p-dioxins (PCDDs), polychlorinated dibenzofurans (PCDFs), polybrominated diphenyl ethers (PBDEs), polycyclic aromatic hydrocarbons (PAHs), and perfluoroalkyls (PFOA). Moreover, plastics are shredded, washed with water, and melted without proper ventilation to obtain pelletized forms for reuse. The unsalvageable parts are often burned or disposed to land and water bodies, causing pollution of atmospheric, aquatic, and terrestrial ecosystems [7]. Table 11.1 shows a list of different contaminants present in e-waste and their associated adverse human health effects.

Bangladesh Case Study Part 1: Rise of Digital Bangladesh, Ship Breaking Industry, and E-Waste Repair and Recycling in Bangladesh
Bangladesh is a small LMIC. Its area is 66 times smaller than the USA, but its population (~170 million) is half that of the USA. The rapid economic development of Bangladesh in the last decade has empowered its people with increased disposable purchasing power, which has led to the increased usage and disposal of electronics. Uniquely, Bangladesh is undergoing a strategic implementation of information and communication technology (ICT) development program named "Digital Bangladesh" as part of the government's "Vision 2021" to rapidly improve the education, health, and the financial situation of the population by infrastructure building and employment of digital technology [8]. Between 2004 and December 2017, the number of mobile and internet network subscribers increased from 5 to 140 million and from <1 to 80 million, respectively [9]. This staggering jump in electronics consumption has led to increased e-waste disposal, repair, and recycling. Moreover, Bangladesh is home to the second largest ship breaking industry in the world that produces significantly more e-waste than consumer electronics, but is not

typically accounted for in the global e-waste mapping [10]. There is no effort by the government to estimate the quantity, sources, or composition of any e-waste recycled in Bangladesh. A recent study by a non-government agency Environment and Social Development Organization (ESDO) showed an increase of e-waste in Bangladesh from 2.7 million metric tons in 2010 to ten million metric tons in 2015 [11].

All the e-waste undergoes informal recycling in different urban pockets, largely in the capital city Dhaka and the second largest, port city of Chittagong. Due to people's low incomes, Bangladesh has an electronics repair culture, which impedes somewhat the rate of e-waste generation. Figure 11.1 shows the examples of e-waste repair and recycling activities in Dhaka city. Both repair and recycling activities occur in informal setting and involve manual processing with little appropriate personal and environmental protection against contaminant release and exposure. While formal large scale industrial facilities in developed countries consists of high-tech machineries, an organized labor force equipped with advanced tools and personal protective equipment, and upper-level management; this informal e-waste repair/recycling industry in Bangladesh and other LMICs consists of many small shops with 1–4 people working in each shop with small tools and without proper personal protection (Fig. 11.1d–g). Some repair facilities are located inside buildings and sometimes in underground markets, where ventilation is poor or lacking, thus resulting in severe inhalation exposure. Repair of electronics often involves the use of organic solvents and additives, contributing to exposure through inhalation and skin (Fig. 11.1c).

In contrast to the repair shops, recycling facilities are more open in nature, in many cases at a roadside shop that is open to outdoors for public engagement for business purposes (Fig. 11.1d). Recycling consists of multiple steps, each potentially occurring in a different location [12, 13]. For example, e-waste may first arrive at a retail shop and then initial recycling activities are performed on them (Fig. 11.1d). E-waste recycling workers are typically called "Bhangaries." They first check if the electronics can be repaired. If not, Bhangaries go through a series of steps, separating the plastics from metals. These metals and plastics are then separately taken by two different facilities/shops each of which performs the metal extraction or plastic processing (depending on their expertise). Figure 11.1e–g shows such an e-waste derived plastic recycling facility. Plastics are sorted (Fig. 11.1e), broken down into small pieces and washed with water (Fig. 11.1f), and melted and extruded to obtain plastic pellets (Fig. 11.1g). All these activities cause direct exposure of contaminants through dermal, inhalation, and ingestion. Also, they can cause contamination of soil/dust (Fig. 11.1e), water (Fig. 11.1f), and air (Fig. 11.1g), thus further endangering the environment and human health.

Fig. 11.1 E-waste repair/recycling in Bangladesh: (**a**) underground market for electronics repair, (**b**) electronics repair with bare hands, (**c**) fume-exposure during electronics repair, (**d**, **e**) separating Fig. 11.1 (continued) and sort of recyclable metals, plastics, wires, etc. using bare hands, (**f**) plastic washing with water, (**g**) plastic melting to produce beads, and (**h**) worker's hand with skin lesions. (Source: Compilation of photos from author: Syed Ishtiaque Ahmed)

Table 11.1 Major e-waste contaminants, their source components of e-wastes, and relevant adverse health effects (adapted with permission from Bakhiyi, Bouchra, Sabrina Gravel, Diana Ceballos, Michael A. Flynn, and Joseph Zayed. 2018. Has the question of e-waste opened a Pandora's box? An overview of unpredictable issues and challenges. *Environment International* 110:173–192. doi: https://doi.org/10.1016/j.envint.2017.10.021)

Contaminants	Example of sources	Main type of toxicity
Metals		
Aluminum	WPCBs, microchips, hard drives, LED monitors, plastic housing; plastics, cables, and wires containing inorganic flame retardants (e.g., aluminum hydroxide and trihydroxide)	Lung irritant, neurotoxic
Americium 241	Ionization smoke detectors	Carcinogenic, in the case of high dose radiation
Antimony	Tin-lead alloys, WPCBs, CRT; LCD TVs; plastics, cables, and wires containing inorganic flame retardants (e.g., antimony trioxide)	Lung, eye, and gastro-intestinal irritant
Arsenic	Dopant for semi-conductors, PTVs, LCD monitors, and TVs	Carcinogenic, hematotoxic, endocrine disrupter
Barium	CRT, fluorescent lamps, LCD TVs, PTVs, gutters in vacuum tubes.	Neurotoxic, cardiotoxic, gastro-intestinal irritant
Beryllium	WPCBs, power supply boxes, ceramic components	Berylliosis, carcinogenic
Cadmium	Batteries, toners, cartridges, plastics, WPCBs, solder, chip resistors, CRT, PTVs, cell phones, infrared detectors	Carcinogenic, cardiotoxic, nephrotoxic, endocrine disrupter
Cobalt	Batteries, hard drives, laptop computers, LCD monitors and TVs, PTVs, CRT	Cardiotoxic, allergen (asthma), possibly carcinogenic to human (IARC)
Copper	Cables, electrical wiring, WPCBs, microprocessors, terminal strips, plugs, PTVs, cell phones	Lung, eye, and gastro-intestinal irritant
Gallium	Data tapes, integrated circuits	Skin and probably eye and mucous membrane irritant
Hexavalent chromium VI	Corrosion resistant coatings, WPCBs, data tapes, floppy disks, pigments, PTVs	Carcinogenic (lung cancer), sensitizer, skin irritant
Indium	LCD TVs, transistors, rectifiers	Probably carcinogenic to human (IARC), eye and lung irritant, indium lung disease

(continued)

Table 11.1 (continued)

Contaminants	Example of sources	Main type of toxicity
Lead	CRT (glass, solder), LCD TVs, PTVs, fluorescent tubes, WPCBs, lead-acid batteries	Probably carcinogenic to human (IARC), neurotoxic, cardiotoxic, nephrotoxic, endocrine disrupter
Lithium	Rechargeable batteries	Skin and eye irritant
Manganese	Cell phones, CRT, PCBs	Cardiotoxic, neurotoxic, lung irritant, endocrine disrupter
Mercury	Fluorescent tubes, compact fluorescent lamps, batteries, switches, thermostats, sensors, monitors, LCD TVs, laptop computers	Neurotoxic, skin, eye, and gastro-intestinal irritant, endocrine disrupter
Nickel	Nickel-cadmium batteries, ceramic components of electronics, computers, LCD monitors and TVs, laptop computers	Carcinogenic, sensitizer
Platinum	WPCBs, LCD and LED notebooks, cell phones	Eye and lung irritant
Selenium Silver	Rectifiers, WPCB, old photocopiers	Eye, skin and lung irritant, selenosis at high concentrations
Thallium	PTVs, laptop computers, LCD and LED monitors	Nephrotoxic, reprotoxic
Tin	Semi-conductors, batteries	Neurotoxic, cardiotoxic, hepatotoxic, birth defects
Tungsten	LCD screens, tin-lead alloys, LED monitors	Skin and eye irritant, hematotoxic, hepatotoxic
Vanadium	LCD and LED monitors	Lung, eye, and skin irritant
Yttrium	PCBs, red phosphor emitters, tablet PCs	Cardiotoxic, nephrotoxic, skin and lung irritant
Zinc	LCDs, superconductors, lasers WPCBs, PTVs, CRTs, batteries, soldering flux, cell phones, plastics, wire and cables containing inorganic flame retardants (e.g., in the form of zinc stannate)	Lung and eye irritant Neurotoxic, hematotoxic, gastric irritant, may induce fume fever (inhalation of large amount of zinc in fume or dusts), probably endocrine disrupter

(continued)

Table 11.1 (continued)

Contaminants	Example of sources	Main type of toxicity
Organic pollutants		
Halogenated flame retardants – Brominated flame retardants (e.g., PBDEs, TBBPA, HBCD) – Novel brominated flame retardants (e.g., DBDPE, TBPH, TBBPA-BGE) – Chlorinated flame retardants (e.g., DPs, DBHCTD)	WPCBs, plastics condensers, transformers, plastics	Endocrine disrupter, neurotoxic, carcinogenic (chlorinated flame retardants)
Halogen-free flame retardants	IT housing, plastics, epoxy resins in WPCBs	Organophosphorus: endocrine disrupter Nitrogen-based: nephrotoxic, neurotoxic
Organophosphorus-based flame retardants (e.g., TCEP, TCIPP, TDCIPP, TPHP/TPP, BPA-BDPP, PBDPP, and DOPO)		
Nitrogen-based flame retardants (e.g., melamine cyanurate, melamine polyphosphate) PCBs	Old capacitors and transformers, fluorescent lamps, electrical motors	Endocrine disrupter, hepatotoxic, carcinogenic
Ozone-depleting substances e.g., CFCs, HCFC, HFC, HCs	Old refrigerators, freezers and air conditioning units, insulation foam	Neurotoxic, lung and eye irritants
Other components		
Phthalates	Plasticizers to soften plastics and rubber	
PFOS/F	Antireflective coating	Endocrine disrupter, reprotoxic, hepatotoxic
PVC	Wiring and computer housing	May effect lipid metabolism (increase in blood cholesterol level)

Outdoor and indoor air pollution are aggravated by e-waste recycling; indoor air quality within recycling workshops is particularly poor. Various atmospheric heavy metals and POPs occur in higher quantities near e-waste recycling facilities than at reference sites. For example, the highest levels of the organic contaminant PCDDs in air ever reported occurred around an e-waste dismantling area in Guiyu, China. Similarly, the same air samples also contained PBDEs at ~300 times the reference level. Guiyu air/dust samples were also shown to contain heavy metals including lead, copper, zinc, and nickel at concentrations several hundred times higher than at reference sites [7, 14, 15].

Direct dumping of e-wastes, acid leaching (i.e., dissolving in acids) for precious metal recovery, and plastic burning can contaminate water, soil, and sediments with heavy metals and POPs [7, 15]. Soil sampling at e-waste recycling and disposal sites, along with nearby roadside and agricultural sites, have shown levels of contamination that often exceed the maximum allowable concentrations for individual contaminants set by the national environmental quality standards. For example, the abundance of organic contaminants and heavy metals in road dust and agricultural soil near e-waste sites were at least two orders of magnitude higher than at reference sites and more than 700 times the maximum allowable limits for cadmium, mercury, nickel, lead, copper, arsenic. Similarly, waters of four Chinese rivers near e-waste sites in Guiyu, Longtang, and Taizhou were highly contaminated with PBDEs and heavy metals at levels exceeding local/national drinking water quality standards.

The environmental pollution by e-waste contaminants can contribute to the uptake and accumulation in plants and animals, which can be biomagnified in species higher up the food chain, finally resulting in significant human exposure [7]. For example, lead and cadmium levels in polished rice and vegetables grown in e-waste contaminated soils contained exceeded maximum allowable concentrations for food [16]. Even more striking result from another study was that food of animal origin (e.g., fresh water and marine fish, shellfish, poultry, eggs, chicken/duck, viscera) contained significantly higher levels of e-waste contaminants (~30 times) compared to plant foods [17]. This happened likely due to biomagnification—a process by which contaminants are transferred and accumulated inside animals that are at higher level in the food chain from plants that are at the lower level in food chain.

Human Exposure to E-Waste

Humans can be exposed to e-waste contaminants through inhalation, dietary intake, soil/dust ingestion, and dermal contact. In addition to occupational exposures of the workers from e-waste recycling activities, environmental contamination can result in exposure to residents in the surrounding areas. E-waste contaminants can bioaccumulate in human tissues and organs, and levels of contaminants in tissues tend to be related to the distance from the source as well as the individual's age. For example, higher contaminant levels were found in children's blood and serum samples compared to adults at or near e-waste recycling sites, indicating the need for special attention to children's health outcomes. E-waste contaminants can also be found in placenta, umbilical cord, and breast milk, suggesting that pregnant women, fetuses, and young infants are also vulnerable to e-waste exposure. Placental transfer of the contaminants during pregnancy to the fetus has been evidenced in literature, thus indicating further risks to newborns of being exposed to e-waste derived contaminants [18–27].

The high body burden of contaminants in children may be explained by the following: (1) a high number of children, although not accurately estimated, are employed in informal e-waste recycling industries; (2) children have comparatively higher food and water intake per body weight than adults during growth; (3) children near e-waste recycling areas may be exposed to pollutants during outdoor play via inhalation or ingestion during crawling and hand-to-mouth behavior [18]. Higher contaminant levels in fetal tissues may have resulted from mothers' involvement in e-waste recycling, their environmental exposures from living near e-waste sites during pregnancy, or as secondary exposure from family members employed in e-waste recycling industry.

Food consumption can be a major exposure pathway for e-waste pollutants. In one study, the estimated consumption of e-waste derived contaminant PBDE by 6-month-old breastfed infants at or near e-waste recycling sites were found to be ~57 times higher than that by children in control areas, possibly due to high maternal exposure levels [25]. Another important route of exposure to e-waste contaminants is inhalation. The severity of exposure will vary by location, being higher inside e-waste recycling workshops than outside. For example, e-waste workers in Guiyu China breathed in five times higher concentrations of e-waste related contaminants than other local residents [28]. Similar to dietary intake, children appear to be more susceptible than adults to contaminant intake through inhalation and dermal contact [21, 22].

Health Effects of E-Waste Exposure

Various studies have found a link between the exposure of e-waste and physical and mental health outcomes, e.g., thyroid function, reproductive and neonatal health, lung function, physical growth, temperament, and behavior [18–20]. Many of these outcomes have significant health implications for both individuals and at the population level. For example, e-waste derived POPs contribute to changes (increased or decreased) in thyroid stimulating hormone levels of workers, and resident populations, pregnant women, and newborn babies [29–32]. Lung function of school children living near an e-waste dismantling site in Guiyu, China was lower compared to the children from a reference town, which was correlated with the exposure to chromium and nickel levels in air. Children exposed to metals from e-waste also were shorter, lighter, and leaner than children from the reference town [33, 34]. Children living near e-waste recycling sites also appear to have poorer neonatal behavioral neurological assessment scores, poorer temperament scores, and lower intelligence quotient (IQ), possibly due to increased blood lead levels. The exposure of pregnant women to different POPs, e.g., PBDEs, PCBs, PFOA, and PAHs, has been associated with spontaneous abortions and stillbirths, premature birth, and lower birth weight and length [26, 35, 36]. Thus, e-waste exposures pose serious health threats, especially for vulnerable population such as pregnant women and children.

Although much work has been done to understand the effects of e-waste pollution on human health, the full extent of effects is not known because very few studies have followed people for long periods of time. Long-term studies can provide evidence on the cumulative effects of exposure to e-waste contaminants over time, as well as the mixtures of chemicals to which humans are exposed. Finally, most studies on the effects of e-waste on human health have been conducted in China. A better understanding of the global disease burden related to e-waste exposure is needed. Recognizing that e-waste recycling is a complex environmental and social phenomenon that exposes people directly and indirectly to chemical contaminants, there is a need to engage with exposome scientists (discussed in Chap. 3) to identify the myriad risk and susceptibility factors that drive individuals and communities to take up e-waste recycling and that put humans and ecosystems at risk (Fig. 11.2).

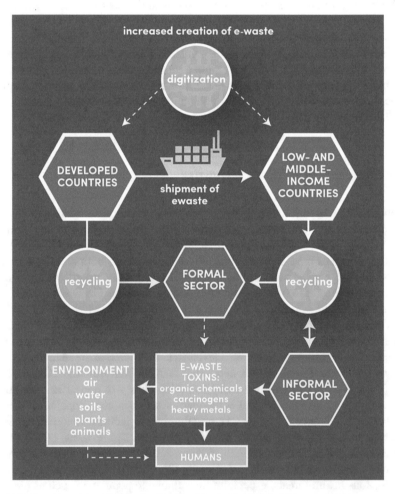

Fig. 11.2 E-waste and health impacts. (Source: Nicole C. Little and Nirapum Aich)

Historical and Recent Developments of E-Waste Issues

During the 1970s and 1980s, industrialized countries started shipping hazardous and e-waste to LMICs [37, 38]. The escalation of e-waste disposal to LMICs and its potential environmental health impact attracted international attention in the late 1980s. As a result, the Basel convention was drafted and initiated as an international treaty in 1992 to prevent transboundary movement of e-waste. Approximately 184 countries and the European Union (EU) are the signatories of this convention, which has reduced global e-waste movement. However, significant transboundary movement of e-waste continues in the form of illegal trade and the transfer of usable second-hand products to LMICs. Furthermore, e-waste generation per capita and per year within LMICs is exponentially increased by the rapid deployment of cellular and internet networks. Political agendas can also motivate the generation of e-waste in-country—as seen in the *"Digital Bangladesh Case Study."*

Solving the E-waste Problem (StEP) initiative was started by the United Nations University (UNU) in 2007 to engage and collaborate among different stakeholders and to strategize solutions for addressing the e-waste problem. The initiative provided a strategic framework to determine and monitor national and global e-waste generation and management, identify and analyze country specific e-waste regulations and policies, develop sustainable e-waste management practices, and capacity building for international collaboration on e-waste management. In 2013, the WHO organized an international cooperation initiative, named the Geneva Declaration, which states that children's health is significantly impacted by informal e-waste recycling. This declaration created WHO's E-waste and Child Health Initiative that calls for engagement of international stakeholders in developing sustainable e-waste management processes, and also for developing intervention strategies to protect the health of e-waste recycling workers and nearby vulnerable populations.

These initiatives have raised awareness regarding environmental and health impacts of e-waste recycling in LMICs [39, 40]. For example, the Chinese government implemented the Management Regulation on the Recycling of Waste Electrical and Electronic Products (WEEE Regulation) in 2011. China has also employed "old-for-new" (or trade-in) policies to reduce e-waste production. Moreover, formal recycling has grown tremendously. In Guiyu, the entire informal recycling industry was removed to create government-subsidized formal recycling facilities elsewhere. These initiatives by China resulted in better material recovery from e-waste along with the reduction of heavy metals, POPs, and greenhouse gas emissions. Few other LMICs, including India and Ghana are in the initial stages of creating policy, regulations, and technology; but China's experience can help other LMICs.

Bangladesh Case Study Part 2: Potential Environmental and Human Health Consequences of Informal E-Waste Recycling in Bangladesh

Surprisingly, there have been no studies to estimate the severity of the contamination and exposure of the Bhangaries and nearby residents from the e-waste recycling in Bangladesh. In 2011, the Bhangaries comprised a workforce of almost 120,000 people and their exposure is deemed more serious [41]. Most Bhangaries have low education levels and live in poverty; they have few employment options. Environment and Social Development Organization (ESDO), a non-government organization in Bangladesh, reported in 2015 that around 50,000 children are involved in informal e-waste collection and recycling [11]. Also, according to ESDO, 15% of these child e-waste workers die each year while 83% live with long-term illness related to e-waste contamination. There is no information on how these data were collected and there is a clear need for systematic investigation of health among worker and their families. To understand the occupational hazards from e-waste recycling in Bangladesh, we conducted an informal survey of 20 Bhangaries in 2016. We found that while no women were involved e-waste repair, women performed e-waste recycling, which pays less than repair. Thus, there is a risk of contaminant exposure and health hazards to vulnerable population such as pregnant women and newborns. Of the respondents in our survey, 16 e-waste workers reported skin problems on hand and feet, 15 reported reduced eyesight and burning eyes, 10 reported breathing difficulties, and 8 discussed experiencing loss of appetite. This survey raises concerns regarding health-related hazards from e-waste pollution in Bangladesh. E-waste is likely to grow with increasing digitization and inadequate regulations/policies to guide e-waste management. Bangladesh represents a unique case where the political agenda ("Digital Bangladesh"), lack of political and economic incentives to protect socially and economically vulnerable populations, and the existence of non-traditional industries ("ship breaking") all contribute to e-waste generation and recycling problems. This complex issue can only be solved through an integrated effort between the government and private sectors, where health and economic parity among individuals, and the sustainability of environmental resources are treated on par with economic gains.

Existing Challenges in E-Waste Issues

E-waste is a complex global environmental and health issue complicated by social, economic, and political factors. Most LMICs facing problems with e-waste do not have regulations or even documentation of the amount of waste that generated, imported, repaired, and recycled. The awareness of e-waste contaminants and their health effects is limited among e-waste workers, the general public, and the government. Moreover, informal recycling in LMICs is driven by the exploitation of marginalized populations (i.e., migrants) with low income and education levels for maximization of profits. Gender-based labor differences are also evident in e-waste facilities—while higher paying e-waste repair jobs are dominated by males with technical training; low paying and riskier e-waste recycling jobs are often performed by non-technical personnel, including women and children. Technologically, formal e-waste recycling faces challenges that stem from the level of difficulty and complexity of individual processes and from the large variety of e-waste types and components. Engineering solutions for protecting environmental and human health from e-waste contaminants in low-resource settings are virtually non-existent. Even if informal recycling is replaced by more formal processes (as in Guiyu), there will be legacy contamination, with extensive remediation needs and large financial inputs. Finally, public interest in repair and reuse is quite low because of lack of sustainable designs, high repair cost, and lack of incentives to improve the repair industry. The ease of purchase and relatively low cost of new products discourages repair.

Integrated Approach for Addressing E-Waste Issues

E-waste problems cannot be solved with single-sector approaches [37, 38]. Rather, an integrated approach is needed to understand the challenges, design, and implement interventions, which are likely to differ depending on setting. This approach will require active participation from a number of disciplines including engineering (e.g., environmental, computer, materials, industrial), environmental chemistry, public health, behavioral and health economics, business, public policy, political science, communication, and social work. The philosophical boundaries need to be extended beyond the assessment of environmental and human health impacts and the need to identify the economic, social, behavioral, and political determinants e-waste production and management. Any research and solutions will need to address the seven major components of "human security"—economic, food, health, environmental, personal, community, and political—that are significantly impacted by e-waste pollution, particularly for the most vulnerable groups. What follows are some recommendations for the prevention and management of e-waste pollution to prevent environmental and human exposure.

In the production of electronics, eco-design should be embraced to include safer materials and improve the reparability of devices. These strategies will reduce the release of contaminants during end-of-life disposal and reduce the rate of e-waste generation. Such strategies will require stringent regulations, affordable and innovative green technology, and effective communication with the public and stakeholders in electronics manufacturing/recycling industry to raise awareness about e-waste safety. Downstream, the illegal trade in e-wastes needs to be minimized through political, legal, and economic interventions, by individual governments and through coordinated regional and global efforts. Policies should be devised to eliminate or decrease informal recycling, by incentivizing stakeholders at different points in the chain rather than punitive measures. Moreover, an integrative approach of merging informal and formal recycling practices ("best of two worlds") can be an option, where the cost-effective but protected manual dismantling of e-waste is complemented by state-of-the-art formal recycling processes for more complex steps of resource recovery.

To protect the human health from e-waste recycling activities, it would be important to ban or disincentivize child labor in this sector, while strengthening health protections for other workers. Furthermore, programs should be created to educate e-waste workers about potential health risks, and to provide additional training on safer e-waste processing techniques. Remediation of contaminated e-waste sites is needed, as are resources for the diagnosis, treatment, and prevention of e-waste derived health issues for the existing cadre of e-waste workers. These potential solutions require coordinated efforts of governments, the health sector, non-governmental organizations, academics, and businesses. Finally, effective and sustainable solutions will be context-specific and require knowledge of local cultures, and the socio-economic and educational backgrounds of e-waste workers.

References

1. WHO. Air pollution causes 1 in 9 deaths worldwide. http://www.who.int/airpollution/en/. Accessed 3 Oct 2018.
2. Hajat A, Hsia C, O'Neill MS. Socioeconomic disparities and air pollution exposure: a global review. Curr Environ Health Rep. 2015;2(4):440–50. https://doi.org/10.1007/s40572-015-0069-5.
3. Townsend TG. Environmental issues and management strategies for waste electronic and electrical equipment. J Air Waste Manag Assoc. 2011;61(6):587–610.
4. Balde CP, Kuehr R, Blumenthal K, Fondeur Gill S, Kern M, Micheli P, Magpantay E, Huisman J. E-waste statistics-guidelines on classification, reporting and indicators. Bonn: United Nations University, IAS-SCYCLE; 2015. p. 51.
5. Balde CP, Forti V, Gray V, Kuehr R, Stegmann P. The global e-waste monitor 2017: quantities, flows and resources: United Nations University, International Telecommunication Union, and International Solid Waste Association; 2017.
6. Chi X, Streicher-Porte M, Wang MY, Reuter MA. Informal electronic waste recycling: a sector review with special focus on China. Waste Manag. 2011;31(4):731–42. https://doi.org/10.1016/j.wasman.2010.11.006.

7. Robinson BH. E-waste: an assessment of global production and environmental impacts. Sci Total Environ. 2009;408(2):183–91. https://doi.org/10.1016/j.scitotenv.2009.09.044.

8. Zaman H, Zaman R. Achieving digital Bangladesh by 2021 and beyond: background paper for the 7th five year plan (7FYP). 2015.

9. bdnews24. Mobile phone users in Bangladesh top 140 million. 2017. https://bdnews24.com/business/2017/11/14/mobile-phone-users-in-bangladesh-top-140-million.

10. Rabbi HR, Rahman A. Ship breaking and recycling industry of Bangladesh; issues and challenges. Procedia Engineering. 2017;194:254–9. https://doi.org/10.1016/j.proeng.2017.08.143.

11. ESDO. Magnitude of the flow of E-waste in Bangladesh. Dhaka: Environmental and Social Development Organization (ESDO); 2015.

12. Rifat MR, Prottoy HM, Ahmed SI. The breaking hand: knowledge, care, and sufferings of the hand of an electronic waste worker in Bangladesh. In: Paper presented at the 37th annual ACM conference on human factors in computing systems (CHI), Glasgow, UK; 2019.

13. Rifat, MR, Aich N, Prottoy HM, Ahmed SI. Understanding the opportunities and challenges in e-waste management practices in Dhaka, Bangladesh. In: Paper presented at the ACM CHI conference on human factors in computing systems, SIGCHI, Montreal, Canada. 2018.

14. Dai Q, Min X, Weng M. A review of polychlorinated biphenyls (PCBs) pollution in indoor air environment. J Air Waste Manage Assoc. 2016;66(10):941–50. https://doi.org/10.1080/10962247.2016.1184193.

15. Song Q, Li J. Environmental effects of heavy metals derived from the e-waste recycling activities in China: a systematic review. Waste Manag. 2014;34(12):2587–94. https://doi.org/10.1016/j.wasman.2014.08.012.

16. Fu J, Zhou Q, Liu J, Liu W, Wang T, Zhang Q, Jiang G. High levels of heavy metals in rice (Oryzasativa L.) from a typical E-waste recycling area in Southeast China and its potential risk to human health. Chemosphere. 2008;71(7):1269–75. https://doi.org/10.1016/j.chemosphere.2007.11.065.

17. Song Y, Wu N, Han J, Shen H, Tan Y, Ding G, Xiang J, He T, Jin S. Levels of PCDD/Fs and DL-PCBs in selected foods and estimated dietary intake for the local residents of Luqiao and Yuhang in Zhejiang, China. Chemosphere. 2011;85(3):329–34. https://doi.org/10.1016/j.chemosphere.2011.06.094.

18. Song Q, Li J. A systematic review of the human body burden of e-waste exposure in China. Environ Int. 2014;68:82–93. https://doi.org/10.1016/j.envint.2014.03.018.

19. Grant K, Goldizen FC, Sly PD, Brune M-N, Neira M, van den Berg M, Norman RE. Health consequences of exposure to e-waste: a systematic review. Lancet Glob Health. 2013;1(6):e350–61. https://doi.org/10.1016/S2214-109X(13)70101-3.

20. Song Q, Li J. A review on human health consequences of metals exposure to e-waste in China. Environ Pollut. 2015;196:450–61. https://doi.org/10.1016/j.envpol.2014.11.004.

21. Wen S, Gong Y, Li JG, Shi TM, Zhao YF, YongNing W. Particle-bound PCDD/Fs in the atmosphere of an electronic waste dismantling area in China. Biomed Environ Sci. 2011;24(2):102–11. https://doi.org/10.3967/0895-3988.2011.02.003.

22. Ma J, Kannan K, Cheng J, Horii Y, Wu Q, Wang W. Concentrations, profiles, and estimated human exposures for polychlorinated dibenzo-p-dioxins and dibenzofurans from electronic waste recycling facilities and a chemical industrial complex in eastern China. Environ Sci Technol. 2008;42(22):8252–9. https://doi.org/10.1021/es8017573.

23. Li H, Yu L, Sheng G, Jiamo F, Peng P'a. Severe PCDD/F and PBDD/F pollution in air around an electronic waste dismantling area in China. Environ Sci Technol. 2007;41(16):5641–6. https://doi.org/10.1021/es0702925.

24. Guo Y, Huo X, Li Y, Wu K, Liu J, Huang J, Zheng G, et al. Monitoring of lead, cadmium, chromium and nickel in placenta from an e-waste recycling town in China. Sci Total Environ. 2010;408(16):3113–7. https://doi.org/10.1016/j.scitotenv.2010.04.018.

25. Leung AOW, Chan JKY, Xing GH, Ying X, Sheng Chun W, Wong CKC, Leung CKM, Wong MH. Body burdens of polybrominated diphenyl ethers in childbearing-aged women at an intensive electronic-waste recycling site in China. J Environ Sci Pollut Res. 2010;17(7):1300–13. https://doi.org/10.1007/s11356-010-0310-6.

26. Wu K, Xu X, Liu J, Guo Y, Li Y, Huo X. Polybrominated diphenyl ethers in umbilical cord blood and relevant factors in neonates from Guiyu, China. Environ Sci Technol. 2010;44(2):813–9. https://doi.org/10.1021/es9024518.

27. Zhao Y, Ruan X, Li Y, Yan M, Qin Z. Polybrominated Diphenyl ethers (PBDEs) in aborted human fetuses and placental transfer during the first trimester of pregnancy. Environ Sci Technol. 2013;47(11):5939–46. https://doi.org/10.1021/es305349x.

28. Xing GH, Chan JKY, Leung AOW, Wu SC, Wong MH. Environmental impact and human exposure to PCBs in Guiyu, an electronic waste recycling site in China. Environ Int. 2009;35(1):76–82. https://doi.org/10.1016/j.envint.2008.07.025.

29. Ju Y, Xu G, Chen L, Jiang Q, Li L, Yang K, Chen X. Analyses of levels of thyroid hormones and its receptor expression in puerperants and newborns from an e-waste dismantling site. J Front Med China. 2008;2(3):276–82. https://doi.org/10.1007/s11684-008-0052-8.

30. Yuan J, Chen L, Chen D, Guo H, Bi X, Ying J, Jiang P, et al. Elevated serum polybrominated diphenyl ethers and thyroid-stimulating hormone associated with lymphocytic micronuclei in Chinese workers from an E-waste dismantling site. Environ Sci Technol. 2008;42(6):2195–200. https://doi.org/10.1021/es702295f.

31. Zhang J, Jiang Y, Zhou J, Bin W, Liang Y, Peng Z, Fang D, et al. Elevated body burdens of PBDEs, dioxins, and PCBs on thyroid hormone homeostasis at an electronic waste recycling site in China. Environ Sci Technol. 2010;44(10):3956–62. https://doi.org/10.1021/es902883a;.

32. Wang H, Zhang Y, Liu Q, Wang F, Nie J, Qian Y. Examining the relationship between brominated flame retardants (BFR) exposure and changes of thyroid hormone levels around e-waste dismantling sites. Int J Hyg Environ Health. 2010;213(5):369–80. https://doi.org/10.1016/j.ijheh.2010.06.004.

33. Zheng G, Xu X, Li B, Wu K, Yekeen TA, Huo X. Association between lung function in school children and exposure to three transition metals from an e-waste recycling area. J Expos Sci Environ Epidemiol. 2012;23:67. https://doi.org/10.1038/jes.2012.84.

34. Huo X, Lin P, Xu X, Zheng L, Qiu B, Qi Z, Bao Z, Han D, Piao Z. Elevated blood lead levels of children in Guiyu, an electronic waste recycling town in China. Environ Health Perspect. 2007;115(7):1113–7. https://doi.org/10.1289/ehp.9697.

35. Wu K, Xu X, Lin P, Liu J, Guo Y, Huo X. Association between maternal exposure to perfluorooctanoic acid (PFOA) from electronic waste recycling and neonatal health outcomes. Environ Int. 2012;48:1–8. https://doi.org/10.1016/j.envint.2012.06.018.

36. Xu X, Yang H, Chen A, Zhou Y, Wu K, Liu J, Zhang Y, Huo X. Birth outcomes related to informal e-waste recycling in Guiyu, China. Reprod Toxicol. 2012;33(1):94–8. https://doi.org/10.1016/j.reprotox.2011.12.006.

37. Bakhiyi B, Gravel S, Ceballos D, Flynn MA, Zayed J. Has the question of e-waste opened a Pandora's box? An overview of unpredictable issues and challenges. Environment International. 2018;110:173–92. https://doi.org/10.1016/j.envint.2017.10.021.

38. Heacock M, Kelly CB, Asante KA, Birnbaum LS, Bergman ÅL, Bruné M-N, Buka I, et al. E-waste and harm to vulnerable populations: a growing global problem. Environ Health Perspect. 2016;124(5):550–5. https://doi.org/10.1289/ehp.1509699.

39. Cao J, Lu B, Chen Y, Zhang X, Zhai G, Zhou G, Jiang B, Schnoor JL. Extended producer responsibility system in China improves e-waste recycling: government policies, enterprise, and public awareness. Renew Sust Energ Rev. 2016;62:882–94. https://doi.org/10.1016/j.rser.2016.04.078.

40. Kumar A, Holuszko M, Espinosa DCR. E-waste: an overview on generation, collection, legislation and recycling practices. Resour Conserv Recycl. 2017;122:32–42. https://doi.org/10.1016/j.resconrec.2017.01.018.

41. Ahmed FRS. E-waste management scenario in Bangladesh: Department of Environment, Government of Bangladesh; 2011.

Chapter 12
Overcoming Shortages of Essential Medicines: Perspectives from Industrial and Systems Engineering and Public Health Practice

Biplab Bhattacharya and Felix Lam

Disparities in Pharmaceutical Supply Chains

The landscape of access to essential medicines changes drastically across national boundaries. Inequity is severe, yet, difficult to capture accurately [1]. Those who manage pharmaceutical supply chains rely on measures like number of pharmacies per capita, availability of essential medicines and number of pharmaceutical personnel per capita to get an estimate of the spread of this inequity. These metrics are very important in making policy level decisions on pharmaceutical spending, pharmaceutical procurement, pharmaceutical distribution, education, infrastructure, and personnel structuring. But what do these metrics mean to the consumer? How do people's health or quality of life compare in locations with starkly different indicators? Let us picture this through the eyes of two people, Sarish and Aden, both born in 1993.

Sarish was born in the state of Maharashtra in India and Aden was born in the state of New York in the USA. According to the World Bank (in ref), India is a low-middle income economy and the USA is a high-income economy. Although Sarish and Aden grew up in completely different parts of the world, both of them had similar interests growing up. As children, they shared a fascination for trains, shapes, and playing outside. However, they experienced education, transportation, food accessibility, and healthcare very differently.

Life through high school followed a rather consistent pattern for Aden. He would wait at the end of his street for a school bus to pick him up. He would go through his routine school day with 40 other students in the class. During snowstorms, Aden's

B. Bhattacharya (✉)
University at Buffalo, State University of New York (SUNY), Buffalo, NY, USA
e-mail: biplabsu@buffalo.edu

F. Lam
Clinton Health Access Initiative, Boston, MA, USA

© Springer Nature Switzerland AG 2020
K. H. Smith, P. K. Ram (eds.), *Transforming Global Health*,
https://doi.org/10.1007/978-3-030-32112-3_12

parents would pick him up from school. Usually after school, he would play soccer, do his homework, and get ready for the next day. The high school experience was different only in a few ways for Sarish. He walked a mile and a half to school every day. Some days he could borrow his older brother's bike, which meant an extra 20 min of sleep. He would go through his school day with 80 other students in the class. Every alternate day, there would be no power at his house for 3 h, so those days, he would rush to finish homework and miss soccer. During heavy monsoons, Sarish would walk home in ankle deep water. He did not mind that much, playing with the odd floating plank along the way. From their own perspective, both of them thought their version of life was the norm. It is not until someone from the outside compares their two experiences that the differences in opportunity and priority become evident.

Now in their twenties, they are responsible for finding medical care when needed. Aden called his doctor when he felt ill one morning. He had an appointment later in the afternoon and picked up his prescriptions that evening. A few weeks later, he had a minor accident and rushed to an emergency room, where a triage nurse classified him as a level 4 severity according to the Emergency Severity Index (ESI) scale. The ESI is a triage system used by the emergency department ranging from 1 (most severe) to 5 (least severe). He received care after waiting for 45 min. Aden picked his prescription up from a pharmacy of his choice that was close to his home after a couple of hours. Occasionally, Aden stops by that pharmacy to pick up generic variants of ibuprofen, amoxicillin, or some common antibiotic as needed.

Access to healthcare services and pharmaceuticals is much more difficult for Sarish. When he was due for a rabies vaccine, he found out that his town and the neighboring towns were out of the immunization. Sarish was forced to travel an hour to the neighboring town to get the vaccine. The number of pharmacists per 10,000 people is just under 6 much lower than the USA which is just under 10. When Sarish needs to see a doctor, unless urgent, he usually gets an appointment for the next day. When he cannot afford that much time or wants to save money, he will often go to the pharmacist for help directly. The pharmacist will listen to his symptoms and give him a small dose of prescription medication. This might not be legal, but is seldom regulated. It saves him time and money while the pharmacist earns the trust and goodwill of a patient. The doctor's office in India is often flooded with patients because the physician density is 0.725 physicians per thousand people in India—much lower compared to the USA, which is at 2.568 physicians per thousand people. On getting a prescription, Sarish has to travel to four pharmacies before finding the medicines he needs. Even though the travel time and cost add up for him, this is a good scenario, because he got his prescription filled on the same day. It is not uncommon to have to wait a week to a month for certain pharmaceuticals, in which case, Sarish would have to arrange to get a pharmaceutical from another city or state while fighting an illness. He would have to either travel himself, which is difficult because it would mean having to take a day off of work or try and find someone coming into town to bring it with them. In the meantime, his condition would deteriorate.

When Sarish gets ill even with something that is not life threatening, it impacts his community. He will lose workdays and probably get worse before he gets better. It is common to come back from spending a day going to different pharmacies to find the pharmaceuticals he is looking for and not getting what he needs. His friends and family spend time searching for the right pharmaceuticals or finding people coming in from out of town who can maybe pick it up. He knows that sometimes, the pharmacist will offer to give him a substitute pharmaceutical which might not be as effective as the one prescribed. But something is better than nothing. This is his normal. Sarish thinks *this* is how access to essential medicines looks like all over the world.

When Aden gets ill with something similar, his immediate community knows to support him by giving him time to rest and recuperate. They know that he will have to go to a doctor, pick up his prescribed pharmaceuticals from a pharmacy nearby and recover. This is his normal. Aden thinks *this* is how access to essential medicines look like all over the world.

Both stories bring into perspective the differences in experiences in access to healthcare and pharmaceuticals between low-income and high-income countries. The stories assume that Sarish and Aden could both afford the prescription pharmaceuticals. However, since co-pays through insurance are uncommon in most of the Indian population (like most other low- and middle-income countries), some pharmaceuticals are unaffordable. Expensive pharmaceuticals are also out of stock more often, as pharmacies work on cash in hand. In this case, the pharmacy will order the pharmaceuticals from a regional distributor or wholesaler. The lead time[1] (time between ordering and receipt) added in procurement will add to a patient's wait time, which will in effect let the illness advance. This is a time when the patient might also take inappropriate substitutes instead of what is necessary. These narratives highlight only the tip of the iceberg of inequity in pharmaceutical access and availability in low-and middle-income countries (LMICs).

> At least one third of the world's population has no regular access to medicines. Inequity in access to essential medicines is part of inequity in health care. Key evidence to document such inequities is rarely collected. 60 countries do not recognize the right to health in their national constitution [2].

The above excerpt from a report by the World Health Organization (WHO) summarizes the depth and spread of the challenge of access to essential medicines. The WHO model essential medicines list comprises of medications that are most effective and safe to meet the most important needs of a health system. This list does not include herbal, local, or natural remedies. Although such remedies might be used frequently in low- and middle-income countries, impacts of drug shortages are still found to be severe. The availability of generic medicines is less than 60% in the world. There exists both inter- and intra-country inequity in access to medicines with rural and low-income communities often getting the short end of the stick. These communities are faced with the challenge of access, availability, and affordability combined. The three challenges are not always considered together in interventions. Acute shortages where 95% of the pharmaceuticals needed are stocked out is common. Stock-outs can last from days to months. Stock-outs have been recognized as an important and complex global challenge [3–5].

The Pharmaceutical Supply Chain Challenge

The difficulty in tackling the pharmaceutical stock-out problem can be largely attributed to inefficiencies and over-complexity in an overburdened pharmaceutical supply chain. This complexity is further increased in LMICs, which characteristically are prone to weak infrastructure, not enough trained personnel, an absence of a standardized ordering and distribution system, inaccurate and insufficient demand data, and a possible dependence on foreign aid. This problem is magnified by a distribution network that is not usually streamlined because of the number of stakeholders (key decision makers) in the system. As compared to a high-income country where external donors and non-governmental organizations (NGO) do not enter the system, LMICs have to accommodate for additional inefficiencies. In most high-income countries, pharmaceutical stocks are managed by a few privately owned national wholesalers who make deliveries to retail pharmacies [6]. Private, public, and NGO distribution networks often work in parallel in many LMICs.

Pharmaceutical Distribution System in LMICs

In this section, we will look at what we mean by a pharmaceutical distribution system and how it looks like in LMICs. A generalized medicine procurement and distribution network for the private and public sector in LMICs is described in Fig. 12.1. The arrows connecting two entities (connectors) indicate the direction of flow of pharmaceuticals. Each level on the left is a stage that represents where the pharmaceuticals are being stored. This network gets more stages, connectors, and stakeholders depending on the country being studied.

Pharmaceuticals change a lot of hands between getting out of the manufacturing plant into the hands of the patient in LMICs, while in high-income countries the distribution system is more streamlined and efficient. Depending on the system, a pharmaceutical might be manufactured in-country or internationally. In the public sector, most LMICs have a central medical storage (CMS) location to store the national stock of pharmaceuticals that arrive from imports or from manufacturing plants in-country. In the private sector, this step is bypassed and pharmaceuticals may be passed directly to wholesalers or retailers. Depending on the stock at the CMS, decisions to distribute regionally based on demand is made in the public sector. Based on the decisions of quantity and timing, pharmaceuticals are pushed into regional medical storage centers. In the private sector, bulk orders are made from regional distributors to an importer/wholesaler who then maintains regional stocks of the pharmaceuticals. The higher level pharmacies which are typically at the

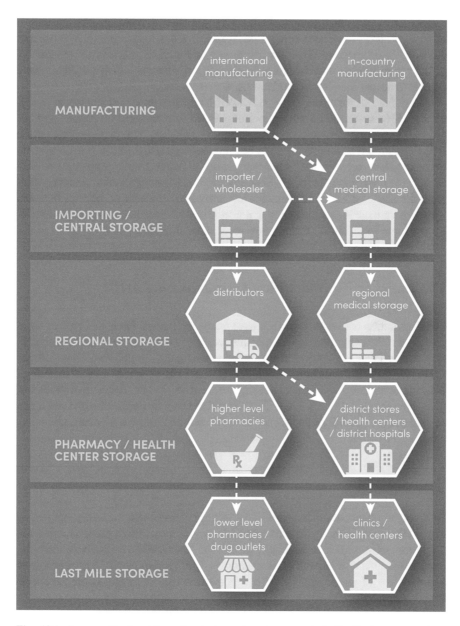

Fig. 12.1 A generalized public and private sector pharmaceutical distribution system in a LMIC. Source: concept: Biplab Bhattacharya; graphic design: Nicole C. Little

district level order pharmaceuticals from distributors as per requirement, cost, profit, and storage availability. District level stores, health centers, and hospitals receive medicines based on availability at the regional medical storage. The available pharmaceuticals at the regional medical storage are rationed as per need and are distributed to these facilities. At the last stage, smaller pharmacies, pharmaceutical outlets, clinics, and health centers place orders from bigger pharmacies district stores or health centers. The last two stages of Fig. 12.1, namely the pharmacy/ health center storage and the last mile storage, are where a patient would typically be able to access the pharmaceuticals needed.

The private sector usually works on a PULL∗ system and the public sector works on a PUSH∗ system. The advantages of a PULL∗ system are that lead times are reduced and the quantities delivered are more accurate. The disadvantage of the PULL∗ system is that orders need to be more frequent. Also, in the PULL∗ system, because orders can be placed whenever needed, a larger inventory needs to be kept in stock at the source. In the PUSH∗ system, the advantage is that delivery decisions are made centrally and a larger inventory does not need to be kept in stock. The disadvantage is that the stores that need the medicines have to wait for deliveries and have less control on when they receive the shipments.

Challenges in Pharmaceutical Distribution in LMICs

The distribution networks in LMICs have many challenges. There can be item-level challenges like expiration and temperature control. There can be facility-level challenges like inventory management, order management, warehouse management, shipment visibility, and shortage avoidance. There can be systems-level challenges like lack of coordination, lack of demand information, human resource dependency, and shipment visibility [4]. We will group and discuss challenges such as too many stages, storage conditions and capacity, funding, logistical inefficiencies, and information transparency.

Excessive Stages

Each stage in the network has time and cost implications, starting with manufacturing. While there are many benefits to producing pharmaceuticals needed locally, the proportion of imports is much higher in LMICs. Most LMICs do not have the capability to manufacture most of these medicines [7]. The continent of Africa produces less than 2% of the pharmaceuticals it consumes [8]. Importing pharmaceuticals lead to additional costs and lead time due to transportation.

Let the lead time between placing an order with the manufacturer and receipt of the order at the central medical store or importer/wholesaler be L1. The lead time here can be in the order of months. Let the lead time for pharmaceuticals to travel from CMS locations to regional storage locations be L2. This is typically in the order of weeks. Let the lead time for pharmaceuticals to travel between regional storage to district level storage and from district level storage to last mile storage be L3 and L4, respectively. This can be in the order of days or weeks.

Hence the total lead time at the last mile compounds to LT = L1 + L2 + L3 + L4.

This number adds up quickly and the more stages added, the more time it usually takes for pharmaceuticals to reach the patient at the last mile, especially in the public sector. Moreover, procurement systems do not always account for lead times accurately, which also contribute to stock-outs [9].

The complexity of the stages of distribution creates the problem of longer lead times in the public sector and higher costs to the consumer in the private sector. The longer the distribution network, the higher the mark-up costs. This is not always true for the public sector, where most pharmaceuticals are often subsidized.

Storage Conditions and Capacity

In the public sector, demand can sometimes never be completely met with the stock available at the CMS; hence, shortages are experienced in the last mile. Decisions regarding quantities of shipment to be sent to a lower stage in the public sector are made based on storage conditions, storage capacity, demand, and transportation available. Depending on the climate of the country, certain pharmaceuticals will have specific temperature and humidity requirements for storage. Limited capital and interrupted electricity supply impose limits on the quantities of supplies that can be stored. In the poorest population groups and rural areas, these storage conditions are more challenging to meet and therefore certain pharmaceuticals are less available.

In the private sector in LMICs, in addition to similar problems of not having the proper storage conditions, drug shops and pharmacies often do not have enough storage space to store the supplies. Hence, they rely on frequent trips and low inventory levels. This leads to additional transportation costs.

Funding Sources

Medical spending per capita in LMICs is much lower compared to high-income countries as seen in Table 12.1. In the public sector, a lot of LMICs depend on donor funding for purchasing pharmaceuticals. As beneficial and critical as donor funding is, it needs to be more streamlined. There are usually several channels of donor funding. The combination of lack of communication between donor organizations

Table 12.1 Total medicine spending by income 2005–2006 [10]

Income group	Number of countries	Total expenditure (million US $)	Per capita expenditure (US $)
High	43	660,609	438
Upper-middle	35	81,235	82
Low-middle	33	76,857	31
Low	19	4123	7

and the public health body, differences in rules and regulations on funding use by the donors, and unsynchronized timings of funding receipt contribute to irregularities in order placement and distribution of pharmaceuticals.

In the private sector, most drug shops and pharmacies in LMICs are small businesses that look to serve the communities they are a part of. They work with small amounts of cash on hand. Because of the limitation of funds, it has hard to maximize the purchase quantities. Only those supplies that are in back-ordered (ordered by a patient but was not available previously), in demand, and most profitable are purchased.

Logistical Inefficiencies and Information Transparency

In-country distributions are typically made by road networks restricted by limited infrastructure in the form of road conditions, weather, number of vehicles, etc. Inadequate infrastructure can also lead to wastage or delays in deliveries [10]. To emphasize that logistical issues can be a deal-breaker even when procurement from manufacturers is timely, a study found that "most drugs are available in South Africa's medicine depots but patients are unable to get their prescriptions filled in local health centers because of local logistical and management problems, ranging from inaccurate forecasting to storage or transport issues [11]. Only a minority of shortages are due to pharmaceutical companies being unable to provide enough drugs worldwide."

At the CMS and regional storage center level, the distribution network includes dedicated vehicles and personnel but in the last mile, pharmacy or drug shop owners often procure supplies themselves and might have to rely on public transportation to do so. This can lead to downtime in having to close the shop while they are procuring medicines. This, in turn, leads to reduction of availability for the patient while the shop is closed and a loss of sale for the pharmacy or drug outlet.

Lack of information transparency in the pharmaceutical distribution system causes other logistical inefficiencies. For example, typically in LMICs there is very little information available about the pharmaceutical inventory levels so that the right amount of pharmaceuticals can be replenished.

In most LMICs, electronic health records (EHRs) are not in place. This makes the transfer of information unclear, untimely, and inaccurate. In the public sector, this can translate to a scenario like the central medical store having to make assumptions of the needs of the regional warehouse and the regional warehouse having to make assumptions of the needs of the district level stores. This can lead to inaccurate demand estimation and insufficient supply in those systems [12]. Logistical and information inefficiencies combined often lead to stock-outs even after interventions to improve inventory levels are implemented [13].

Interventions to Improve the Pharmaceutical Supply Chain

Supply chain and distribution system improvements have been the focus of many industrial engineering and operations management studies, offering a vast number of strategies for addressing stock-outs. However, these strategies are usually limited to a small number of entities in the supply chain with control over the actions of each entity. In pharmaceutical supply chains, the number of stakeholders is high and there is little control over each entity. The products are perishable and sometimes need special storage conditions. Interventions in the field are usually specific to a particular case being studied. Some interventions that have worked are shown in the next paragraph.

Supply chain experts have shown that reducing the number of stages in the distribution network and improving the information flow increases efficiency [14]. Pharmaceutical experts call for reducing the number of stages and partnering with the private sector to boost transportation time and distribution strength [15]. Public health professionals have shown that technologies like Short Messaging Services (SMS) when used to obtain stock counts for certain essential medicines in LMICs, improve information transparency and exchange. Using SMS to exchange stock count information helped better timely restocking and reduced stock-outs in Tanzania [16]. It should be stressed that most interventions use basic technologies to address core challenges. The state-of-the-art technologies and interventions often fail to work in these settings because they are not compatible with the infrastructure, personnel, and maintenance levels available at the region of intervention. Operations researchers have recognized the types of dynamic routing algorithms that can be implemented to accommodate the complex nature of the pharmaceutical supply chain [17]. Dynamic routing algorithms are methods of developing routes based on time and need (among other requirements) for transportation of a commodity. This is particularly useful because demands fluctuate so often that vehicles used to distribute pharmaceuticals cannot function on the same route all the time.

Since every country has a unique system and characteristics, they also have unique challenges. Different disciplines like management sciences, pharmaceutical sciences, public health experts, industrial engineers, health policy experts, and systems engineers have tackled the problem from different angles. The pharmaceutical stock-out problem is one that needs combined action from different disciplines with simplistic and systemic solutions.

Key Actors Locally and Globally

In-country, there are several cogs in the machine that make a pharmaceutical supply chain work efficiently. Decision makers exist from central medical stores nationally to public health ministries where orders are placed or shipments are dispatched to pharmacy owners at the last mile. Logistics and distribution companies work on the

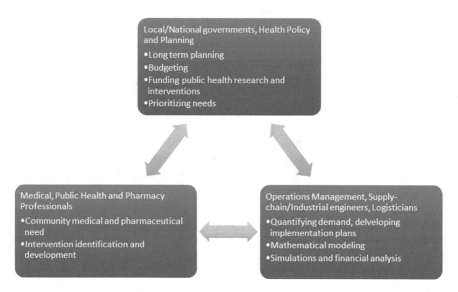

Fig. 12.2 Interdisciplinary relationships in pharmaceutical supply chain functioning. Source: Biplab Bhattacharya

ground to efficiently distribute pharmaceuticals, maintain information, inventories, and warehouses.

Figure 12.2 describes how these different professionals complement each other to tackle pharmaceutical supply chain challenges. Government policy and planning professionals work at higher level with a long-term scope where they prioritize areas of improvement, maintenance, pressing issues and allocate budgets. Public health, medical, and pharmacy professionals work together with local and national government policy and planning professionals to measure and evaluate medical and pharmaceutical needs along with developing intervention plans. Sometimes, they implement these projects themselves and other times they then work with operations management, supply chain/industrial engineering, logistical professionals to develop detailed implantation strategies, mathematical models, simulations, and financial analyses. These groups communicate and collaborate at different levels to make possible a smooth functioning pharmaceutical supply system.

Globally, a variety of organizations like the United States Agency for International Development, Clinton Health Access Initiative, Bill and Melinda Gates Foundation, World Health Organization, and United Nations Children's Fund have ongoing projects and material on improving pharmaceutical supply chains. They partner with health ministries and other local organizations to help strengthen the pharmaceutical supply chain.

As barriers to timely information exchange reduce and the pharmaceutical supply chain infrastructure improves, combining expertise from various disciplines will 1 day enable people like Aden and Sarish to have similar experiences when they purchase essential medicines.

Practitioner Notes (Felix Lam: Clinton Health Access Initiative)
The Pharmacist Head of Bishoftu General Hospital has an enormous responsibility. She and her team manage all essential medicines, medical equipment, and medical supplies at the hospital so that patients can receive lifesaving treatments whenever needed. However, she is not alone in this responsibility. About 25 miles away, a warehouse in Addis Ababa is busy in overseeing deliveries to over 1200 facilities including Bishoftu Hospital. Both the Bishoftu Hospital pharmacy department and the warehouse team are important actors in implementing of Ethiopia's Integrated Pharmaceutical Logistic System (IPLS).

Availability of safe, effective, and affordable medicine and supplies are a critical component in the delivery of quality healthcare. In Ethiopia, essential lifesaving medicines are often stocked out as a result of a weak and fragmented supply chain. One of the main challenges contributing to this issue was that medicines were managed in several vertical supply chains depending on the disease area or population served. Different agencies within the government were responsible for ordering, storing, and delivering medicines in their own disease areas. This resulted in huge inefficiencies and wastage of resources.

To try and solve these problems, the Pharmaceuticals Fund and Supply Agency (PFSA) was established in 2007 to be the central authority for sustainably ensuring the availability and affordability of high-quality pharmaceuticals to all public health facilities and ensure their rational use. Furthermore in 2010, PFSA instituted a system for integrating all medical commodities into a single supply chain system. This system was called IPLS.

Today under IPLS, the Bishoftu Hospital pharmacist's job is to coordinate directly with the Addis Ababa PFSA hub to ensure patients at her hospital will have the medicines that they need. Every 2 months, the head pharmacist works closely with her team to forecast all medicines and supplies patients will need for the following 2 months. She sends an order form to the regional PFSA hub, and the PFSA hub delivers those medicines to the hospital. The Addis Ababa PFSA hub also manages supply orders from the facilities in their region and coordinates with the central PFSA warehouse to ensure that there are sufficient supplies at the hub to meet the orders.

Despite the vast improvements that IPLS has made to simplify the supply chain, healthcare facilities throughout Ethiopia still experience stock-outs of essential drugs. There are many reasons why this happens. At times, demand for the medicine is higher than initially forecasted by the hospital. In other cases, the PFSA hub may also be stocked out of the medicine that facilities need. Often, it is not clear whether stock-outs at the PFSA hub are due to inaccurate forecasting by the hub, central warehouse, or other facilities under the hub's management. And lastly, low-resource countries like Ethiopia have enormous budget constraints that make it difficult to meet every need, including health supplies, infrastructure, and staffing.

These challenges notwithstanding, the government of Ethiopia continues to forge ahead in partnership with donors, non-governmental organizations, and private industry. For example, the government is expanding the flexibility for hospitals to order medicines directly from the open market when PFSA is stocked out. Donors are helping to fill some of the government's gap in funding, and NGOs are providing technical assistance to strengthen the system through capacity building, such as developing improved forecasting and supply chain management tools. These coordinated and innovative solutions hold great promise towards ensuring that every patient in Ethiopia gets the treatment they need.

Glossary of Terms and Supplemental Reading
- Lead time—The time between placing an order to a supplier or manufacturer and receipt of the order.
- PUSH system—A PUSH system is a commodity distribution system where the commodities are supplied based on demands forecasted to meet the customer demand. The advantage of this system is that the supplier has control on how to release orders and can plan on release based on stock available. The disadvantages of this system are that demand forecasts are usually unreliable and too much inventory is held at the supplier which increases the cost to hold inventory. This system is typically seen in the public sector.
- PULL system—A PULL system is a commodity distribution system where the commodities are supplied based on orders placed by the customer. The advantage of this system is that the customer can order supplies when needed and not have to wait for the supplier to determine the time and quantity of delivery. One disadvantage of this system is that the suppliers face inconsistent orders from the customers in terms of time and quantity. This makes it harder for the supplier to manage his own inventory. This system is typically seen in the private sector.
- Inventory levels—Inventory levels refer to the quantity of supplies in store.
- Routing algorithm—A routing algorithm is a set of instructions that help develop routes that a vehicle needs to take. Such an algorithm usually takes into consideration constraints like vehicle capacity, time of delivery, demand, and distance between locations to be visited.

References

1. Bhattacharya B, Lin L, Batta R, Ram PK. Stock-out severity index: tool for evaluating inequity in drug stock-outs. Central European Journal of Operations Research, 2019;1–21.
2. Hogerzeil HV, Mirza Z. The world medicines situation 2011: access to essential medicines as part of the right to health. Geneva: World Health Organization, 680, 689; 2011.
3. Masina L. Voice of America. Malawi Public Hospitals face acute drug shortage. 2013. https://www.voanews.com/africa/malawi-public-hospitals-face-acute-drug-shortage
4. Privett N, Gonsalvez D. The top ten global health supply chain issues: perspectives from the field. Oper Res Health Care. 2014;3(4):226–30.
5. Gray A, Manasse Jr, HR. Shortages of medicines: a complex global challenge. 2012.
6. Yadav P, Stapleton O, Van Wassenhove LN. Always cola, rarely essential medicines: comparing medicine and consumer product supply chains in the developing world. 2010.
7. Kaplan W. Local production and access to medicines in low-and middle-income countries: a literature review and critical analysis. Geneva: World Health Organization; 2011.
8. Pheage T. Dying from lack of medicines. 2016. http://www.un.org/africarenewal/magazine/december-2016-march-2017/dying-lack-medicines. Accessed 2017.
9. Schouten EJ, Jahn A, Ben-Smith A, Makombe SD, Harries AD, Aboagye-Nyame F, Chimbwandira F. Antiretroviral drug supply challenges in the era of scaling up ART in Malawi. J Int AIDS Soc. 2011;14(Suppl 1):S4.
10. Seoane-Vazquez E, Rodriguez-Monguio R. Access to essential drugs in Guyana: a public health challenge. Int J Health Plann Manag. 2010;25(1):2–16.
11. South Africa: drug shortages threaten progress made in the world's largest HIV programme. 2015. http://www.msf.org/article/south-africa-drug-shortages-threaten-progress-made-world%E2%80%99s-largest-hiv-programme. Accessed 30 Aug 2015.
12. Rais A, Viana A. Operations research in healthcare: a survey. Int Trans Oper Res. 2011; 18(1):1–31.
13. Kangwana BB, Njogu J, Wasunna B, Kedenge SV, Memusi DN, Goodman CA, Zurovac D, Snow RW. Malaria drug shortages in Kenya: a major failure to provide access to effective treatment. Am J Trop Med Hyg. 2009;80(5):737–8.
14. Vledder M, Friedman J, Sjöblom M, Brown T, Yadav P. Optimal supply chain structure for distributing essential drugs in low income countries: results from a randomized experiment. 2015.
15. Tetteh E. Creating reliable pharmaceutical distribution networks and supply chains in African countries: implications for access to medicines. Res Soc Adm Pharm. 2009;5(3):286–97.
16. Barrington J, Wereko-Brobby O, Ward P, Mwafongo W, Kungulwe S. SMS for life: a pilot project to improve anti-malarial drug supply management in rural Tanzania using standard technology. Malar J. 2010;9(1):298.
17. De M, Miguel J, De Sousa JP. Dynamic VRP in pharmaceutical distribution—a case study. Cent Eur J Oper Res. 2006;14(2):177–92.

Chapter 13
Communicating Taboo Health Subjects: Perspectives from Organizational Leadership, Clinical Psychology, and Social Work

Dorothy Siaw-Asamoah, Erica Danfrekua Dickson, Emma Seyram Hamenoo, and Deborah Waldrop

Introduction

Worldwide, improving health literacy is a compelling need. Health literacy is the degree to which individuals have the capacity to obtain, synthesize, and understand basic health information and processes needed to make appropriate health decisions [1]. The use of code speak[1] and inadequate communication about taboo subjects undermines efforts to understand basic health information and services. The inter-relationship between serious illness, dying, death, and fear influences how diagnoses and prognoses are understood, whether or not and how treatment is delivered. Clear communication taking cognizance of cultural and societal characteristics will optimize understanding about death/dying/illness. It is essential for helping people who have serious illnesses to make meaningful choices when they are nearing death. It is imperative to have a clear understanding of the best practices in different cultural settings to meet the needs of the dying in addressing death at individuals' level of comfort.

---

[1] Code speak refers to issues or topics that are not easily discussed in certain cultures, so we use encrypted ways of conveying the message.

D. Siaw-Asamoah (✉)
University at Buffalo, State University of New York (SUNY), Buffalo, NY, USA
e-mail: dasamoah@buffalo.edu

E. D. Dickson
37 Military Hospital, Accra, Ghana

E. S. Hamenoo
University of Ghana, Accra, Ghana

D. Waldrop
University at Buffalo, State University of New York (SUNY), Buffalo, NY, USA

© Springer Nature Switzerland AG 2020
K. H. Smith, P. K. Ram (eds.), *Transforming Global Health*,
https://doi.org/10.1007/978-3-030-32112-3_13

What Is the Problem/Issue? Death and Dying Are Taboo Subjects

Death has been considered "the last great taboo" [2]. Stigma and code speak contribute to a pervasive lack of openness and discussion about death and dying [3]. Addressing death with the right balance of frankness and sensitivity can be a challenge. Yet, failure to acknowledge the elephant in the room can mean that dying people feel shut out of important and meaningful social interaction and distanced by others. It is only by discussing our wishes openly and honestly, making such conversations a part of everyday life that we can alter this sense of isolation and offset the difficulties that may arise in the final stages of life. Thulesius et al. [2] indicate that de-tabooing dying has begun with open awareness of death and dying, and is seen in the growth of palliative care.

Is the fear of death universal? The idea of death, the fear of it haunts human beings like nothing else; it is a driving force of human activity—that which aims to avoid the fatality of death, to overcome it by denying that it is the final destiny. The path to death is for all but a few accompanied by pain. Death can be a lonely and isolating experience. Given that human beings are biopsychosocial in nature (e.g., involving the interaction of biological and psychosocial factors), our interactions with others complete our existence and give our lives meaning. Death is a final separation from everything that gives our life form and everything we hold dear. The death of a loved one has been identified as one of the most emotionally painful and stressful human experiences. Fear has been one of the most commonly expressed human responses to death [4, 5].

Fear creates anxiety. Death anxiety describes the amorphous set of feelings that thinking about death can arouse [4]. The fear of death has been conceptualized with varying dimensions. Hoelter and Hoelter (1978) [6] distinguish eight dimensions of death as fear of the dying process, premature death, (for) significant others, being destroyed, the body after death, the unknown, the dead and phobic fear of death. Florian and Mikulincer (1993) [7] suggest three components of death fear to be: *intrapersonal*—the impact of death on the mind and the body which include fears of loss of fulfillment of personal goals, and of the body's annihilation; *interpersonal*—the effect of death on interpersonal relationships; and *transpersonal*—fears about the transcendental self, the hereafter and punishment after death. Recognizing death-related fears can enhance the quality of an individual's life whereas ignoring them may lead to self-deception [8, 9].

Distance from and the avoidance of confrontation with death have marked it as a taboo subject in global health. The further death can be removed from common experience the more of an abstraction it becomes. The abstract nature of death makes the fear of it vulnerable to social manipulation. Modern societies have created a variety of mechanisms for removing the actual experience of death from everyday life. In addition to traditional mechanisms (such as religion), hospices, drugs, death education, psychotherapy, philosophical belief systems serve to remove, sanitize, and ease the pain of the transition from life to death making it easier not to fear such abstraction. Euphemisms such as he has "passed," "gone," or "we lost him" become code speak for the reality of death and dying. The use of euphemisms in death and dying suggests a strategy for relegating death to the sub-conscious by creating labels that are more tolerable. The practice of using euphemisms for death is likely to have originated with the unfounded belief that to speak the word "death" invited death; to "draw death's attention" is the ultimate bad fortune. The need to buffer or soften the impact of death demonstrates the reality of futile attempts to evade death and can influence how grieving people adapt to bereavement [10].

Cultures vary in terms of the extent to which they deny the reality of death as a function of the contexts in which people are born, grow, mature, and die. Death is socially and culturally constructed; feelings and attitudes about death and dying are reflected in the use of language, religious or funeral rituals, faith talk, and the values placed on younger versus older lives. A culture's stance toward death—its death ethos—affects everyday behavior (e.g., willingness to engage in risky behavior, the likelihood of taking out an insurance policy) as well as attitudes toward a variety of issues, such as the justifiable loss of life through war, euthanasia, organ donation, reincarnation, the death penalty, abortion, and the possibility of an afterlife resurrection. Cultures can be classified as *death accepting*, *death denying*, or *death defying*; it follows that tolerance of death varies, dictating responses to natural disasters, deaths of public figures, and the loss of life through violence. Individual responses to death are intertwined with the character of death within cultures [8].

Death anxiety's alter ego is the concept of a "good death." The idea that death can be a good experience may be considered a romanticized notion that white-washes the experience of dying. A good death is highly individual, influenced by faith and culture. Some feel that death while sleeping is preferable while others prefer to be awake and alert [11]. Descriptions of a good death in contemporary western culture have been characterized as physical comfort, social support, acceptance, and appropriate medical care, while minimizing psychological distress for a person who is dying and his or her family [12]. However, contrasting opinions of a good death echo research findings that what one person considers a good death may be completely opposite another's [11].

Death with dignity has emerged as an important goal in care for people who are dying. Yet, the concept of dignity is culturally bound and understood differently in Asian and Western contexts [13]. Psychological symptom distress, heightened dependency needs, and loss of will to live diminish the sense of dignity and are closely associated with certain types of distress in terminal illness [14]. A person's self-worth can be enhanced or threatened by extrinsic factors creating perceptions of being valued or devalued.

What Are the Key Historical and Recent Developments of This Problem: Death and Culture?

Historically, death was a family matter and relatives of the deceased handled the details of processing the dead and death. Death is a ubiquitous event in farming cultures, assimilated into the fabric of social life and accepted as a matter of inevitability and the natural order. During the nineteenth century the advent of embalming and increasing industrialization and urbanization in the USA shifted dying and death from homes to hospitals and funeral homes. Intimate familiarity with death faded and, the USA became a death-denying society—making death a taboo and according to some, "pornographic" topic—ultimately shielding it from public attention. After World War II a new "death awareness" movement surfaced [15]. Many countries including the USA and UK have increased the focus on home-based models of hospice and palliative care. Trends of increasing home deaths followed [16]. The wider use of palliative care and home death has been promoting more open dialogue about death and dying and diminishing the taboo of death.

Cultural practices surrounding death combined with ideas about what happens after death to form the basis of religion is one of the cornerstones of all civilizations. Religion provides expectations and guidelines for human behavior by setting up a series of taboos, as well as sacred objects and rites. Humans still attempt to reduce the shock of death by confronting and understanding it. Early societies developed religious systems, including ancestor worship that bridged the divide between the dead and the living and portrayed death not as an end but as a transition to another world [4]. Recent developments in death and dying include national movements that are aimed at encouraging patient-family-provider conversations. Movements such as *National Healthcare Decisions Day, Dying Matters, Speak Up, The Conversation Project* have been launched to provide materials and education about healthcare communication. The emergence of Death Cafes, which are scheduled non-profit events where people come together, eat cake, drink coffee or tea, and talk about death, suggests the increasing public desire to communicate about death. *Hello* (formerly *My Gift of Grace*) is a conversation game that is played in public and private spaces to start meaningful conversations about life and death. *The Conversation Project* provides starter kits to help people begin conversations about their wishes for end-of-life care. *Dying Matters* (UK), *National Healthcare Decisions Day* (USA), and *Speak Up* (Canada) are public awareness campaigns that urge individual-family-provider conversations about healthcare decisions on identified dates. Each of these developments contributes to a more open dialogue about death and coping with death thus reducing the use of code speak. A partial listing of websites for these movements can be found in Appendix 1 (Figs. 13.1 and 13.2).

Fig. 13.1 Graveyard in Ghana. Source: Emma Seyram Hamenoo

Fig. 13.2 Cemetery in the USA. Source: Emma Seyram Hamenoo

What Are the Current Gaps? Why Does the Problem/Issue Persist? Individual-Family-Provider Communication and Concerns As Death Approaches

Family communication is often fraught with conflict which can lead to a discord about end-of-life decisions and care. Healthcare professionals' attitudes and skills are influenced by the level of personal awareness and discomfort with dying which is embedded in their own cultural leanings of death and dying, a prescribed focus on curative treatment, and lack of education, mentorship, and models for initiating conversations can all influence communication.

Communication

Ethnic and cultural patterns of patient populations have changed as a result of the increasing migration into Western countries. The principles of truth telling and patient autonomy are embedded in the framework of Anglo-American medical ethics. In contrast, in many countries, the cultural norm is to protect the patient from the truth, decisions are made by the family with respect for the tradition of filial piety where it is dishonorable not to do everything possible for one's parents. While such attitudes can be explained partly by the Asian philosophy of filial piety, a similar sense of duty, labeled role obligation can be found in Western cultures [17]. The challenge for health care professionals is to understand the influence that culture has on patients' responses to decision-making, healing, and suffering and the physician–patient relationship [18].

Surrogate decision-making involves healthcare decisions that are made by a person who is chosen in advance to make decisions when an ill or injured person is unable to voice his or her wishes. Surrogate decision-making is often difficult and emotionally charged, isolating and overwhelming. Firstborn sons in Asian cultures naturally assume responsibility in decision-making resulting from traditional expectations of filial duty. Discussion of death is taboo in different cultures including Asian, African-American and often leads to significant stress. Firstborn surrogates have described how they feel alone in the process without the support of others which further compounds their stress. Many younger siblings describe feeling angry and powerless when day-to-day caregiving and carefully chosen care plans are changed and overruled not by an individual with more knowledge of the parent's preferences but solely because of birth order. For firstborns in the context of death as taboo, attempts at decision-making as a family may create conflict, power struggles, and differences of opinion about what is in a parent's best interest [19].

There is a reticence to talk openly about issues surrounding end-of-life care in many cultures because of cultural taboos and fear that doing so will destroy hope. Ng et al. [20] found that surrogate decision-makers were willing to consider conversations about their own wishes but some were ambivalent about discussing loved ones' wishes, especially when they were from an older generation. Cultural sensitivity about open conversations and the importance of closely involving the family unit in the process is critically important [20]. Additionally, the acceptability of truth telling about diagnosis and prognosis is an important consideration. Cultural aversion to discussions of end-of-life issues and planning for the future are strong. Superstitions positing that discussions about death and dying are ominous and bring bad luck need to be considered [21]. The disclosure of terminal status and family involvement in the process have been important issues in end-of-life decision-making in many cultures [17].

Wook Shin et al. [17] surveyed patients and caregivers about their expectations about disclosure of a terminal diagnosis. Most patients and family caregivers expressed the expectation the patient should be informed of a terminal prognosis while a small number of caregivers believed that it was better for patients not to be informed if families decide not to. Over half of patients responded that their physician should inform them directly but family caregivers most frequently answered that a physician should inform the family first and then inform the patient if the family agreed. Patients preferred to be informed of their terminal status more than family caregivers and they preferred to be informed directly by their physicians [17].

Filial expectations and family dynamics related to birth order and surrogate decision-making were explored with surrogate decision-makers and the majority offered unprompted discussions about birth order and family dynamics. When death is taboo, surrogate decision-makers experienced communication difficulty, unspoken expectations, emotional stress, loneliness, and family conflict. Birth order and family dynamics can have profound effects on surrogate stress and coping [19]. Unspoken expectations are often accompanied by the assumption that the firstborn will be the primary decision-maker for the family. Oldest children do not need to be formally asked to take on their decision-making responsibility or have their role as surrogate discussed because it is implied. Death as a taboo subject compounds the firstborn's lack of knowledge about a loved one's wishes. An individual who takes on the caregiving role may feel they cannot broach the topic for fear others would perceive them as trying to hasten their loved one's death (Fig. 13.3).

What Are the Traditional Disciplines/Approaches Related to the Issue/Problem?

Advances in medical technology that have made it possible to prolong and sustain life have also begun to transform communication about and care for people with serious illnesses. Palliative medicine is a relatively new discipline that has become a central element of contemporary care for people who are nearing death—and has been influential in shaping dialogue about the taboo subjects of death and dying. The care that is possible is not always desired; conversations about such choices have become essential. *Palliative care* is derived from the word palliate which means to make a disease or its symptoms less severe or unpleasant without removing the cause. Palliative care is interdisciplinary, holistic care that aims to improve the quality of life for people with serious illness, and their families at any stage on the disease trajectory. Ideally, palliative care begins at diagnosis and is provided concordantly with other disease directed treatments. Palliative care is both a philosophy of care and an organized system for delivering care. The World Health Organization (WHO) stipulates that palliative care is applicable early in the course

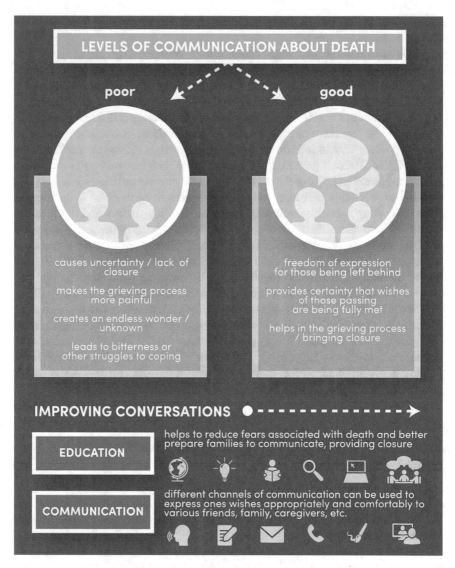

Fig. 13.3 Levels of communication about death. Source: concept: Dorothy Siaw-Asamoah; graphic design: Nicole C. Little

of illness, in conjunction with other therapies that are intended to prolong life such as chemotherapy or radiation therapy. *Supportive care* is defined as improving the quality of life of people who have serious or life-threatening disease, often including psychosocial support in end-of-life care and is often used interchangeably and synonymously with palliative care.

Over a million people die each week worldwide yet, 75% do so without access to essential medicines necessary to relieve suffering. The International Covenant of

Economic, Social, and Cultural Rights calls for the "right of everyone to the enjoyment of the highest attainable standard of physical and mental health" (article 12:1, 2000). Palliative care is not specifically codified as a human right; however, it can be inferred from important covenants and is supported by a number of international declarations.

The term palliative care carries a stigma for physicians, patients, and their caregivers who regard it as synonymous with death and dying, loss of control, hopelessness, and abandonment. The use of the term supportive care has been suggested as more favorable because it is associated with better understanding, more favorable impressions, higher future perceived need, and earlier referrals [22–24]. Unfortunately, the stigma and taboo that surround the delivery of palliative care may become a barrier for those who would benefit from receiving it.

Clinicians have expressed difficulty with the term "palliative care" that does not exist for the term "supportive care." Palliative care carries the taboo of being associated with only the very end of life while supportive care may seem more palatable to clinicians who are responsible for initiating conversations about end-of-life care and choices. This difficulty or dissonance arises because the meaning most widely accepted and promoted by the WHO and National Consensus Project is not consistent with the meaning derived from common association. Palliative care became the preferred term because hospice was seen as too limited in scope. Palliative care was adopted and defined by the WHO and it became the dominant term in public health discourse and documentation.

The delivery of palliative care is team based, interdisciplinary and at its best, the whole is far more than the sum of its parts. Each team member's contribution adds dimension to this complex type of whole person care and strengthens discussions about end-of-life care. Physicians, nurses, social workers, spiritual care providers, volunteers, and professionals in alternative therapies (e.g., music and massage therapy) each bring specific scopes of practice that contribute to and enhance the physical, psychosocial, spiritual, and social care of the patient and family as a unit. Each discipline brings a unique perspective to strengthen discussions about death, dying, and end-of-life care, using transparent and clear language instead of code speak and diminishing the taboo of palliative care. The Appendix presents Voices from the Field illustrating differential perspectives.

Global efforts to universalize understanding and utilization of palliative care continue. The International Association for Hospice and Palliative Care (IAHPC) is a global non-profit membership organization that is dedicated to the development and advancement of palliative care. IAHPC's vision is universal access to high-quality palliative care, integrated into all levels of healthcare systems in a continuum of care with disease prevention, early diagnosis, and treatment, to assure that any patient's or family caregiver's suffering is relieved to the greatest extent possible. IAHPC's mission is to improve the quality of life of adults and children with life-threatening conditions and their families worldwide. (IAHPC; www.hospicecare.com).

The European Society for Medical Oncology (ESMO) has adopted pragmatic definitions of supportive and palliative care based on the common understanding of the terms and a strategy to recognize and encourage a critical shift in emphasis to a palliative care paradigm when patients no longer have curable disease. ESMO has strongly supported a comprehensive program of policies, education, research, and incentives to encourage the integration of supportive and palliative care into cancer centers and programs. The American Society for Oncology (ASCO) has endorsed the integration of palliative care and thrown their weight and credibility behind it in an effort to educate clinicians and consumers about palliative care and to reduce stigma associated with the term [25].

Gomes et al. [26] conducted a systematic review and found moderate evidence that the majority of people prefer dying at home and most patients did not change preference as their illness progressed. This aligned with the direction by current end-of-life care strategies to target a home setting. Notwithstanding, even in countries where these resources to support a home death exist, the majority of people still do not die at home. This highlights the need for stronger action on factors previously found to influence death at home so that more are able to have their preferences met. There is also the need for further research to understand what factors influence death at home for people dying from non-malignant conditions, here the evidence is thinner and the chances of dying at home are generally lower than for cancer patients. The evidence that the majority prefers home to other settings is not as strong as it once was and there is a substantial minority of patients and caregivers for whom home is not the first choice or who change their mind as death approaches. The findings highlight the importance of allowing for a degree of diversity and flexibility in service planning, alongside the focus on home care [26].

Culture is the mechanism through which people learn how to be in the world, how to behave, what to value, and what gives meaning to existence. Cultural congruence is a process of effective interaction between the provider and patient or client. Effective provider communication involves sensitivity to cultural diversity, awareness, and competence. Diversity exists in many domains including but not limited to race, ethnicity, culture, age, gender, sexual orientation, socioeconomic status, education, individual backgrounds, and experiences. Exposures to diverse human cultural patterns vary widely and change over time. As patients, families, and communities grapple with the realities of human mortality, issues of culture come into focus in ways that heighten the relevance of cultural concerns at the end of life [27].

What Disciplines/Approaches Might Be Leveraged to Speed Progress?

Reluctance to participate in end-of-life discussions is common all over the world. If we do not know how to communicate what we want and those around us do not know how to listen, it is almost impossible to express a clear choice. Provider education and training—for all disciplines can focus attention on addressing barriers, addressing the stigma of palliative care, raising public awareness, and using social media. There is the distinct possibility that patients are deprived of care because of stigma.

Enabling health professionals to have clear and informed discussions on death and dying is a major step on the road to improving end-of-life care. Specific barriers to provider communication about end-of-life issues have been found to include physician, institutional/organizational and social factors [28–30].

Physician Factors

- Discomfort with discussing death and dying in the treatment context
- Fear of diminishing or undermining hope
- Fear of causing harm to the patient
- Lack of good models and mentorship
- Uncertainty about who is responsible (e.g., primary vs. specialist)
- Working with interdisciplinary professionals (e.g., who initiates?)

Societal Barriers

- Social and cultural taboos.

Institutional Barriers

- Stigma about palliative care being associated with imminent death; lack of protocols for introduction and referral.

Organizational/Institutional Barriers

- Divergent perspectives and goals (e.g., providers, payers, surveyors).

An example of educational and training program that aims to address barriers is *The Sidney Project in Spirituality and Medicine and Compassionate Care*™ which was created to provide physician residents an opportunity to receive training and awareness of the significance of spirituality and compassionate care in the medical encounter with people who have serious illnesses. The project emphasized making connections with patients; the importance of self-care strategies; opportunities to discuss real feelings; common myths about physicians. Medical residents became more aware of compassionate medical practices, which they were able to integrate in patient care despite the diversity of beliefs among them. The themes that emerged offer validation of the beneficial aspects of this program and other curricula in spirituality and medicine including "deeper connection to patients," importance of self-care strategies, and a consciousness of the connection between physician and patient [31].

The online world is another timely venue for increasing awareness and knowledge; it is alive with discussions, comments, and anecdotes about death and dying. Taubert et al. focus on how the subject of death and dying has influenced the world of microblogging, how it is fast becoming a focus of research, and how this may impact the creation of digital legacy building for people who are approaching death. They address what social media can provide for palliative care and bereavement workers including information, discussion forums, feedback, opinion gathering, areas of controversy, representing organizations and research. Social network sites have brought grieving and dying out of the private domain and into the everyday life of the online world. A careful discussion about their limitations but also an open mind towards the social change they are bringing are called for [32].

Conclusion

Improving health literacy is a powerful means for transforming global health across the life course. Yet, communication about death, dying, and end-of-life care remains taboo topics worldwide, stigmatized by the fear and anxiety that people experience when confronting their inevitability. There is perhaps no topic more personal nor intimate than considering one's own death or that of a loved one and no time in life that is more important for understanding and communicating than when death is approaching. Without open, honest conversation about what is hoped for and desired at the end of life, tremendous opportunities for resolution, meaningful conversations, and life-closure are missed. Absent knowledge about a person's goals, values, and end-of-life wishes decisions can be made that bring unwanted and burdensome treatment. Death is a fundamental part of life and our interaction with it bringing deeper connection with religious, cultural, and personal value considerations into its communication can transform our global humanity. The barriers faced by the various disciplines that are confronted with death and dying must be addressed to equip them with capacity to address this taboo subject hence de-tabooing death and dying (Table 13.1).

Best Practices
1. Do not judge. What we think is the problem may not be at all.
2. Respect people's wishes.
3. Respect people's boundaries.

4. Respect people who do not want to hear "it."
5. Do not go in talking about God.
6. Listen to hear not to answer.
7. Never lose hope.
8. Be sensitive to people's faith.
9. Help me know what you need at this time.
10. Use a structured protocol for guiding a conversation (e.g., the SPIKES protocol or The Conversation Project).
11. We should not shy away from talking about death.
12. Talk about it freely unless you perceive discomfort.

Spikes Protocol [33]

S-Setting up the interview: (arrange for privacy; involve significant others; sit down; make connection with the patient; manage time constraints and interruptions)

P-Patient's perception: before you tell, ask.

I-Invitation: How would you like me to give you information? Would you like all the information?

K-Knowledge: Preface bad news with a warning, "Unfortunately I've got some bad news to tell you."

E-Exploring/Empathy: Respond to emotions with offers of support and solidarity

S-Strategy/Summary: Check patient's understanding and establish a clear plan for the future

Appendix 1: A Partial List of Movements

Canadian Virtual Hospice: http://www.virtualhospice.ca/en_US/Main+Site+Navigation/Home/For+Professionals/For+Professionals/Tools+for+Practice/Education/End+of+Life+_+Palliative+Education+Resource+Center+(EPERC).aspx.

ELNEC: https://elnec.academy.reliaslearning.com/.

Death Cafes (international): http://deathcafe.com/.

Green Burial Council: https://greenburialcouncil.org/home/what-is-green-burial/.

Dying Matters (UK): https://www.dyingmatters.org/.

National Healthcare Decisions Day (US): https://www.nhdd.org/.

Speak UP (Canada): http://www.advancecareplanning.ca/.

The Conversation Project: https://theconversationproject.org/.

Respecting Choices: http://www.gundersenhealth.org/respecting-choices/.

VitalTalk: http://vitaltalk.org/.

My Gift of Grace/Hello: https://www.mygiftofgrace.com/.

World Health Organization: http://www.who.int/cancer/palliative/definition/en/.

Appendix 2: Other Examples of Taboo Topics in Health Communication—A Start List of Future Reading

There are many other health-related topics that are taboo and often addressed in code speak. Here is a beginning list of possibilities.

Table 13.1 Voices from the field: Cross Cultural—Cross-Disciplinary Perspectives from interviews with a Chaplain, a Priest and a Palliative Care physician—The elements of communicating about the Taboo of Death, Dying, and Palliative Care

Quotes from an interview with a hospital chaplain at a county hospital in the USA	Quotes from an interview with a Catholic Priest from Ghana	Quotes from an interview with a Palliative Care physician in Ghana
Difficult communication		
People don't want to mention the "D" word. They will say "Don't tell me anything about the medical. My mom was up and strong before and she will get up again. They are afraid that if you talk about death it will happen right away	So it gets to a stage where I tell them no! the end is coming, let's face it, let's prepare ourselves for it, don't let us think there is some miracle going to come from somewhere. Even though I am a priest, I know miracles can happen, but I don't want to depend on miracles. Let's face the reality, if you face it and it doesn't come, when he is alive we are happy. But when he is going we should let him or her go in peace and we will prepare ourselves so we are not taken by surprise	I realized that it was very likely I would never see the patient again because he had deteriorated so rapidly. I had to explain the pathological processes that he had gone through because I wanted them to understand and grasp. I didn't use the "D" word. I said, "we may not have a lot of time" but they looked distress They want answers. But the initial resistance usually comes from the family who feel like if you tell "xyz" that you are going to die immediately
Avoiding talk about the death		
I have actually seen people experience relief when they are invited to talk about death. People don't realize how much relief they can give their families if they talk about it	Some people don't want to talk about death, but they know it. So, I will use the word death and say this person doesn't have much time. He could die at any time. I will say it, and when l say it, sometimes they are silent or they will ask me how do I know?	A lot of us feel that the Ghanaian doesn't want to know and that if you tell them it would be an emotional mess. I've found out that in fact a lot of patients do know what is going on and they just want confirmation

(continued)

Table 13.1 (continued)

Quotes from an interview with a hospital chaplain at a county hospital in the USA	Quotes from an interview with a Catholic Priest from Ghana	Quotes from an interview with a Palliative Care physician in Ghana
The stigma of palliative care		
I used to wear a collar until one day when I came onto the unit and one of the residents was laying on the floor putting her hands up in the air saying "Oh Father." She thought she was going to die. I learned that the collar was an anxiety thing that I wasn't aware of. One man said, "Please don't wear that collar. My wife thinks that you are here to give her last rites." Then, I realized that I was causing something to happen so I don't wear it to work anymore. That collar means "it's worse than I thought." We want them to hear palliative and think life—not death	Like I had a case where they called me, a very good friend of mine called me and said Rev., we need your expertise in palliative care, because our father has been diagnosed and so I rush out and journeyed for about 3 months. We had to send the man to America, which I thought was not necessary, but, they had the money so they did that and when they got to America the doctor said this man will not last for the next 2 weeks. So they brought the person back and we journeyed and they all knew it will happen so that one, the acceptance was there	No more elephant in the corner. This is it, it is an elephant and it is sitting in the middle of our sitting room. Let's deal with it There was a burden lifted. I've found that if you don't think it is so much of a taboo topic but it still is. The manner in which we communicate about it is what makes the difference
Culture—religion		
I don't start out with God and I don't start out with death. The way I might bring it up is if I see a cross. I might comment on a family picture if I see one. I do my best work when I don't say a thing and practice the ministry of presence. When you hit someone over the head with something—that's abuse. I never challenge when I hear someone say, "God's going to heal me."	I went out and the woman said what kind of song is this? Did you come to pray that the man will get well or for the man to die. I don't understand you So when I sang my song, I said okay. I mean if you don't like it that is what I feel like doing. I went home and then they called me around 3 am that the man is gone. So this woman was now telling me that when they call me I am going to forecast death, so they should be careful about me	Often its religion against culture. So, how much respect do you give to each? You try to tread cautiously so that you don't offend anyone. Within the same family you may have Christians and then you may have the traditionalists or a Christian who sees spiritual and the others who think that tradition is tradition and it must be followed by all means so you still have to honor the family traditions

(continued)

Table 13.1 (continued)

Quotes from an interview with a hospital chaplain at a county hospital in the USA	Quotes from an interview with a Catholic Priest from Ghana	Quotes from an interview with a Palliative Care physician in Ghana
Guilt—regret		
Families agonize when they reach the point where they can no longer take care of someone and have to place someone in long term care. I deal with guilt all the time. I work to help them change guilt to regret. The hardest person to forgive is yourself. It's not your fault that she got old or got a disease. You can regret that it is happening but you are not responsible. Family members make other members feel guilty. When you love someone for real you put them in the right place. I talk about self-forgiveness	Sometimes they know it, only they don't want to say it. They have seen the development and have noticed that they reached a stage where there is no hope. Only they don't want to say it themselves. They don't want to say it. They are hoping against hope	I remember a patient who had been living in Nigeria so her kids had grown up there and not in Ghana. When she fell ill, she asked to move to Ghana. Her son was trying to get a job in Ghana and her daughter was managing but they were refusing to get the family here in the Volta Region involved. We had to explain that sometimes cultural issues override a lot of things so if your mother says you must do this then take her at her wishes so that you can have peace because you don't want the family to descend on you when that eventually happens. Plus, your father is not here and they would descend on him and ask him why he wasn't here when their princess was sick. Sometimes it's the cultural issues that cause difficulty in discussion

Sexuality, Menstruation, Contraception

Aizenman, N. (2015). People are finally talking about the thing nobody wants to talk about. In *Goats and Soca*. National Public Radio: All Things Considered.

Kragelund Nielsen, K., Malue Nielsen, S., Buttler, R., & Lazarus, J. V. (2012). Key barriers to the use of modern contraceptives among women in Albania: A qualitative study. *Reproductive Health Matters, 20*(40), 158–165.

Mellor, R. M., Greenfield, S. M., Dowswell, G., Sheppard, J. P., Quinn, T., & McManus, R. J. (2013). Health care professionals' views on discussing sexual wellbeing with patients who have had a stroke: A qualitative study. *PLoS ONE [Electronic Resource]*.

Nambambi, N. M., & Mufune, P. (2011). What is talked about when parents discuss sex with children: Family based sex education in Windhoek, Namibia. *African Journal of Reproductive Health, 15*(4), 120–129.

Rani, A., Kumar Sharma, M., & Singh, A. (2016). Practices and perceptions of adolescent girls regarding the impact of dysmenorrhea on their routine life: A comparative study in the urban, rural, and slum areas of Chandigarh. *International Journal of Adolescent Medical Health, 28*(1), 3–9.

Redelman, M. J. (2008). Is there a place for sexuality in the holistic care of patients in the palliative care phase of life? *American Journal of Hospice & Palliative Medicine, 25*(5), 366–371.

Warriner, J., & Power, L. (2009). Let's talk about sex. *British Journal of Nursing, 18*(22), 1356.

Vaccines in Pregnancy

de Martino, M. (2016). Dismantling the taboo against vaccines in pregnancy. *International Journal of Molecular Sciences, 17*(894). doi: 10:3390/ijms17060894

Cancer Screening

Hicks, E. M., Litwin, M. S., & Maliski, S. L. (2014). Latino men and familial risk communication about prostate cancer. *Oncology Nursing Society, 41*(5), 509–516.

Rohan, E. A., Boehm, J. E., DeGroff, A., Glover-Kudon, R., & Preissie, J. (2013). Implementing the CDC's colorectal cancer screening demonstration program: Wisdom from the field. *Cancer* (August 1, 2013).

Mental Health, Dementia

Kaduszkiewicz, H., Bachmann, C., & van den Bussche, H. (2008). Telling "the truth" in dementia--Do attitude and approach of general practitioners and specialists differ? *Patient Education and Counseling70*, 220–226.

Steinberg, D. S., & C.T., W. (2017). OCD Taboo thoughts and stigmatizing attitudes in clinicians. *Journal of Community Mental Health, 53*, 275–280.

Obesity

Katz, A. (2014). The last taboo? *Oncology Nursing Forum, 41*(5), 455.

Pregnancy (de Martino, 2016).

References

1. Ratzan SC, Parker RM. Introduction. In: Selden CR, Zorn M, Ratzan SC, Parker RM, editors. National Library of Medicine current bibliographies in medicine: health literacy. Bethesda: National Institutes of Health, U.S. Department of Health and Human Services; 2000. NLM Pub. No. CBM 2000-1.
2. Thulesius HO, Scott H, Helgesson G, Lynöe N. De-tabooing dying control - a grounded theory study. BMC Palliat Care. 2013;12:13.
3. Sutton L. We can't let crucial end of life choices be impaired by silence. Nurs Times. 2010; 106(10):24.
4. Conzelus Moore C, Williamson JB. The universal fear of death and the cultural response. In: Bryant CD, editor. Handbook of death and dying, vol. 1. Thousand Oaks: Sage; 2003.
5. Folkman S. Stress, coping, and hope. Psycho-Oncology. 2010;19(9):901–8.
6. Hoelter JW, Hoelter JA. The Relationship between Fear of Death and Anxiety. The Journal of Psychology Interdisciplinary and Applied. 1978;99(2):222–226.
7. Florian V, Mikulincer M. The Impact of Death-Risk Experiences and Religiosity on the Fear of Personal Death: The Case of Israeli Soldiers in Lebanon. Journal of Death and Dying. 1993;26(2):101–111.
8. Hayslip B. Death denial: hiding and camouflaging death. In: Bryant CD, editor. Handbook of death and dying, vol. 1. Thousand Oaks: Sage; 2003. p. 34–42.
9. Nuland SB. How we die: reflections on life's final chapters. New York: A.A. Knopf; 1994.
10. Nyatanga B. Talking openly about death. Int J Palliat Nurs. 2010;16(6):263.
11. Dempsey L, Dowling M, Larkin P, Murphy K. The unmet palliative care needs of those dying with dementia. Int J Palliat Nurs. 2015;21(3):126–33.
12. Carr D. A "good death" for whom? Quality of spouse's death and psychological distress among older widowed persons. J Health Soc Behav. 2003;44(2):215–32.
13. Li HC, Richardson A, Speck P, Armes J. Conceptualizations of dignity at the end of life: Exploring theoretical and cultural congruence with dignity therapy. J Adv Nurs. 2014; 70:2920–31.
14. Chochinov HM, Hack T, Hassard T, Kristjanson LJ, McClement S, Harlos M. Dignity in the terminally ill: a cross-sectional, cohort study. Lancet. 2002;360(9350):2026–30.
15. Bryant CD. In: Bryant CD, editor. Handbook of death and dying, vol. 1. Thousand Oaks: Sage Publications; 2003.
16. Bryant CD, editor. Introduction, vol. 1. Thousand Oaks: Sage; 2003.
17. Wook Shin D, Cho J, SYoung Kim S, Joo Chung I, Soo Kim S, Kook Yang H, Park JH. Discordance among patient preferences, caregiver preferences, and caregiver predictions of patient preferences regarding disclosure of terminal status and end-of-life choices. Psycho-Oncology. 2015;24:212–5.
18. Brown EA, Bekker HL, Davison SN, Koffman J, Schell JO. Supportive care: communication strategies to improve cultural competence in shared decision making. Clin J Am Soc Nephrol. 2016;11(10):1902–8.
19. Su CT, McMahan RD, Williams BA, Sharma RK, Sudore RL. Family matters: effects of birth order, culture, and family dynamics on surrogate decision-making. J Am Geriatrics Soc. 2014;62(1):175–82.
20. Ng RW, Chan S, Wee Ng T, Ling Chiam A, Lim S. An exploratory study of the knowledge, attitudes and perceptions of advance care planning in family caregivers of patients with advanced illness in Singapore. Support Palliat Care. 2012;3:343–8. https://doi.org/10.1136/bmjspcare-2012-000243.
21. Braun KL, Nichols R. Death and dying in four Asian American cultures: A descriptive study. Death Studies. 1997;21:327–59.
22. Dai YX, Chen TJ, Lin MH. Branding palliative care units by avoiding the terms "palliative" and "hospice": a nationwide study in Taiwan. J Health Care Organ Provision Financ. 2017; 54:1–6.

23. Ziehm J, Farin E, Shafer J, Woitha K, Becker G, Koberich S. Palliative care for patients with heart failure: facilitators and barriers --a cross sectional survey of German health care professionals. BMC Health Serv Res. 2016;16:361.
24. Zimmerman C, Swami N, Krzyzanowska MK, Leihl N, Rydall A, Rodin G, Hannon B. Perceptions of palliative care among patients with advanced cancer and their caregivers. CMAJ. 2016;188(10):E217–27.
25. Cherny N. Stigma associated with palliative care. Cancer. 2009;115:1808–12.
26. Gomes B, Canlanzani N, Gysels M, Hall S, Higginson IJ. Heterogeneity and changes in preferences for dying at home: a systematic review. BMC Palliat Care. 2013;12(1):7.
27. Schim SM, Doorenbos AZ. A three-dimensional model of cultural congruence: framework for intervention. J Soc Work End-Of-Life Palliat Care. 2010;6(3–4):256–70.
28. Dong F, Zheng R, Chen X, Wang Y, Zhou H, Sun R. Caring for dying cancer patients in the Chinese cultural context: a qualitative study from the perspectives of physicians and nurses. Eur J Oncol Nurs. 2015;21:189–96.
29. Granek L, Krzyzanowska MK, Tazer R, Mazzotta P. Oncologists' strategies and barriers to effective communication about the end of life. J Oncol Pract. 2013;9(4):129–35.
30. Roscoe LA, Schonwetter RS. Improving access to hospice and palliative care for patients near the end of life: present status and future direction. J Palliat Care. 2006;22:46–50.
31. Roseman JL. Reflections on the Sydney Project: Can we talk? Can we give voice to the taboo topics that are usually not embraced in residency medical education? J Pain Symptom Manage. 2014;48(3):478–82.
32. Taubert M, Watts G, Boland J, Radbruch L. Palliative social media. Support Palliat Care. 2014;4:13–8.
33. Baile WF, Buckman R, Lenzi R, Glober G, Beale EA, Kudelka AP. SPIKES- a six-step protocol for delivering bad news: application to the patient with cancer. The Oncologist. 2000; 5:302–11.

Chapter 14
Interpreting the Meaning in Our Genomes: Perspectives from Biochemistry, Genetics, Infectious Disease, and Dance

Jennifer A. Surtees, Thomas Russo, and Anne H. Burnidge

Our understanding of the information contained within our genomes (our DNA) is revolutionizing the way we consider personal and public health and well-being, as well as our concepts of race, ancestry, and identity. At the same time, we are at risk of exacerbating existing health and socioeconomic disparities by limiting the voices that are part of the conversations addressing the impact of genomic data. This chapter is a call to action—to recognize the importance of genomic information across disciplines and to promote an understanding of that importance across societies—a call for broad-based genomic literacy.

Within the human genome resides information about the genetic factors that impact human health and disease—what remains is the daunting task of decoding that information for the benefit of all. Multiple variants can contribute to disease; different versions of genes can impact age of onset or affect the severity of symptoms. Interactions with the environment are likely critical. By understanding the contribution of different genetic factors to health and disease, we can improve treatment and develop preventative approaches that are personalized to one's genomic information [1–3].

J. A. Surtees (✉)
Genome, Environment and Microbiome Community of Excellence, University at Buffalo, State University of New York (SUNY), Buffalo, NY, USA
e-mail: jsurtees@buffalo.edu

T. Russo · A. H. Burnidge
University at Buffalo, State University of New York (SUNY), Buffalo, NY, USA

© Springer Nature Switzerland AG 2020   213
K. H. Smith, P. K. Ram (eds.), *Transforming Global Health*,
https://doi.org/10.1007/978-3-030-32112-3_14

The human genome also contains information about ancestry and migration patterns throughout history. In conjunction with archeological and historical records, our collective genomes help tell the story of the human race—and related hominid species [4–6]. Thus, our genomes not only provide practical health indicators, but also influence the less tangible sense of our personal identities.

What Is a Genome?

One's genome is the complete set of genetic material—all of one's DNA. The genome contains the information required to create, manage, and run an organism, from single-celled microbes to multicellular animals, including humans. It is often referred to as the "instruction manual" for an organism. Its genes encode all the proteins required for a cell to function and to generate different cell types and tissues. There must also be careful gene regulation; the appropriate genes must be expressed, or "turned on," at the right time and in the right place. Thus, in addition to providing the blueprint for the building materials, the genome also encodes intricate regulatory information. Changes in either genes or gene regulation can alter genome function—sometimes for the better, improving the chances of passing one's genome on to the next generation, and sometimes significantly less so.

Understanding the Sources and Importance of Human Genetic Diversity

The DNA sequences of humans are 99.9% identical. The remaining ~0.1% is what makes us all different from each other (except identical twins) and the major source of that diversity among humans is derived from single base changes, i.e. single nucleotide polymorphisms, or SNPs (pronounced snips). SNPs are defined as nucleotide positions within the genome where there is more than one possible version of a DNA sequence within a population. There are about 10 million SNPs in the pairs in our ~3 billion base pair human genome. Therefore, every gene almost certainly contains nucleotide positions that differ between individuals. Furthermore, blocks of SNPs are typically co-inherited, which helps determine ancestry and related-ness of individuals.

Inherited variations arise from changes in the germline DNA, i.e., they are passed on from one generation to the next, as a result of replication errors or genetic recombination. SNPs, or other variants, can change the sequence of the gene product, can change gene expression, turning a gene on or off, or have no discernible effect. Sometimes a SNP is linked to (or nearby) a gene or DNA sequence that encodes a trait, or phenotype, without causing the phenotype itself. Understanding how this works is the basis of genome-wide association studies (GWAS), which we will discuss later in this chapter.

Changes During DNA Replication

Prior to cell division, when a single cell grows and divides to become two cells, the entire genome is copied to create two complete copies of the genome. During this process, called DNA replication, the genome is susceptible to mutations, although multiple mechanisms for DNA repair and genome stability are present in our cells.

The iconic DNA helix reveals the double-stranded structure of DNA. The two strands are held together by base pairing, fitting together in a complementary fashion (Fig. 14.1a). There are four bases that make up the DNA sequence: adenine (A), cytosine (C), guanine (G), and thymine (T). A pairs with T; G pairs with C. When DNA is replicated, the elegant double-stranded helix is separated to reveal two single-stranded DNA templates (Fig. 14.1b). The replication machinery, a team of proteins, synthesizes new DNA *polymers* by connecting the deoxyribonucleotides (dNTPs), the building blocks of DNA. To create this new DNA strand, a free dNTP is inserted opposite a dNTP base in the template (old strand) via A-T and G-C base pairing. The dNTPs are added one at a time along the length of the template strand, and connected together, until there are two complete copies of the genome where there was once only one—each with one "old" strand and one newly synthesized strand of DNA (Fig. 14.1c).

There are a number of safeguards that ensure the correct base is incorporated. As a result, the error rate of DNA replication is one for every ~1 billion nucleotides replicated! But the human genome is about three billion base pairs and some regions of the genome are more prone to errors than others—mistakes do happen, leading to changes in the DNA (Fig. 14.1d). Changes in most cell types (skin, heart) may lead to changes in function and can ultimately lead to genetic diseases such as cancer. These changes in body cells do not get passed on to offspring. However, unrepaired errors in DNA replication that occur in germline cells, i.e., sperm or egg cells, can be inherited in the next generation.

Following DNA replication, there are four copies of every chromosome (rather than two), so that when the cell divides, each daughter cell will inherit two of each chromosome. The chromosomes are separated in a process called mitosis. The replicated chromosome pairs remain attached at their centromeres, forming the familiar X shape that we often think of when we think of chromosomes (Fig. 14.2a). These pairs line up along the equator of the cell. Spindle protein fibers extend from each cell pole, attach to either side of a chromosome pair and pull apart each replicated pair. New cell membrane forms at mid-cell, resulting in two new cells, each with an entire complement of chromosomes.

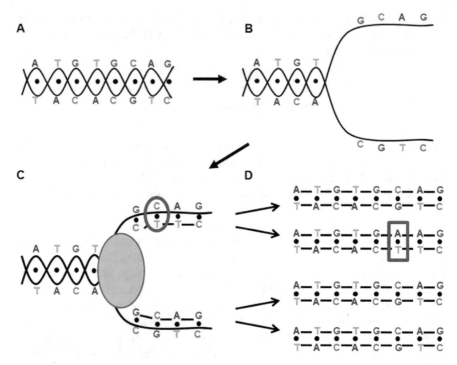

Fig. 14.1 (**a**) The DNA double helix is formed by base pairing between two strands of DNA. (**b**) The strands are separated at origins to initiate DNA replication. (**c**) The replication machinery (turquoise oval) sometimes makes mistakes, resulting in a misincorporation and a mispair (red circle). (**d**) If not corrected, this leads to a sequence change in subsequent rounds of DNA replication (red box). (Source: Jennifer A. Surtees)

Genetic Rearrangements (Recombination) During Meiosis

In contrast to potential errors in replication, which are accidental and typically very low in frequency, genetic diversity is deliberately generated during meiosis, the process that generates the sperm and egg cells that generate the next generation. Prior to meiosis, DNA replication duplicates the genome. Any changes in the DNA sequence that are not corrected have the potential to be inherited. And while this happens infrequently in an individual cell, over the course of human history, millions of inherited changes have occurred—generating SNPs.

Following DNA replication, there are two meiotic cell divisions that reduce the chromosome number to generate germ cells with one set of 23 chromosomes. Upon fertilization, the egg and sperm each contribute 1 set of chromosomes—23 from the mother and 23 from the father in humans, for a total of 46 chromosomes. But before the chromosomes are segregated into the germ cells, DNA recombination generates genetic diversity.

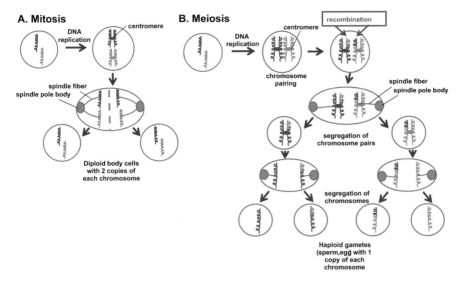

Fig. 14.2 (**a**) Mitosis: following DNA replication, duplicated chromosomes line up along mid-cell. Spindle fibers (curved lines) come from the poles (blue ovals), attach to the centromeres, and pull chromosomes to opposite poles; new cell membrane forms at mid-cell. (**b**) Meiosis: following DNA replication, pairs of replicated chromosomes align at mid-cell, i.e., both replicated copies of chromosome 1. DNA breaks are induced and portions of the chromosome arms are exchanged (red arrows). The spindles separate the pairs; a new membrane forms at mid-cell. A second chromosome segregation reduces the chromosome number by half to generate haploid cells. (Source: Jennifer A. Surtees)

In early meiosis, pairs of chromosomes (both replicated copies of chromosome 1, for example) are aligned and tethered. Double-strand DNA breaks are *purposely* generated along the length of each chromosome. Some of those breaks are repaired by swapping genetic information between the two different versions of the chromosome (Fig. 14.2b). This exchange, or recombination, is reciprocal; there is no loss or gain of genetic material nor is the order of the DNA changed. Rather, there is a mixing and matching of the DNA, including SNPs, along the chromosome. This means that the egg cells in your mother contain chromosomes that are a combination of her mother's and father's chromosomes and the germ cells in you are created from a mash-up of your own parents. With each successive generation, there is more mixing and matching of DNA sequences and SNPs to create tremendous genetic diversity. More closely related individuals have more similar SNP patterns, which is part of the underlying science for direct-to-consumer ancestry DNA services.

How Does Understanding Our Genome Impact Human and Global Health?

How can we connect genomic information (and genetic diversity) with disease and/or health and wellness? We use SNPs as markers, identifying SNPs that are "linked" to a phenotype such as a disease or trait. If a SNP contributes to a trait of interest or is located very close to the gene that encodes that trait, people with that trait are more likely to have that SNP than those individuals without the trait. Using a detailed map of the human genome and databases of genetic variants, we can perform genome-wide association studies (or GWAS) to identify SNPs that are associated with a trait of interest. The basic approach is to take a population with a particular trait (test group) and a population that lacks the trait (control group) and compare SNPs across the genome of each individual within each population. SNPs that are clearly present in both populations are unlikely to contribute to the trait being studied. Some SNPs may be present more frequently in the test group than in the control group. This correlation *indicates* a genetic link between the SNP and the trait of interest. For example, SNPs of *APOL1* have been correlated with chronic kidney disease, have been shown to be associated with increased disease progression, and are more common in African and African-American populations [7]. Of course, this knowledge is of little use if treatment options are unavailable, hence the debate about resource allocation in low and middle income countries—medical facilities versus genomic infrastructure.

Importantly, it is not possible to make statistically meaningful correlations with only a handful of individuals. Similarly, genomic information from more distantly related or diverse populations is important in establishing these critical correlations. We will return to this point shortly. It is important to remember that the genetic factors identified through GWAS are statistical correlations—that is, they *predict* a relationship that must be confirmed with directed studies to establish causation. The presence of a SNP associated with a trait is not *deterministic*—possessing that SNP does not guarantee that trait will be expressed in a particular individual. Environmental factors (such as diet or chemical exposure) may play a role, or there may be protective genetic factors to mitigate the predicted effect of a SNP.

Consider the human *MSH3* gene, which encodes a DNA repair protein. A *MSH3* SNP, rs26279, has been associated with a predisposition to certain cancers and to chemotherapy sensitivity. This SNP means that *MSH3* nucleotide position 3108 can be either an A or a G [8]. Two copies of each gene, including *MSH3*, are present in the genome. Therefore, there are three possible allele combinations at this position in *MSH3*: G/G, G/A (or A/G), and A/A. Notably, this SNP distribution is distinct in different populations (Fig. 14.3), altering the probability of cancer in different groups. With about 180 SNPs within *MSH3*, there is a huge number of possible SNP combinations within this single gene that represents only about 0.0001% of the genome. Imagine the diversity across the entire genome!

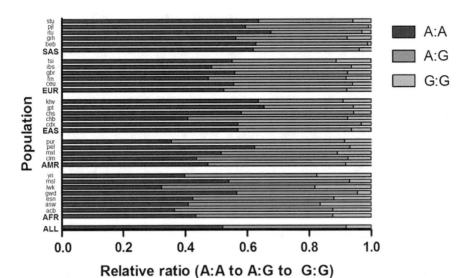

Fig. 14.3 *MSH3* rs26279 genotypes in different populations. *ALL* all populations, *AFR* African, *AMR* Americas, *EAS* East Asian, *EUR* European, *SAS* South Asian. Sub-populations are indicated with lower case letters. Figure has been created by chapter author Jennifer A. Surtees through the adaptation of datasets presented in publicly available databases through the National Center for Biotechnology Information. (NCBI). Datasets located at: https://www.ncbi.nlm.nih.gov/projects/SNP/snp_ref.cgi?do_not_redirect&rs=rs26279 and https://www.ncbi.nlm.nih.gov/snp/rs26279

In addition to predisposing us to particular traits or diseases, SNPs can influence our response to various drugs, which is the foundation of the growing field of pharmacogenomics. Perhaps the most famous example is the response to the anticoagulant warfarin. Two key genes influence a person's metabolism of and sensitivity to this drug—*CYP2C9* and *VKORC1* [9]. Specific *CYP2C9* SNPs are correlated with rapid metabolism of warfarin, leading to rapid removal from the body through the liver. Individuals with these SNPs may need higher doses for more effective treatment. In contrast, a *VKORC1* SNP increases sensitivity to warfarin, requiring lower doses. Obtaining this genetic information from a patient *prior* to prescribing can help healthcare professionals choose an appropriate warfarin dose.

Another example is reaction to the anti-viral drug abacavir, part of a combination therapy for HIV-infected patients. In North America, 5–8% of patients are hypersensitive to abacavir and suffer from fever, skin rash, gastrointestinal symptoms, and eosinophilia within 4–6 weeks of initial exposure and more severe symptoms if re-exposed to the drug. Hypersensitivity is highly correlated with the HLA-B*5701 SNP, therefore genetic testing for this variant prior to treatment is clinically indicated and valid. However, this SNP frequency varies by population (high among Caucasian, Indian, and Thai populations, low among Africans, almost non-existent among Koreans) and therefore the *need* for testing may vary globally [10–12].

Similarly, ~20% of Zimbabweans carry two copies of a *CYP2B6* SNP that severely reduces their ability to metabolize efavirenz, part of a three-in-one HIV combination therapy. In these patients, efavirenz accumulates in the bloodstream, causing hallucinations, depression, and suicidal tendencies, all of which led patients to stop taking the drug and precipitated a health crisis in 2015. Testing for this variant could allow for formulations with lower efavirenz doses. In 2016 Botswana took a different approach, adopting the alternative, albeit more expensive, drug dolutegravir, because ~13.5% of that country's population carries two copies of the *CYP2B6* SNP [7, 13]. In these examples, and many more, it is clear that understanding the connections between genomic information and health outcomes can facilitate and inform public health policy around the world.

Microbial Genomics

We can track human migration and human variation to develop an understanding of how our genomes contribute to traits and disease. The same principles apply to the pathogens that infect us. An important, recent example of this is a massive study of cholera around the world. By comparing the SNPs and other genetic variants within the hundreds of *Vibrio cholera* (the bacterium that causes cholera) genomes from Asia, Africa, and the Americas over the past 50 years, scientists traced the migration patterns of different *V. cholera* strains. Notably, strains that caused pandemics in both Africa and the Americas originated in Asia. Local strains caused disease, but rarely a full-blown epidemic. Therefore, efforts in eradicating local reservoirs of the microbe in Africa and America may be less effective than eradicating Asian strains to prevent outbreaks [14–16].

Similarly, the unprecedented 2018 outbreak of Lassa virus in Nigeria is being tracked with genomics. Lassa virus is borne by Mastomys rats and human infection usually is a result of exposure to the urine or feces of infected animals. Lassa virus can cause hemorrhagic fever, but the disease has a wide range of severity and a variable clinical course. Genomic data suggest extensive diversity of Lassa virus strains circulating in Nigeria, likely partially accounting for the diverse clinical outcomes. No evidence was found for hypervirulence. Furthermore, despite concerns of extensive human-to-human transmission, analysis of the genomic structure and variation obtained from dozens of samples suggest that there have been multiple cross-species (rat-to-human) transmissions, rather than human-to-human transmission [17–20].

Genomics and Diversity

Our ability to decode genomic information, for health or for ancestry, is limited by the lack of diversity in genomic databases. The vast majority (currently about 80%) of individuals represented in the current databases around the world are from

European ancestry. This is a significant improvement over the past decade; in 2009, 96% of collected genomic information was derived from European descendants [8, 21]. Nonetheless, this dearth in diversity within the databases translates to disparities in our understanding of how genetic variation affects health in different populations. More broadly, it compromises research designed to understand both genetic predictors of disease and the underlying mechanisms [22].

Notably, recent studies of skin pigmentation and schizophrenia, among others, have demonstrated that there are a huge number of genetic variants that are only observed in African populations [5, 22–26]. Because humans originally evolved in Africa, there is much more genetic diversity represented across that continent. Smaller groups migrated from Africa to Europe, through the Middle East and into Asia—and eventually the Americas. These small groups carried limited genetic diversity to new colonies around the globe, restricting the genetic variants that are observed in different populations today (Fig. 14.4). Tracking these variants is how we can determine likely ancestry from genomic information. But we are missing a tremendous amount of genomic information because of a lack of diversity in genomic databases, particularly with respect to populations of African descent. Only about 3% of GWAS have been performed with African populations; 81% involve individuals of European descent. As a result, we are behind in understanding the genetic factors that contribute to various disease traits, or drug responses, in these underrepresented populations [8, 24, 27, 28].

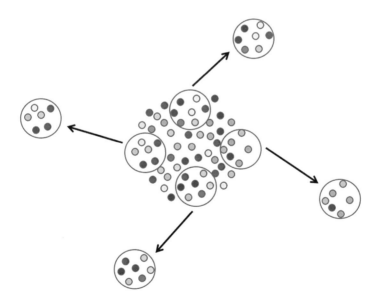

Fig. 14.4 Genetic diversity is highest across African populations (represented in the center of the cartoon), where humans originated. Diversity is indicated by different color circles. Smaller subpopulations migrated out of Africa (encircled) decreasing the genetic diversity. Over time, novel genetic variants arose in these smaller groups. These patterns of mutation allow us to track migration. (Source: Jennifer A. Surtees)

Beyond the impact on health, differences in the frequency of SNPs and genetic variation within and between populations can help us understand human evolution and diversity, as well as the migration of different human populations around the globe. The more similar the patterns of SNPs between individuals, the more likely they are to be related. DNA regions containing SNPs rearrange through generations, making more distantly related individuals less similar with respect to these SNPs. Thus this diversity can be traced back through generations by examining and analyzing DNA sequences and SNP patterns. Understanding genomic diversity can therefore fundamentally change how we think our own identity and about so-called differences among us.

There is an ongoing concerted effort to broaden the diversity of genomic databases. In 2015, the National Institutes of Health (NIH) in the USA and the Wellcome Trust in the United Kingdom funded the Human Heredity and Health in Africa (H3 Africa) Initiative [29, 30]. H3 Africa funds African scientists, has helped set up collaborative research centers, including biobanks and tissue repositories for genomic research, and has developed training programs to promote genomic literacy in Africa through peer training. Similarly, the Sanger Institute in the UK participates in the African Genome Variation Project, a collaborative network of scientists across Africa and the UK seeking to understand the genetic variation among African populations [9]. In the USA, the NIH launched *All of Us* in 2018, a broad-based effort to recruit participation of all Americans in genomic databases, with a particular emphasis on populations that are currently underrepresented. Many other initiatives are underway: Global Gene Corp in India is focused on populations from South Asia; South Korea launched its own genomics project in 2015; the GenomeAsia 100K Initiative will sequence 100,000 genomes in south, north, and east Asia; 23andMe launched its African Genetics Project in 2016 and recently launched its Global Genetics Project; the Population Genomics and Global Health team at Stanford University is focusing on individuals from the Caribbean, Mexico, and South America [31–33]. All seek to expand the diversity of available genomic information to improve health and treatments for all populations around the globe [18, 34, 35]. Broader genomic literacy that creates an understanding of the benefits of broad participation in these efforts will be critical for their success.

Beyond precision medicine based on one's personal genomic information, a better understanding of *all* human genetic diversity will improve its ultimate implementation. And "precision public health," basing public health decisions on populations and communities instead of individuals, may well precede personalized precision medicine. For example, in 2015, Ethiopia banned the use of codeine because ~30% of Ethiopians are ultra-metabolizers (a result of a *CYP2D6* SNP) who convert codeine to morphine extremely efficiently leading to excessive morphine formation [36].

Challenges in Genomic Literacy

The term "genome" is used all around us in many different contexts and is therefore familiar in that sense. But what the genome actually "is" and what we can learn from it is often misunderstood. A Wellcome Trust survey performed in the United Kingdom in 2012 found that only about 25% of respondents (ages 17 and up) self-reported "some understanding" or "better understanding" of the term "human genome." In a national genomic literacy survey that the University at Buffalo, The State University of New York deployed in the USA in the spring of 2017, researchers found that about 42% of respondents ages 18 and older self-reported a "good" or "better" understanding of the term "human genome." While this is an improvement, it is notable that in the same survey, about 75% of respondents reported good to excellent understanding of "human DNA," which means that almost half did not make the connection between human genome and human DNA.

Why does it even matter if people know about their genome? It may seem important but somewhat esoteric—primarily the purview of doctors and scientists. In fact, there are compelling reasons why broad genomic literacy is an important goal. First, genomic literacy empowers patients, which is pretty much everyone, to ask questions and to request information about genomic testing. Second, genomic literacy empowers everyone to make better informed choices about their own health and lifestyle. Everyone should strive for a healthy lifestyle, but some choices may be more important to the patient with genomic makers indicating heart disease or diabetes. Third, in this age of direct-to-consumer genetic testing, there is a significant gap in the genetic information provided and people's ability to interpret and apply that potentially vital information—and understand its limitations. Fourth, genomic information informs questions of identity and self. Finally, it is critical that we have an informed electorate to ensure that many voices are heard when decisions are being made about genomic public policy. Whether related to legal issues, issues of informed consent, privacy and security issues as well as patients' rights, a diversity of voices should be part of the conversation to ensure public policy that mirrors society's values.

There are several barriers to genomic literacy. One is a lack of access to the relevant information—genomics, in general, is not taught in schools or other formal settings. Another is the perception that a lot of background information in required in order to get people to understand what their genome is and what it means. Lack of interest and/or a sense of intimidation is related to this latter issue. But sitting in a lecture hall, building on years of biology content, is not the only way to learn about the genome. In developing genomic literacy across a spectrum of audiences, it is important to *partner* and *engage* with people, to make the information interesting and relevant to them.

Reaching a broad audience for genomic literacy will require new, innovative ways of engaging diverse audiences and means of communicating this information.

An interdisciplinary approach is essential and research is needed to continue to develop these approaches. In the meantime, we know that promoting genomic literacy broadly requires educating healthcare professionals who can then inform their patients. It requires genomically literate teachers to prime the next generation for the genomic age—and to shift educational standards in this direction. It requires scientists and researchers to initiate conversations with a broad range of lay people in our communities—to discuss genomics and the complexity surrounding genomic information. It requires lawyers to explore legal issues ranging from informed consent for testing to the privacy and security of our genomic information. It requires social science and humanities researchers to reflect on our evolving ideas about identity in this genomic age and their impact on our society. It requires artists to embody and reflect the processes and the ethical questions inherent in genomic information. It requires entire communities to recognize that knowing our genome is paradigm shifting in ways that we cannot predict. It requires all of us to start and continue these very important conversations (Fig. 14.5).

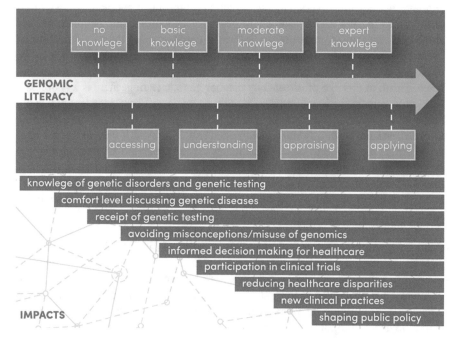

Fig. 14.5 Levels of genomic literacy and impacts. (Source: concept: Jennifer A. Surtees; graphic design: Nicole C. Little)

Case Study #1 Using Microbial Genomics to Understand Pathogen Colonization: Implications for Public Health by Thomas Russo

With the advent of DNA sequencing it is now possible to define the microbiome (the genomic content of the microbes that colonize the body). Although this has not yet become part of personalized medicine practice for individual patients, it is being done for research purposes. These data have the potential to define the presence of microbes capable of causing infection (pathogens) within a population, which in turn can be used to estimate the risk of subsequent infection. This approach can be exemplified by an evolving pathotype of Klebsiella pneumoniae.

In the mid-1980s and 1990s, reports from Taiwan described a unique clinical syndrome of community-acquired, invasive K. pneumoniae infection in otherwise healthy individuals that often presented with multiple sites of infection. The offending pathogen has been termed hypervirulent K. pneumoniae (hvKp). Although these observations were initially made in the Asian Pacific Rim where the majority of cases have been described, an increasing number of such cases are being reported worldwide. However, the risk of infections due to hvKp is presently unknown for most countries around the world.

Colonization with hvKp is the first step that may lead to infection. Determination of colonization rates would assist in predicting the risk of subsequent infection. Mining the ever-increasing fecal microbiome sequence data generated from individuals around the world would be an efficient means to measure the prevalence of hvKp colonization. To accomplish this goal, our group generated a DNA database from fecal microbiome sequences in the public domain. This database was segregated by geographic location and ethnicity of the donor and subsequently interrogated for a combination of genes that have been shown to accurately identify hvKp. This process enables an accurate determination of the presence of hvKp and reflects the prevalence of colonization. The use of this information across communities, ethnic groups, and countries would lend insights into who would be at risk for hvKp infection and would assist in defining the overall public health risk.

Mining the microbiome likewise could help in our battle against antibiotic resistant bacteria by determining the prevalence of antibiotic resistance genes at the local, national, and international level. If, at presentation, the doctor is aware that an individual is colonized with antibiotic resistant bacteria or even the prevalence of these bacteria at the local community level, then this would greatly assist physicians in making optimal treatment choices pending final culture results, which can take hours to days. It is possible to imagine that in the near future individuals at risk for infection, and perhaps everyone, will have their microbiome analyzed as part of their personalized preventative health program and that this information, in turn, would be used to assist doctors in the diagnosis and treatment of infections if and when they occur [37, 38].

Case Study #2 Dancing the Science: An Embodied Approach to Genomic (Scientific) Literacy by Anne H. Burnidge

I hear and I forget. I see and I remember. I do and I understand.

—Confucius

Dance is a uniquely human endeavor that is created with the human body, a complex art form that engages the body's knowledge to tell personal narratives, express emotions, spark new ideas, create dialog, connect with others, and even explore scientific discoveries such as the human genome. Investigating scientific concepts through dance brings science to a human level, allowing observers and practitioners to develop a visceral, first-person understanding of multifaceted theories [39–41].

Dance and movement are particularly well suited to exploring the inner workings of the body. Processes and systems at the molecular level exhibit the same characteristics as macro-scale human movement, including action/dynamics, shape/structure, spatial intent/directionality, duration/phrasing, and function/expression. Thus, one can create dance projects that embody genomic concepts in order to facilitate a deeper understanding of both science and dance—for both the dancer/choreographer and the audience.

The first step in the process is developing questions and gathering information. What structures are involved? What processes are happening, in what order and with what characteristics? What relationships exist and what kinds of interactions occur? Then the scientific information is translated into movement by making decisions about what the dancers represent (base pairs, enzymes), how they move (dynamics, spatial pathways), what they are doing (communicating, breaking bonds, replacing one another), how they relate to one another (through touch, focus), and what else is needed (costumes, props, music, lighting). The choreographer ensures that the artistic elements are clear and that the piece communicates an overarching message that is true to the scientific foundation. Now, the dance project is ready to be performed for an audience.

Participating in creative inquiry through embodied exploration and expression allows individuals of all ages and abilities to experience scientific enquiry. The performance brings science into a new context, providing a rich, metaphoric landscape from which observers can gain a holistic understanding and appreciation of the nuances of the science. Ultimately, dancing is a form of inquiry; the body is a site for knowledge creation; and investigating, experiencing, and observing with and through dance has the ability to engage participants and the public in embodied learning that promotes genomic literacy.

Acknowledgements Thomas Russo, MD, professor of infectious disease, wrote Case Study #1. Anne H. Burnidge, MFA, associate professor of theatre and dance, wrote Case Study #2.

References

1. Akhmetov I, Bubnov RV. Assessing value of innovative molecular diagnostic tests in the concept of predictive, preventive, and personalized medicine. EPMA J. 2015;6:19. https://doi.org/10.1186/s13167-015-0041-3.
2. Orchard C. Genomic medicine in the real world: 'hope' and 'hype'. 2015. Accessed 29 Mar 2019.
3. Shendure J, Findlay GM, Snyder MW. Genomic medicine–progress, pitfalls, and promise. Cell. 2019;177(1):45–57. https://doi.org/10.1016/j.cell.2019.02.003.
4. Dannemann M, Racimo F. Something old, something borrowed: admixture and adaptation in human evolution. Curr Opinion Genet Dev. 2018;53:1–8. https://doi.org/10.1016/j.gde.2018.05.009.
5. Montinaro F, Capelli C. The evolutionary history of Southern Africa. Curr Opinion Genet Dev. 2018;53:157–64. https://doi.org/10.1016/j.gde.2018.11.003.
6. Metspalu M, Mondal M, Chaubey G. The genetic makings of South Asia. Curr Opinion Genet Dev. 2018;53:128–33. https://doi.org/10.1016/j.gde.2018.09.003.
7. Popejoy AB, Fullerton SM. Genomics is failing on diversity. Nature. 2016;538:161–4.
8. Ensembl. 2018. Accessed 29 Mar 2019.
9. Wellcome Sanger Institute. 2018. African genome variation project. Accessed 29 Mar 2019.
10. Ma JD, Lee KC, Kuo GM. HLA-B∗5701 testing to predict abacavir hypersensitivity. PLoS Curr. 2010;2:RRN1203. https://doi.org/10.1371/currents.RRN1203.
11. Rodriguez-Nóvoa S, Soriano V. Current trends in screening across ethnicities for hypersensitivity to abacavir. Pharmacogenomics. 2008;9(10):1531–41. https://doi.org/10.2217/14622416.9.10.1531.
12. Nyakutira C, Röshammar D, Chigutsa E, Chonzi P, Ashton M, Nhachi C, Masimirembwa C. High prevalence of the CYP2B6 516G→T(∗6) variant and effect on the population pharmacokinetics of efavirenz in HIV/AIDS outpatients in Zimbabwe. Eur J Clin Pharmacol. 2008;64(4):357–65. https://doi.org/10.1007/s00228-007-0412-3.
13. Bernard S, Neville KA, Nguyen AT, Flockhart DA. Interethnic differences in genetic polymorphisms of CYP2D6 in the U.S. population: clinical implications. The Oncologist. 2006;11(2):126–35. https://doi.org/10.1634/theoncologist.11-2-126.
14. Kupferschmidt K. Genomes rewrite cholera's global story. Science. 2017;358(6364):706–7. https://doi.org/10.1126/science.358.6364.706.
15. Weill F-X, Domman D, Njamkepo E, Tarr C, Rauzier J, Fawal N, Keddy KH, et al. Genomic history of the seventh pandemic of cholera in Africa. Science. 2017;358(6364):785–9. https://doi.org/10.1126/science.aad5901.
16. Domman D, Quilici M-L, Dorman MJ, Njamkepo E, Mutreja A, Mather AE, Delgado G, et al. Integrated view of *Vibrio cholerae* in the Americas. Science. 2017;358(6364):789–93. https://doi.org/10.1126/science.aao2136.
17. Siddle KJ. New Lassa Virus genomes from Nigeria 2015–2016. 2018. Accessed 29 Mar 2019.
18. Beaubien J. Nigeria faces mystifying spike in deadly Lassa fever. 2018. Accessed 29 Mar 2019.
19. Maxmen A. Deadly Lassa-fever outbreak tests Nigeria's revamped health agency. Nature. 2018;555:421–2.
20. Siddle KJ, Eromon P, Barnes KG, Mehta S, Oguzie JU, Odia I, Schaffner SF, et al. Genomic analysis of Lassa virus during an increase in cases in Nigeria in 2018. New England J Med. 2018;379(18):1745–53. https://doi.org/10.1056/NEJMoa1804498.

21. Gibbons A. How Africans evolved a palette of skin tones. Science. 2017;358(6360):157–8. https://doi.org/10.1126/science.358.6360.157.
22. Munung NS, Mayosi BM, de Vries J. Genomics research in Africa and its impact on global health: insights from African researchers. Global Health Epidemiol Genomics. 2018;3:e12. https://doi.org/10.1017/gheg.2018.3.
23. McClellan JM, Lehner T, King M-C. Gene discovery for complex traits: lessons from Africa. Cell. 2017;171(2):261–4. https://doi.org/10.1016/j.cell.2017.09.037.
24. Callaway E. Ancient genomes expose Africa's past. Nature. 2017;547:149.
25. Crawford NG, Kelly DE, Hansen MEB, Beltrame MH, Fan S, Bowman SL, Jewett E, et al. Loci associated with skin pigmentation identified in African populations. Science. 2017;358(6365):eaan8433. https://doi.org/10.1126/science.aan8433.
26. Schleit J, Johnson SC, Bennett CF, Simko M, Trongtham N, Castanza A, Hsieh EJ, et al. Molecular mechanisms underlying genotype-dependent responses to dietary restriction. Aging Cell. 2013;12(6):1050–61. https://doi.org/10.1111/acel.12130.
27. Browning SR, Browning BL, Zhou Y, Tucci S, Akey JM. Analysis of human sequence data reveals two pulses of archaic denisovan admixture. Cell. 2018;173:53–61. https://doi.org/10.1016/j.cell.2018.02.031.
28. van de Loosdrecht M, Bouzouggar A, Humphrey L, Posth C, Barton N, Aximu-Petri A, Nickel B, et al. Pleistocene North African genomes link Near Eastern and sub-Saharan African human populations. Science. 2018;360:548–52. https://doi.org/10.1126/science.aar8380.
29. Africa, H3. 2019. https://h3africa.org/. Accessed 29 Mar 2019.
30. NHGRI. Genomics in Africa. 2015. Accessed 29 Mar 2019.
31. Arney K. The power of a billion: India's genomic revolution. 2017. Accessed /29 Mar 2019.
32. Newswire PR. GenomeAsia 100K initiative announced to sequence 100,000 genomes in south, north and east Asia. 2016. Accessed 29 Mar 2019.
33. McGonigle I, Schuster SC. Global science meets ethnic diversity: Ian McGonigle interviews GenomeAsia100K Scientific Chairman Stephan Schuster. Genet Res. 2019;101:e5. https://doi.org/10.1017/S001667231800006X.
34. Mlotshwa BC, Mwesigwa S, Mboowa G, Williams L, Retshabile G, Kekitiinwa A, Wayengera M, et al. The collaborative African genomics network training program: a trainee perspective on training the next generation of African scientists. Genet Med. 2017;19:826. https://doi.org/10.1038/gim.2016.177.
35. Qian W, Deng L, Lu D, Xu S. Genome-wide landscapes of human local adaptation in Asia. PLoS One. 2013;8(1):e54224. https://doi.org/10.1371/journal.pone.0054224.
36. Parsa A, Kao WHL, Xie D, Astor BC, Li M, Hsu C-y, Feldman HI, et al. APOL1 risk variants, race, and progression of chronic kidney disease. N Engl J Med. 2013;369(23):2183–96. https://doi.org/10.1056/NEJMoa1310345.
37. Shon AS, Bajwa RPS, Russo TA. Hypervirulent (hypermucoviscous) Klebsiella pneumoniae. Virulence. 2013;4(2):107–18. https://doi.org/10.4161/viru.22718.
38. Sellick JA, Russo TA. Getting hypervirulent Klebsiella pneumoniae on the radar screen. Curr Opinion Infect Dis. 2018;31(4):341–6. https://doi.org/10.1097/qco.0000000000000464.
39. Batson G, Wilson M. Body and mind in motion: dance and neuroscience in conversation. Chicago: University of Chicago Press/Intellect ltd.; 2014.
40. Foster SL. Choreographing empathy: kinesthesia in performance. London: Taylor & Francis; 2010.
41. Sklar D. Can bodylore be brought to its senses? J Am Folklore. 1994;107(423):9–22. https://doi.org/10.2307/541070.

Chapter 15
Understanding Violence Against Civilians by Government and Rebel Forces: Perspectives from Political Science and Health Behavior

Sarahmona M. Przybyla and Jacob D. Kathman

Figure Lord's Resistance Army, area of operations. The shaded region marks the approximate areas of operation for the LRA following Operation Lightning Thunder. (Source: Director of National Intelligence, 2018, The National Counter Terrorism Center, Counter Terror Guide: The Lord's Resistance Army (LRA), https://www.dni.gov/nctc/groups/lra.html)

S. M. Przybyla (✉) · J. D. Kathman
University at Buffalo, State University of New York (SUNY), Buffalo, NY, USA
e-mail: mona@buffalo.edu

© Springer Nature Switzerland AG 2020
K. H. Smith, P. K. Ram (eds.), *Transforming Global Health*,
https://doi.org/10.1007/978-3-030-32112-3_15

What Is the Problem?

In January 1999, in a last ditch effort to turn the tide of Sierra Leone's civil war in its own favor, the Revolutionary United Front (RUF) rebel organization embarked upon an all-out assault on the nation's capital, Freetown. The notoriously violent group aptly termed the offensive "Operation No Living Thing," as rebel soldiers engaged in a wave of unseemly violence toward the civilian population. In its advance on Freetown, the RUF escalated the brutality that had become its calling card over the previous decade of war: indiscriminately killing innocent civilians, engaging in systematic sexual predation, mutilating bystanders by amputating arms and sex organs, and forcibly recruiting children to serve as soldiers [1]. Assessments of the brief offensive estimated the killing of approximately 7000 people, with scores of others suffering gross human rights abuses [2].

More recently, groups like Boko Haram in Nigeria and the Islamic State of Iraq and Syria (ISIS) have built their own grisly reputations for indiscriminately targeting civilians, conducting extrajudicial killings, engaging in medieval torture practices, and directing murderous terror campaigns [3, 4].

Stories such as these seem all too common in media coverage of contemporary civil wars, and the public health consequences of such extreme behaviors are dire. As a result of civil wars and the violence they engender, public health systems degrade, access to basic needs such as food and clean water declines, and people are forced from their homes and into refugee camps where infectious diseases easily spread. These consequences of civil war significantly affect the lives of noncombatant populations, degrading their health and reducing their lifespans at both individual and population levels [5].

Yet, the most direct and often brutal consequence of civil conflict for civilian populations comes in the form of directly targeting noncombatant populations with atrocious violence, as described in the examples above. In this, the targeting of civilians speaks to the most basic issue in public health: life and death.

Recent Developments in Civil War and Political Violence

Considering the difficult decision on whether or not to fight for the would-be rebel, the very occurrence of civil war is rather extraordinary. Think about it: what would possess a person to take up arms against his/her government? The proposition is so fraught with danger that the idea of accepting such risks is mindboggling. In making the decision to rebel, a potential insurgent likely recognizes that their chances of being killed by the government increase if they decide to pick up arms. Indeed, rebel organizations face long odds, as they rarely defeat the government to win their wars [6].

And in the process of this likely futile challenge to the sitting regime, their loved ones are more likely to be brutalized and their communities are likely to suffer. As these costs mount, the relative value of victory declines given the severe price to be paid in the effort to achieve victory. It is difficult enough to imagine the perspective of an individual making the choice to rebel. Now iterate that decision across hundreds or thousands of individuals who must also make that same decision to take up arms in order for a rebellion to become viable.

A basic assumption in the field of economics is that individuals commonly behave in manners driven by self-interest, seeking to minimize the burden they bear for any goal they seek [7]. Thus, rebel groups are faced with a severe collective action problem. In other words, how can rebels motivate individuals to betray their self-interest and instead choose put themselves at significantly greater risk in support of the rebel group's collective goals. This collective action problem must be overcome in order to coordinate their challenge to the government. Where individuals face such great risks to their wellbeing and survival by virtue of joining such an organization, one might presume that mass rebellions rarely get off the ground, as the impediments to a collective challenge against militarily superior government forces are too great. And yet, as Fig. 15.1 shows, the number of civil wars throughout the world has grown since the end of World War II [8]. For comparison, Fig. 15.1

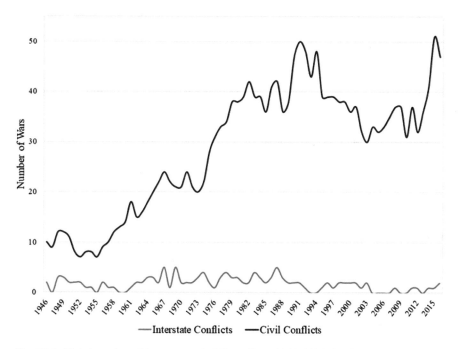

Fig. 15.1 Global number of interstate and civil conflicts, 1946–2016. Conflicts are defined as a contested incompatibility between two states (interstate) or between the government of a state and a non-state actor (civil) in which battlefield fighting occurred and at least 25 casualties were sustained in a given calendar year, and where both sides shared some number of those casualties. Figure created by chapter authors from open data-sets presented in source article. Website: https://ucdp.uu.se/downloads/#d3. Data sources: Uppsala Conflict Data Program, Peace Research Institute, Oslo Armed Conflict Dataset. Gleditsch, Nils Petter, Peter Wallensteen, Mikael Eriksson, Margareta Sollenberg, and Havard Strand. 2002. Armed Conflict 1946-2001 A New Dataset. Journal of Peace Research 39(5): 615–637. https://doi.org/10.1177/0022343302039005007

displays the number of ongoing civil and interstate wars throughout the world in the post-WWII era. While wars between countries have been flat and limited in number, the number of civil conflicts have exploded in recent decades, peaking at 51 in 2015.

So why fight? Why engage in such a destructive activity for such a poor chance at success? One reason may be that people believe that the political and socio-economic status quo is unsustainable in a growing number of countries [9]. For instance, in less developed countries in particular, populations may have grievances against their incumbent governments that they seek to remedy. It may be that these grievances are so strongly held that the political distance between rebels and their government foes helps to explain the brutality of conflict. Indeed, genocidal violence often occurs in tandem with ongoing civil conflicts, when seemingly intractable differences, like race or religion, separate rebels from government. For instance, recent wars like those in Rwanda and Bosnia-Herzegovina were defined by ethnic slaughter [10].

It may also be that the opportunity for civil war has grown. Historical approaches to civil conflict often point to processes of decolonization, which left many newly independent states with limited administrative control over their populations, particularly in Africa and Asia [11]. Further, the availability of assistance to rebel movements was in grand supply during the Cold War era, as Western and Eastern blocs sought to support their respective surrogate rebel movements around the world. Thus, to the extent that increased access to the implements of war, such as weapons and funding, affects the ability of rebellions to challenge governments in limited capacity states, civil wars may become more common and more violent [12].

However, these trends do not necessarily speak to the relative brutality of civil conflicts like those described above. Whereas participants in interstate wars often attempt to abide by formal and informal rules of war, seeking where possible to treat the civilian population humanely, civil wars in recent decades have been defined by extraordinary destruction, the ruthless abuse of civilians, and cycles of violence that negatively affect the health and livelihoods of populations for generations [13]. Below, we offer rationales the violence committed by rebel and government factions in an effort to improve our understanding of the plight of civilians in civil conflict and the population health implications that result.

Why Does Anti-Civilian Violence in Civil War Persist?

Popular sentiments of civil war often view violence against civilians in one of two predominant ways. In one, civilian deaths are viewed as collateral damage. In other words, civilians suffer the incidental consequence of being caught in the crossfire between the combatants. As an old African proverb puts it, "When elephants fight, it is the grass that suffers."

In an alternative view, civilian suffering at the hands of the combatants in war is a product of barbarism. A unique feature endemic to the war context is a destruction of informal social norms and formal institutions that enforce the humane treatment of civilians. In other words, "war is hell." This perspective supports the idea that where men must engage in barbaric violence on the battlefield to fulfill their conflict goals, this barbarism spills into their treatment of civilians off the battlefield [14].

However, popular views of violence, particularly in civil war, often miss the underlying strategic nature of targeting noncombatants. Civil war belligerents view the civilian population not simply as a passive element of the conflict environment that is subject to their predatory whims. Rather, the rebels' need to secure the support of the civilian population, often through fear or force, is a central component of its conflict strategy, and this in turn makes the civilian population a battleground for both rebel and government forces. For both sides, securing civilian support improves access to such resources as recruits, weapons, and information, and these resources improve a combatant's ability to pursue its conflict goals.

Thus, to the extent that violence can be used to shape civilian support, combatants may use the targeting of civilians as a *rational* conflict strategy, where violence is not simply a product of barbarism or collateral damage. In this sense, abusing civilians is a purposeful, strategic element of civil war, and only by recognizing this relationship can the health consequences of civil war be more fully understood. Below, we consider motivations of both rebel organizations and government forces for targeting civilians in the course of civil war (Fig. 15.2).

The Rebel Calculus

To say the least, rebel groups in their early stages of organization are faced with a distinct disadvantage relative to the government in terms of military capability. In order to justify rebellion, insurgents must have an expectation that they will prove successful in their challenge to the regime. Otherwise, there is little reason to rebel in the first place.

Thus, a newly developing rebel organization must be able to manufacture a strategy for confronting a government that has significant warfighting advantages. Generally speaking, governments have near monopolies on the capacity for committing violence. Relative to rebel forces, government militaries often have more troops, better training, more secure strongholds, more and better weapons, and tactical advantages like air superiority and military expertise.

As the literature on military tactics reveals, rebel groups thus most often choose guerrilla warfare as a conflict strategy that seeks to adjust for this asymmetry in power [15]. Guerrilla tactics rely on the use of small bands of soldiers that largely operate independently of one another, such that the capture of any one leader or group does not directly threaten the others. These small groups generally engage in hit-and-run attacks that are meant more to disrupt government operations than they are to inflict significant casualties. Insurgents turn to this strategy when conventional fighting would almost certainly lead to failure. Given that governments often

Fig. 15.2 Conceptual diagram of civil conflict. (Source: Nicole C. Little and Jacob D. Kathman)

have superiority in troops and armored capabilities, engaging in pitched battlefield fighting with massed armies would be futile.

Guerrilla operations, like Castro's insurgency in Cuba and the Viet Cong's resistance in South Vietnam, are a shadowy form of fighting, in which insurgent attacks occur often and seemingly out of nowhere, as fighters attempt to appear both ubiquitous and invisible all at once. Military victories are not the goal. Instead, insurgent objectives are psychological in nature, aiming to affect the government's will to continue fighting against an enemy that the government cannot effectively engage. As such, an insurgent faction that is militarily weak relative to its government opponent seeks to achieve victory via death by a thousand paper cuts rather than by striking one momentous blow.

In order to fight in this way, however, insurgent forces are highly dependent upon the civilian population. It is the civilians that provide guerrillas with the resources that are necessary for their continued survival. From the population, guerrilla fighters can recruit additional soldiers to enhance their ranks. Rebels also depend on civilians for securing supplies such as food, shelter, weapons, and ammunition. Guerrillas live and fight among the civilian population. They are thus reliant upon the population for their continued survival via the provision of these resources [14].

Two resources in particular are critically important. The first is camouflage. Unlike the conventional use of camouflage in which soldiers wear uniforms with patterning and coloration that allow them to blend into the surrounding natural environment, guerrillas seek to blend into the demographic environment. Insurgent soldiers thus dress and act like civilians. Much of their time is spent living civilian lives. Though occasionally brandishing weapons in attacks on government forces, what makes the guerrilla soldier a hard target is the fact that s/he looks like everyone else and blends into the community. In a sense, they are hidden in plain sight.

Thus, the second critical resource that guerrillas require from the civilian population is access to information. It is the population that holds information about government operations to resist the rebellion. Insurgents are dependent upon civilians to share this information such that the rebels can effectively carry out their guerrilla tactics and inflict losses upon government forces. Maybe more importantly, insurgents are reliant upon the population for *withholding* information from the government. It is the civilian population that knows the locations of rebel strongholds and weapons caches. Most importantly, the civilians have information about which individuals are active rebel soldiers, which civilians collaborate with those rebel soldiers, and which civilians that have not aligned themselves with the insurgents. Should members of the population become informants to the government, the value of guerrilla camouflage evaporates, and the ability of the rebels to evade capture and death diminishes [16, 17].

Thus, guerrilla rebel groups are highly dependent on the support of the civilian population and are therefore incentivized to motivate civilian collaboration with their cause. Unfortunately for insurgents, most civilians tend to be apolitical. This does not mean that civilians do not have their own political preferences that they seek to secure. Rather, civilians generally do not hold these preferences so dearly that they are willing to die for them, and collaborating with the rebels puts civilians in the government's crosshairs.

Rebel organizations must convince civilians that collaboration is economical and preferable to the status quo of remaining nonaligned. Rebel groups might do this through the provision of present or future benefits. For instance, rebels may provide an occupation and a wage to recruits. They provide security against government attacks. Rebels also make promises about future governance, like democratization or Marxist redistribution, that the civilians prefer over the existing political regime.

However, civilians tend to devalue these positive inducements for collaboration. The occupational benefit is only attractive to the extent that it exceeds the opportunity cost of normal employment and accounts for the riskier nature of the work. And the individualized benefit to a civilian of grand political regime changes tends to yield limited payoffs for those who will not directly wield power.

Notably, all other things being equal, civilians wish to support the side that ultimately wins the war. By doing so, civilians get to enjoy the spoils when siding with the winner while avoiding the post-war punishments at the hands of the victor for having supported the vanquished. It is thus rather difficult for rebels to induce collaboration given their resource weaknesses relative to the government [18].

However, there is one preference that civilians hold more strongly than any other: civilians wish to survive the war. It is this preference that guerrilla rebel organizations can most directly manipulate. The clear advantage that rebel groups hold over civilians is violence. And the advantage that rebels have over government forces is the rebels' easy access to the population. For rebels, civilians are soft targets. They cannot easily defend themselves against rebel forces. Given that rebels often live among the civilians, and since government forces cannot be present everywhere at all times to defend civilians from the rebels, the ability of rebels to threaten civilians with violent reprisals is great. It is this threat of violence that rebels wield in convincing civilians that it is costly to remain nonaligned and safer to commit their support to the insurgents [19]. When rebels fear that the population may be turning away from them, violence serves as a means of reviving and bolstering support. It is for this reason that rebel losses in battle, territorial retreat, and negative changes in relative rebel power have all been associated with an escalating violence against civilians by rebel organizations [14, 20, 21].

For instance, ISIS used extreme violence in an effort to maintain order in their areas of control as the group began to retreat [22]. Further, public displays of torture and death, like the beheadings for which ISIS became so notorious, serve as billboards that advertise the violent sanction that comes with civilian rejection of ISIS rule [23]. Similarly brutal activities have been carried out by the Janjaweed in Darfur, the RUF, and Boko Haram. Notably, these activities are purposefully made public. They serve as reminders to civilians of the punishments for resistance. In this way, violence serves as a means of manipulating the population at large into collaboration even when the violence targets individuals.

Finally, to the extent that support cannot be obtained through violence, violence may at least serve to intimidate the population into submission, punishing civilians into recognizing that the government cannot secure their welfare. As an example, the Mozambican National Resistance rebel organization engaged in massive anti-civilian campaigns in the Mozambican Civil War, including striking public health facilities like hospitals and health clinics, in an effort to punish civilians indiscriminately for their tacit support of the government [24].

Thus, while it is true that noncombatants may be caught in the crossfire and suffer as a result of collateral damage, and while it may also be true that the simple brutality of war threatens noncombatants, these explanations miss a primary motivation of rebel violence. Rebels threaten the health and wellbeing of individual civilians and their communities in a *rational* effort to improve their chances of winning the war. Thus, the most fundamental elements of public health (i.e., issues of life and death for members of the civilian population) in civil war are intricately related to rebel war strategy.

The Government Calculus

In much the same way that rebels employ violence, government forces also have strong incentives to use cruelty against civilians during civil war as a rational conflict strategy that negatively affect the population's health outcomes. Indeed, it is the insurgents' use of guerrilla tactics that helps us to understand the government's use of mass violence against noncombatants. Government forces recognize that their guerrilla opponent is critically dependent upon the civilian population for its sustenance. In recognition of this, Mao Tse-tung famously noted that "the guerrillas are as the fish and the people the sea in which they swim" [16].

However, government forces have a limited ability to confront insurgents effectively. The government needs to know a great deal more about the rebels beyond simply realizing that the rebels depend upon civilians. Given the civilian camouflage used by the insurgents, government forces have a difficult time determining which individuals are active members of the insurgency and which are not. Even more difficult is attempting to distinguish among civilians themselves, delineating by those who do and do not support the rebels.

Adding yet another layer of complexity to this is the fact that the type of information that the government must obtain from civilian collaborators is the very type of information that rebel forces attempt to induce civilians to withhold. This is one important reason why civil wars tend to endure for such long periods [25]. Even though governments often have substantial military advantages over the rebels, these capabilities are not easily brought to bear when it is difficult to effectively identify the enemy. As such, guerrilla conflicts tend to be long and extraordinarily costly wars for the government. Indeed, this is an inherent part of the guerrilla design [17].

Still, like rebels, government forces have options for overcoming this information asymmetry. Government forces seek to induce civilians into becoming informants and collaborators. These inducements may include the provision of benefits, like monetary payments, deputation of civilians as new authorities in the security apparatus, or policy reforms to satisfy civilian political interests.

Again, like the rebel experience, the provision of benefits by the government may not be a strong motivator of collaboration. This is especially the case in government efforts to induce collaboration because the rebels often have more direct contact with the population, living and working among the civilians in their daily lives. Thus, the relative proximity of rebels and distance of government forces often means that promises of future benefit provision fall on deaf civilian ears.

Given the information limitations as to which individuals are rebels, rebel supporters, and or nonaligned civilians, governments can choose to engage in massive indiscriminate violence. Drawing on Tse-tung's analogy, Guatemala's General Efrain Rios Montt argued that an effective counter-guerrilla strategy must "dry up the human sea in which the guerrilla fish swim" [26]. In other words, to defeat the insurgents, the civilians upon whom they depend must be removed. This may be achieved in different ways. For instance, a government may forcibly relocate large civilian populations to different locations in an attempt to separate civilians from insurgents. Such policies were implemented by the Ethiopian government during the Tigre insurgency and by the British government in the Boer War [27]. Relocations often produce unseemly living conditions, as populations are commonly quarantined in camps with little in the way of health considerations. Even the process of relocation itself can be so fraught with danger and poor conditions that mass death results. This was the case in the Ottoman relocation of its Armenian population in which Armenians were deprived of food and water in deportation "death marches."

However, when these policies fail or are ill-considered due to expected ineffectiveness, governments may turn to mass violence [28]. Popular perceptions of mass killing (also commonly referred to as genocide or politicide) often consider the perpetrators to be bloodthirsty, motivated by ethnic or religious hatred, and otherwise psychologically deranged to engage in extermination policies. Indeed, instances of mass killing are so imbued with immorality that defining the perpetrators as irrational zealots may be a satisfying way to make sense of the brutality. Yet, considered in the counter-guerrilla context, mass killing can be seen by governments as a rational tool for quashing a costly insurgency. While such a view requires the relegation of morality to the trash bin, to the extent that mass killing serves as a tool for achieving the notable benefit of ending a guerrilla insurrection at "acceptable cost," then genocidal policies can be seen as strategic in nature. Once the sea is drained, the fish cannot survive. It is for this reason that statistical research has found that governments are more than three times more likely to engage in mass killing when faced with rebels that employ guerrilla tactics relative to any other rebel military strategy [27].

Given this perspective, it is somewhat less surprising that when governments become increasingly desperate to win in war due to the high costs associated with continued fighting, they also become more likely to unleash massive violence against the civilian population [29]. Thus, like the rebels, government forces impose themselves on civilians in ways that are unrelated to popular accounts of civilians being caught in the crossfire or being subject to the innate brutalities of man. Instead, much of the life and death risk to public health in civil conflict is a product of rational calculation on the part of the government and rebel forces.

Foreign Interventions and Their Health Consequences

The rational motivations for brutality against civilians in civil war have implications for the ways in which the international community can respond to such crises. In an effort to ameliorate the violence and health consequences of civil conflict, debates often center around how best to address these problems. Two broad intervention types are prominent in these debates: interventions that are biased in support of one side and interventions that are neutral. Arguments for the former often argue that defending civilians can only be effective if the perpetrators of violence are confronted and defeated. In defense of the latter, neutral interventions often seek negotiated resolutions and the distribution of humanitarian aid to affected populations. Both forms of intervention ostensibly seek an end to violence and an improvement in the public's health outcomes, which includes a reduction in civilian deaths, reduction in the spread of disease, and improvement in the population's access to basic needs. All of these health outcomes thus contribute to a generally improved standard of health for the population.

One of the theoretical foundations of this chapter is that civilians are a critical component of rebel and government conflict strategies. Motivating civilian loyalty is beneficial to each. However, since civilians wish to support the winning side in order to avoid persecution at the hands of the victor, scholars have argued that civilians generally wish to give their loyalty to the more powerful combatant.

With this as a starting point, research has shown that when the balance of power between the sides shifts in one's favor due to combat successes, the weakened side faces an existential crisis. Not only have they become weaker through combat, but as a product of becoming weaker, they face an additional, potentially catastrophic, reduction in power that results from diminishing civilian support. The weakened faction may then turn to violence against civilians as means of recouping resources and coercively enforcing allegiance from a potentially disloyal population [21].

These dynamics raise important ethical questions for how the international community can respond to civil wars without exacerbating the violence and negative health outcomes of those wars. Indeed, any biased intervention that supports one faction, thus weakening the other, will incentivize the weakened side to engage in greater brutality. In fact, recent statistical modeling has revealed that a foreign intervention with a strength of just over 1000 soldiers favoring the government increases rebel killings of civilians by 25%. An equivalent intervention favoring the rebels increases government killings by 40% [18].

Considering this dynamic, it can be difficult for international actors to intervene in civil conflict in favor of one side without inflaming the violent tendencies of at least one combatant party. Yet, neutral interventions have shown greater promise in protecting civilians from harm and improving national health outcomes [30, 31].

Given the rationalist logic of violence detailed here, this may be because these interventions do not disrupt the power balance and may thus be more capable of reaching civilians with humanitarian aid without motivating abuses by the combatants. In this sense, understanding the strategic motivations for combatant abuses is important for structuring international responses meant to defend the wellbeing of civilians. Indeed, short of such an understanding, intervention may serve to exacerbate the negative health consequences of civil war.

Traditional and New Disciplinary Approaches to Civil War Violence

Traditionally, the study of violence in civil war has been the purview of political science and history. Only recently has the political science field engaged in a more rigorous accounting of violence motivations with a rationalist approach, which is the perspective applied here.

Future work would benefit from public health contributions. We know from past research that civil war is prevalent where certain societal-level indicators of suffering are also present, such as poverty, disease, and oppression, and these are also key social determinants of individual and community health. Research also indicates that the experience of civil war further worsens this suffering in a vicious cycle, including population-level morbidity and mortality. As such, those areas that experience civil war at one point in time are highly likely to experience it again in the future. The health consequences of civil war are thus likely to experience spiraling downward trends.

Beyond the violence described above, civil war generates a variety of negative public health outcomes, including famine, disease, and infant mortality, that have both individual-level and societal-level significance. While improving our understanding of these phenomena is important, providing a strategic logic for direct violence against civilians by combatants is critically important, as it is the driver of these other deleterious phenomena. Beyond this, the suffering engendered by civil war notably appears to follow an epidemiological process of infecting nearby regions and countries with its violence. Therefore, the methods of study used in public health would assist in our understanding of violence processes. In much the same way that the discipline explains health outcomes, the study of civil war violence as an extreme health outcome would benefit from greater attention from the public health discipline. Efforts in this regard could inform policy seeking to improve public health during and after war, and recent research suggests that effective intervention options are available to this end [32].

Case Study: Foreign Intervention and the Lord's Resistance Army

The dangers of biased intervention for the purposes of confronting abusive war factions are revealed by the US intervention into Uganda's civil war. In December 2008, the USA aided the Ugandan government in its preparations for confronting the Lord's Resistance Army. The LRA, headed by Joseph Kony, is one of the longest surviving rebel organizations in the world. Over its decades-long effort to topple the Ugandan government, the LRA has also been extraordinarily brutal toward the civilian population. Having long ago lost the voluntary support of the Acholi population in northern Uganda, the LRA's tactics turned extremely violent as a means to coerce support. In an effort to confront Kony, the USA began to support the Ugandan government with equipment, intelligence, and military advisors, viewing this support as a form of humanitarian intervention to protect civilians from future atrocities.

Unfortunately, Operation Lightning Thunder was ultimately unsuccessful. Kony survived the attack, and the LRA, while weakened, remained constituted. Immediately following the operation from December 2008 through May 2009, the LRA set off on a series of coordinated attacks. In this period, the group targeted and killed over 1000 innocent civilians, abducted scores of children to serve as slave soldiers, displaced hundreds of thousands from their homes, and pillaged the goods left behind. In conducting these atrocities, the LRA sought to recoup its losses by spreading terror throughout the countryside, plundering resources, and victimizing the local population. In its bid for survival, the LRA inflicted terrible suffering on innocents, which many viewed as seemingly senseless and irrational violence [33, 34].

Taken in the context of this chapter, however, the LRA's violence fits with a *rational* understanding of insurgent hostilities. The strategic incentives for violence must be considered carefully when developing the means by which to manage civil war factions. If the policy goal is to protect civilians, then a policy tool that simply weakens the perpetrator of human rights abuses is likely to backfire in the form of a substantial escalation of those same abuses. In the LRA's case, the group continued to persist, finding refuge in the border regions between, Uganda, the Democratic Republic of the Congo, South Sudan, and the Central African Republic, and preying on the local population there.

References

1. Gberie L. A Dirty War in West Africa: the RUF and the destruction of Sierra Leone. Bloomington: Indiana University Press; 2005.
2. Human Rights Watch. Sierra Leone: getting away with murder (New Testimony from Sierra Leone. 1999. HRW Report, July. https://www.hrw.org/legacy/reports/1999/sierra/index.htm# TopOfPage
3. Wood G. What ISIS really wants. The Atlantic. 2015;315:78–94.
4. Smith M. Boko Haram: inside Nigeria's Unholy War. London: I.B. Tauris & Company; 2015.
5. Ghobarah HA, Huth P, Russett B. Civil Wars kill and Maim people-long after the shooting stops. Am Polit Sci Rev. 2003;97(2):189–202.
6. Kreutz J. How and when armed conflicts end: introducing the UCDP conflict termination dataset. J Peace Res. 2010;47(2):243–50.
7. Smith A. The wealth of Nations. New York: Random House Publishing, Bantam Classics; 2003.
8. Gleditsch NP, Wallensteen P, Eriksson M, Sollenberg M, Strand H. Armed Conflict 1946-2001 a new dataset. J Peace Res. 2002;39(5):615–37. Data for this figure are taken from the Uppsala Conflict Data Program, Peace Research Institute, Oslo Armed Conflict Dataset.
9. Collier P, Hoeffler A. Greed and Grievance in Civil War. Oxf Econ Papers. 2004;56:563–95.
10. Totten S, Parsons WS. Century of genocide: critical essays and eyewitness accounts. 3rd ed. New York: Routledge; 2009.
11. Birmingham D. The decolonization of Africa. London: Routledge; 1995.
12. Kalyvas S, Balcells L. International system and technologies of rebellion: how the end of the Cold War shaped international conflict. Am Polit Sci Rev. 2010;104(3):415–29.
13. Collier P, Elliott VL, Hegre H, Hoeffler A, Reynal-Querol M, Sambanis N. Breaking the Conflict Trap: Civil War and development policy. Washington, DC: The World Bank and Oxford University Press; 2003.
14. Kalyvas S. The logic of Violence in Civil War. Cambridge: Cambridge University Press; 2006. See Chapters 3, 4.
15. Wimberly S. Special forces Guerrilla Warfare manual. Boulder: Paladin Press; 1997.
16. Tse-tung M. On Guerrilla Warfare (trans: Samuel B. Griffith). New York: Praeger; 1961. For further work on guerrilla warfare, see treatises by Tse-tung (1961) and Guevara (1961). Guerrilla Warfare. BN Publishing.
17. Guevara E. Guerrilla warfare. New York: Monthly Review Press; 1961. For further work on guerrilla warfare, see treatises by Tse-tung (1961) and Guevara (1961). Guerrilla Warfare. BN Publishing.
18. Wood R, Kathman J, Gent S. Armed intervention and Civilian victimization in intrastate conflicts. J Peace Res. 2012;49(5):647–60.
19. Kalyvas S, Kocher M. How free is "Free-Riding" in Civil Wars? Violence, insurgency, and the collective action problem. World Polit. 2007;59(2):177–216.
20. Hultman L. Battle Losses and Rebel violence: raising the cost of fighting. Terrorism Polit Violence. 2007;19:205–22.
21. Wood R. Rebel capability and strategic violence against civilians. J Peace Res. 2010;47:601–14.
22. Gladstone R. U.N. says Islamic State executed hundreds during Siege of Mosul. The New York Times; 2017 November 2.
23. Hultman L. The power to Hurt in Civil War: the strategic aim of RENAMO violence. J Southern Afr Stud. 2009;35(4):821–34.
24. McCoy T. ISIS, beheadings and the success of horrifying violence. The Washington Post; 2014 June 13. https://www.washingtonpost.com/news/morning-mix/wp/2014/06/13/isis-beheadings-and-the-success-of-horrifying-violence/?utm_term=.59912becbf0d.
25. Fearon JD. Why do some civil wars last so much longer than others? J Peace Res. 2004;41(3):275–301. By one estimate, the average civil war in 1999 had endured for 16 years.

26. Richards M. Cosmopolitan World view and counter insurgency in Guatemala. Anthropol Q. 1985;58(3):95.
27. Valentino BA. Final solutions: mass killing and genocide in the 20th century. Ithaca: Cornell University Press; 2004. Chapter 6.
28. Valentino B, Huth P, Balch-Lindsay D. Draining the Sea: mass killing and guerrilla warfare. Int Organ. 2004;58:375–407.
29. Downes AB. Targeting civilians in War. Ithaca: Cornell University Press; 2008.
30. Hultman L, Kathman J, Shannon M. Peacekeeping in the midst of War. Oxford: Oxford University Press; 2019.
31. Kathman J, Wood R. Managing threat, costs, and incentives to kill? The short- and long-term effects of intervention in mass killing. J Confl Resolut. 2011;55(5):735–60.
32. Reeder B. The Spatial concentration of peacekeeping personnel and public health during intrastate conflict. Int Peacekeeping. 2018;25(3):394–419.
33. Johnson S. Hard target: the hunt for Africa's last warlord. Newsweek; 2009 May 16.
34. Gettleman J, Schmitt E. U.S. aided a failed plan to rout Ugandan rebels. New York Times; 2009 February 6, A1.

Chapter 16
Uneven Geographies and Treatment Challenges of People Living with HIV and AIDS: Perspectives from Geography

Torsten Schunder, Sharmistha Bagchi-Sen, and Michael Canty

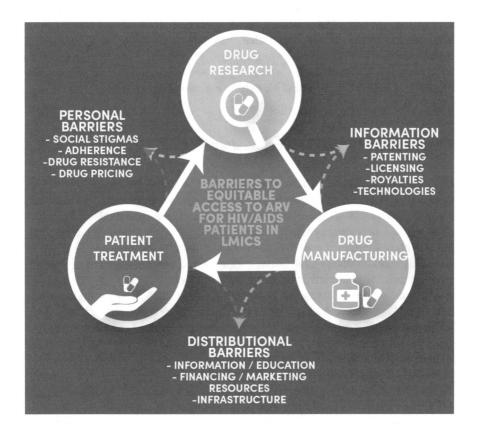

T. Schunder (✉) · S. Bagchi-Sen · M. Canty
University at Buffalo, State University of New York (SUNY), Buffalo, NY, USA
e-mail: torstens@buffalo.edu

© Springer Nature Switzerland AG 2020
K. H. Smith, P. K. Ram (eds.), *Transforming Global Health*,
https://doi.org/10.1007/978-3-030-32112-3_16

Introduction

Approximately 36.7 million people are living with human immunodeficiency virus (HIV) and acquired immune deficiency syndrome (AIDS) and another 1.8 million are infected every year [1]. As one of the biggest global health challenges, HIV attracts attention from researchers, government and non-government organizations (NGOs), and the pharmaceutical industry to discover better drugs/vaccines, stop the spread of the disease, and promote adherence to treatment. Progress in fighting HIV/AIDS has been geographically uneven with a huge burden on low- and middle-income countries (LMICs) in terms of morbidity and mortality. Antiretroviral drugs (ARV), the only known way to prevent AIDS-related death, can be costly and legally challenging to produce and distribute in LMICs. While cost of treatment has decreased, challenges in countering HIV include access, the quality of care, and prevention. In this chapter, we examine the geographic aspects of HIV, ARV production, and the role of international stakeholders (e.g., international organizations and pharmaceuticals). We argue that inclusive innovation involving multi-stakeholder and multidisciplinary (e.g., drug development, economics, epidemiology, geography, international business, social work, patients, or community agencies) collaboration is critical for halting and reversing the trends.

Disease Burden and Geographic Patterns

HIV, a retrovirus, reproduces itself by overtaking "T-Cells" and using them to produce copies of itself while rendering the cell useless in the process [2]. These cells are an important part of the human immune system and the virus weakens the ability of an infected individual to fight infections. The virus infects an individual by entering the body through the exchange of bodily fluids with someone living with HIV. This includes sexual intercourse, breast feeding, sharing injection equipment, blood transfusion, and vertical transmission from mother to child. If untreated, the HIV virus develops a condition known as AIDS characterized by opportunistic infections (e.g., tuberculosis). There is no known cure or vaccine for the virus. ARVs are effective in suppressing the virus and may reduce the viral load (viral count per milliliter of blood) of an infected person to undetectable levels. Reaching an undetectable load makes it highly unlikely that an infected person will transfer the virus to a non-infected person [3].

The overall societal burden of HIV includes loss of income, high medical cost, and AIDS orphans. For example, the loss of income decreases food security; children dropping out of school to work for food further aggravates economic conditions. Furthermore, poor households have limited resources to cope with HIV/AIDS [4]. At the national level, it has been estimated that HIV/AIDS decreases annual GDP growth rates by 2–4% across Africa due to lower labor productivity of infected persons, higher public expenditure to deal with the disease, and the need to import drugs [5, 6].

Disability-adjusted life years (DALY) represent the loss of life years that would have been free from disease and disability—it represents the difference between the current health state and the ideal disease and disability free situation and is a measure of reduced life quality. In 2004, HIV was the fifth largest contributor to DALYs, accounting for 58.5 million disability-adjusted life years. This burden was disproportionately levied on low-income countries, which account for 42.9 million of the DALYs attributed to HIV/AIDS [7]. The overall impact and geographic unevenness persists in 2010—the age group 30–44 HIV/AIDS is the leading cause of DALYs in 21 countries in Eastern and Southern Africa, Central Africa, the Caribbean, and Thailand [8].

Geographic Patterns

Table 16.1 shows the geographic distribution of HIV cases, new infections, and mortality. Sub-Saharan Africa combining Eastern and Southern Africa as well as Western and Central Africa represents 69.5% of all HIV cases, 64.44% of new infections, and 73% of the annual HIV related deaths. HIV/AIDS-related illnesses are the leading cause of death among women of reproductive age in the world [9]. According to the U.S. President's Emergency Plan for AIDS Relief (PEPFAR), adolescent girls and young women account for 74% of new infections in sub-Saharan Africa [10].

According to UNAIDS, the number of deaths from HIV related causes has been declining with a peak of 1.9 million in 2005 to one million in 2016—this is mainly attributed to the increased availability of ARVs. Figures for women and children show improvements—they tend to have higher ARV coverage and are inclined to follow

Table 16.1 Regional HIV and AIDS statistics with confidence intervals, 2016

World regions	Persons living with HIV, all ages	Persons newly infected, all ages	Death due to AIDS, all ages
Eastern and Southern Africa	19.4 million (17.8–21.1 million)	790,000 (710,000–970,000)	420,000 (350,000–510,000)
Western and Central Africa	6.1 million (4.9–7.6 million)	370,000 (270,000–490,000)	310,000 (220,000–400,000)
Middle East and North Africa	230,000 (160,000–390,000)	18,000 (11,000–39,000)	11,000 (7,700–19,000)
Asia and the Pacific	5.1 million (3.9–7.2 million)	270,000 (190,000–370,000)	170,000 (130,000–220,000)
Latin America	1.8 million (1.4–2.1 million)	97,000 (79,000–120,000)	36,000 (28,000–45,000)
Caribbean	310,000 (280,000–350,000)	18,000 (15,000–22,000)	9400 (7300–12,000)
Eastern Europe and Central Asia	1.6 million (1.4–1.7 million)	190,000 (160,000–220,000)	40,000 (32,000–49,000)
Western and Central Europe and North America	2.1 million (2–2.3 million)	73,000 (68,000–79,000)	18,000 (15,000–20,000)
World Total	36.7 million (30.8–42.9 million)	1.8 million (1.6–2.1 million)	1.0 million (830,000–1.2 million)
Share Sub-Saharan Africa	69.5%	64.44%	73%

Source: UNAIDS [1]; and calculation by authors

their treatment regimen resulting in 27% lower deaths compared to men [1]. Reduction of deaths in Eastern and Southern Africa influences the overall pattern—1.1 million in 2004 to 0.42 million in 2016. New infections are still a problem—the UN planned to reduce new infections to 0.5 million a year by 2020 but 1.8 million occurred in 2016. The decline in new infections is only 16% between 2010 and 2018. New infections among children have declined in the same timeframe by 47% as a result of supplying pregnant women with ARVs preventing virus transmission.

Over time, there has been a price decrease in the annual cost of treatment per patient. Figure 16.1 displays the global average annual cost of treatment per patient of imported ARV medication based on total exports and prices reported by the Global Price Reporting Mechanism (GPRM) of the WHO [11]. While the average annual treatment cost well exceeded $200 per patient in 2004, it dropped in 2010 to under $100 per patient, and has continued to decline.

Figure 16.2 shows prevalence rates by countries and changes in prevalence rates between 2000 and 2016 based on data from the WHO Global Health Observatory [12]. Prevalence rates in 2016 are under 1% among the adult population (age 15–49) in most regions. A cluster can be identified in sub-Saharan Africa where rates are higher with a distinctive hotspot in the South. South Africa, Zimbabwe, Botswana, Namibia, Mozambique, and Zambia have the highest rates in the world with 10–27.2% (South Africa) of the population. Most countries in Africa show a decrease or stagnation in prevalence rates with the exception of Angola, South Africa, Sierra Leone, and Sudan (Fig. 16.3)—these countries experienced increases between 30% and 100%. In Asia, Kazakhstan, Vietnam, and Kyrgyzstan experienced similar increases while increases up to 400% can be found in Indonesia, Laos, and Tajikistan. Latin America and South America experienced increases in prevalence rates as well with the exception of the western part of South America. Eastern Europe (east of Poland) notes strong increases in prevalence rates. Interestingly, France, Ireland,

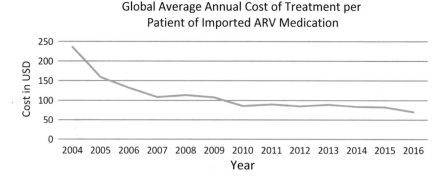

Fig. 16.1 Global average annual cost of treatment per patient of imported ARV medication. (Source: Chapter authors created this figure. The data presented are results of chapter author calculations using the individual ARV drug transaction data "Global Price Reporting Mechanism for HIV, tuberculosis and malaria" provided by the World Health Organization. Link to web presence of data set: www.who.int/hiv/amds/gprm/en/)

Fig. 16.2 HIV prevalence rate among adults aged 15–49 in 2016. (Source: Map produced by authors using data from Global Health Observatory, "Global Health Observatory data HIV/AIDS" provided by the World Health Organization. Link to web presence of data set: www.who.int/gho/hiv/en/.

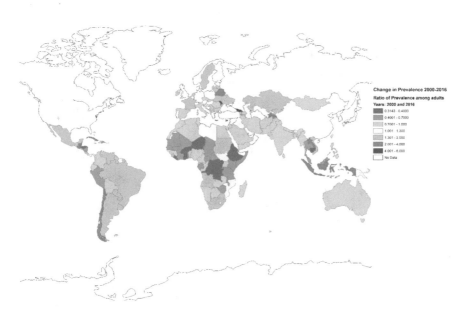

Fig. 16.3 Change in HIV prevalence rates among adults aged 15–49 between 2000 and 2016. (Source: Map produced by authors. Authors calculated change in prevalence rates based on country level data using data from Global Health Observatory, "Global Health Observatory data HIV/AIDS" provided by the World Health Organization. Link to web presence of data set: www.who.int/gho/hiv/en/.

Netherlands, and Sweden—the only developed European countries with available WHO data—show increases in prevalence by 30–100% as well. Reasons of change may vary from country to country and may relate to better education/prevention as a result of increasing awareness, ARV availability affecting life expectancy of infected, changes in testing procedures identifying more/less infected or changes in the cultural perception of HIV.

Regional variation exists also in drug import pricing measured by the annual cost of treatment per patient weighted by units of ARVs by country. Figures 16.4 and 16.5 show the cost at the country level in 2015 and changes in treatment cost by country between 2005 and 2015. Prices in Africa and in developing countries in Asia are low—the estimate is under $150 per year. Drugs in South Africa, where prevalence rates are highest, are under $100 per year per patient. Angola, Libya, and Tunisia import the priciest drugs in Africa. Overall, the cost of imported treatment has decreased since 2005 with a few exceptions. Niger, Angola, Tanzania, and Mongolia experienced slight increases; drugs in Mongolia and Iraq are up to five times more expensive.

Based on GPRM data, cost and market share leaders in global ARV exports in 2015 are India and South Africa. India is responsible for nearly 79.6% of exports with an average annual cost of treatment of the exported drugs at $81, while South Africa is the second largest exporter of ARVs with approximately 15.4% priced on average at $56. In contrast, exports from the USA make up approximately 3.3% and are priced on average at $154.

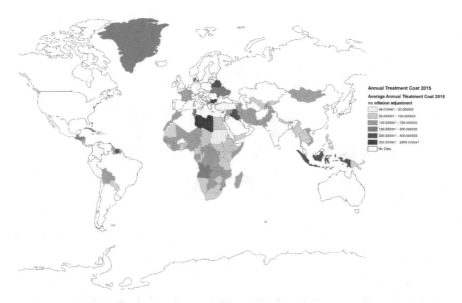

Fig. 16.4 Average annual treatment cost in 2016 weighted by total units imported. (Source: map produced by authors based on data from "Global Price Reporting Mechanism for HIV, tuberculosis and malaria." The data presented are results of chapter author calculations using the individual ARV drug transaction data provided by the World Health Organization. Link to web presence of data set: www.who.int/hiv/amds/gprm/en/.

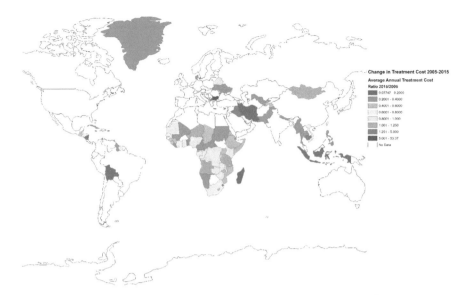

Fig. 16.5 Change in average annual treatment cost between 2005 and 2015. (Source: map produced by authors based on data from World Health Organization, "Global Price Reporting Mechanism for HIV, Tuberculosis and Malaria." The data presented are results of chapter authors' calculations of annual cost of treatment per patient by country) using the individual ARV drug transaction data provided by the World Health Organization. Link to web presence of data set: www.who.int/hiv/amds/gprm/en/.

The Role of Governments, NGOs, and Multinational Enterprises (MNEs)

The U.S. President's Emergency Plan for AIDS Relief (PEPFAR) and the United Nations (e.g., various programs including the World Health Organization (WHO) and UNAIDS) are the leading organizations. UNAIDS is actively involved in shaping policy to achieve the 90-90-90 goal by 2020. This includes diagnosing 90% of those living with HIV, supplying 90% of those diagnosed with ARVs, and suppressing the virus in 90% of those on therapy. UNAIDS supports governments in formulating goals and strategies to achieve these goals [1]. The WHO sets up general frameworks to coordinate efforts: research support, data collection and monitoring, formulating ethical and evidence-based policies. Both organizations provide technical support to help build institutional capabilities. UNAIDS and WHO collaborate with a variety of organizations, including PEPFAR. The role of a government organization (PEPFAR), a non-government organization (Mothers2mothers), and a multinational enterprise (Cipla) is shown in Table 16.2. The data sources are publicly available program documentation, program websites, and newspaper articles.

PEPFAR is a U.S. government program founded in 2003 to address global HIV/AIDS crisis and was endowed with $48 billion (2008–2013). PEPFAR is active in 50 countries and supplies 13.3 million people with ARVs [14]. The majority of these

Table 16.2 Organizations involved in the HIV/AIDS response

The U.S. President's Emergency Plan For AIDS Relief (PEPFAR)	
Description	• U.S. government program • Founded in 2003 as response to global HIV/AIDS crisis • From 2008 to 2013 endowed with up to $48 billion • Active in 50 countries and supplies 13.3 million with ARVs
Mission	• Access to ARVs • Prevention (abstinence promotion) • Palliative care • Support of orphans and children affected by HIV • Scientific support (case study database) and innovation challenges
How	• Runs country based drug purchase programs • Cost reduction deals with drug companies • Large market power, economies of scale influence pricing → spent 7 billion on ARVs in 2017 • Purchase drugs (HIV/AIDS) and co-morbidities (tuberculosis and malaria)
Cooperation	• Cooperation with government and large international organizations and NGOs who subcontract smaller NGOs to distribute drugs • Cooperation with industry and NGOs (innovation challenges)
Other	• Critique: • Prohibit support for organizations offering needle exchange programs, promoting abortion, or providing support to sex workers • Program is inflexible preventing adjustment to local context • Drugs purchased require FDA approval resulting in longer approval times and higher prices
Mother2mothers	
Description	• South African NGO, operates in several sub-Saharan countries • $18 million budget in 2016
Mission	• Policy making and complementary patient care • Creation of national mentoring programs • Aims to reduce mother–child transmission of HIV
How	• Seeks to establish partnerships with governments to establish national mentoring programs. Currently national programs exist in Uganda and South Africa • Links pregnant women living with HIV with mentor mothers also living with HIV who provide supportive services and education • Mentor mother paid fulltime and located in understaffed clinics • Supplies program participants with ARVs (duration country specific: Either during pregnancy or permanently) • M2M's methodology incorporates stigma-fighting mechanisms
Cooperation	• Governments • Clinics • HIV positive mothers and mentors
Other	• Successes: • m2m mothers in Lesotho had a child transmission rate of 1.6%, compared with the national average of 11% • $1 spent on m2m's mentor mother model yields $11.40 in averted treatment costs • 93% of participants are treatment adherent after 3 months compared to 73% outside the program • Higher disclosure rates in relationships increased

(continued)

Table 16.2   (continued)

Cipla	
Description	• Multinational corporation from India • Specializes in ARV therapy
Mission	• Generic drug producer • Low pricing (initially 96% reduction from the cost of treatment in the USA) • New drug development
How	• Initially exploited process patents (patent granted for producing a drug in a novel way)—a practice unique to India and internationally heavily criticized as seen as patent violation • Takes now advantage of TRIPS exemptions for LDCs supplying LDCs with ARVs while forming strategic partnerships with global generic manufacturers
Cooperation	• Has FDA approved drugs participating in PEPFAR • Cooperates with NGO: Offered Medicines Sans Frontieres (MSF) preferential ARV pricing for $350/year and patient • Offers government receiving backing from MSF preferential pricing

Source: Table prepared by authors based on various sources [13–23]

funds were allocated to treatment of HIV and co-morbidities (e.g., tuberculosis), followed by prevention such as promoting abstinence. Smaller portions are allocated to palliative care and the support of children, including orphans, affected by the HIV crisis. PEPFAR runs country based drug procurement programs and has cost reduction deals with drug companies. PEPFAR provides low cost generics to countries or international organizations who in turn subcontract smaller organizations to distribute the drugs. PEPFAR supports the development of a multidimensional understanding (e.g., the role of poverty, gender inequality, sexual violence, and a lack of education) of HIV prevention with its DREAM (Determined, Resilient, Empowered, AIDS-free, Mentored, and Safe women) programs focusing on prevention of HIV in adolescent girls and young women, who make up 74% of newly infected in Africa [10, 21]. The six challenge areas are retaining girls in secondary education, strengthening the capacity of service delivery, linking men to services related to HIV, supporting pre-exposure prophylaxis, providing a bridge to employment, and applying data to increase impact. The challenge is conducted together with multinational enterprises (e.g., ViiV Healthcare, Johnson & Johnson) providing overall $85 million in funding. The majority of 63 communities or districts in 10 African countries in which the initiative is run experienced a 25–40% decline in new infections among adolescent women.

Mothers2mothers is a South African NGO, operating in several sub-Saharan countries, with an annual budget of US$18 million. It aims to reduce mother–child transmission of HIV. In Uganda and South Africa national mentoring programs have been established linking pregnant women living with HIV with mentor mothers also living with HIV—they provide supportive services and education to increase treatment adherence and the removal of stigma. Mentor mothers have access to ARVs. Successes claimed include lower mother to child transmission rates than the national average of participants, higher drug adherence than non-participants, and higher HIV disclosure rates to partners [15, 19, 20].

Mother2mother bridges policy and treatment combining ARV delivery to its customer base with education and extensive mentoring efforts, while at the same time being involved in shaping public policy. However, the program does not seem to be involved in discussions with drug companies (at least beyond price). The examples show that there is a disconnect between people needing the drug in LMICs and drug manufacturers, with little interaction between the two even with mediation from other stakeholders.

The case of Cipla shows that the emphasis on reducing cost is given to generic drugs. But drugs exclusively developed for the developing world with input of stakeholders acknowledging local conditions and production opportunities is not an explicit goal. In this case, the interaction between drug manufacturers and countries seems to end with governments or larger humanitarian organizations (e.g., PEPFAR). Cipla Pharmaceuticals is an Indian company headquartered in Mumbai and specializes in generic HIV antiretroviral therapy, along with other treatments, and has been an industry leader for the past two decades [13]. Their first generic ARV Zidovudine could be sold at 1/10th of the price of GlaxoSmithKline's version of the drug and Cipla received international attention, when it began offering ARVs for $1/day in low-income countries in 2001 to the France-based nonprofit Medecins Sans Frontieres (MSF) [16]. The pricing was a 96% reduction from the cost of treatment in the USA. Cipla operated initially under the Indian Patent Act of 1970 allowing for process based patents giving patent rights to firms who could produce an existing drug in a new way. After joining the WTO, India's patent law became WTO compliant requiring companies to follow WTO rules granting the patent holder of the substance exploitation rights over a 20 year period forcing Cipla to adjust its business and patenting practice. While the WTO does not allow for process patents, it grants exceptions to supply low- and medium-income countries with generics and CIPLA focusses on this market while at the same time forming strategic partnerships with global generic manufacturers.

Approaches in Countering HIV

In response to growing pressure to supply ARVs to the poorest and most affected countries, MNEs use donations, licensing, and access pricing in order to lower the cost barrier to treatment, while international organizations have allowed for compulsory licenses. Table 16.3 highlights the different approaches and their shortcomings. While prices of ARVs have dropped, a considerable number of people continue to lack access to ARVs and the number of new infections has not gone down. This indicates that current price based approaches may not be sufficient to alleviate the problem HIV poses. Donation programs, while beneficial to patients and effective to an extent, are not sustainable as a long-term solution [24].

Preferential, also called differential pricing, was first introduced by UNAIDS' Accelerating Access Initiative in the early 2000s. The program was a large collaboration between governments, non-governmental organizations, and pharmaceutical firms. One of the major outcomes was a surge in "access pricing," or reduced pricing of branded drugs for sale in less developed countries (LDCs) [25]. Under preferential pricing a tiered price structure offers lower prices in LMICs. The framework

Table 16.3 HIV responses

Donation	
Who	• Multinational corporations and international organizations
What	• Donate ARVs to LMICs
Anticipated effects	• Supply of ARVs in resource scarce environments
Critique	• Unsustainable and not able to fulfill demand for affordable ARVs
Preferential pricing	
Who	• Multinational corporations
What	• Price differential between developed (high price) and developing countries (low price)
Anticipated effects	• Ensures revenue of firm in high price markets while increasing access and affordability in LMICs
Critique	• Perception of exploitation in developed markets due to higher prices. • Fear of arbitrage opportunities and black-market trade of drugs bought in LMICs and sold in developed countries
Compulsory licenses	
Who	• World Trade Organization
What	• Allows members to issue a "compulsory license" if a license cannot be obtained from the right holder • Grants LMICs right to export ARVs to LDCs • LDCs exempt from drug patent enforcement
Anticipated effects	• Supply of affordable ARVs in LDCs while compensating right holder • Development of local pharmaceutical industry • Generate interest in developing countries for sustainable development in LDCs
Critique	• Disincentivizes the development of new drugs as large research investments can be only recouped through strict patent enforcement
Voluntary licenses	
Who	• Multinational corporations
What	• Practice of extending non-exclusive licenses to manufacturers in least-developed countries by drug originator
Anticipated effects	• Ensures revenue of originator in high price markets • Compensates originator via licensing fees in LMICs • Non-exclusivity creates competition and reduces price
Pooled purchasing	
Who	• Global Fund (Non-government organization)
What	• Organizes pooled purchasing to achieve economies of scale
Anticipated effects	• Reduces drug price

Source: Table prepared by authors based on various sources [17, 24–29]

allows firms to generate revenue from drug sales in industrialized countries while meeting some of the large demand for cheaper drugs in LMICs. This approach can inadvertently make the firm the subject of criticism from consumers in industrialized markets who may believe the preferential pricing is the fair price, and the pricing in industrialized nations seems exploitative of patients and healthcare systems. Furthermore, preferential pricing creates arbitrage as firms fear that black-market pharmaceutical dealers may buy large stocks of ARVs in LMICs to later sell in industrialized countries.

The 1995 Agreement on Trade-Related Aspects of Intellectual Property (TRIPS) by the World Trade Organization (WTO) established intellectual property rules requiring member states to make patents available and set a framework for the multilateral enforcement of international patent rights. LDCs are exempt from TRIPS compliance till 2021 to incentivize technology transfer to LDCs increasing their competitiveness [26]. Member states were granted the right to issue a "compulsory license" to domestic firms seeking to use patented intellectual property of a foreign firm if a license could not be obtained from the right holder under certain conditions: production must be for domestic consumption, unless the compulsory license was used to address anticompetitive behavior and the right holding firm must be paid reasonable royalties or remuneration for the use of its intellectual property [17]. In 2001, TRIPS was amended allowing for exports under compulsory licensing for middle-income countries to supply the countries most affected by HIV and other diseases. WTO members extended drug patent exemption to LDCs until January 2033, or until they cease to be identified as a LDC [27].

Proponents of compulsory licensing for LDCs argue that it encourages developed nations to have interest in long-term sustainability of LDC markets, while allowing the supply of pharmaceuticals and emergence of a local pharmaceutical industry partners [26]. Critics argue that compulsory licensing disincentivizes the development of new drugs as large research investments can be only recouped through strict patent enforcement.

Voluntary licensing takes place when an originator firm extends production rights of a drug to another firm, usually located in a LMICs. Firms, such as Gilead and GlaxoSmithKline, adopted as response to patent infringements, the practice of extending non-exclusive licenses to manufacturers in least-developed countries requiring the manufacturer to pay reasonable royalties. Voluntary licensing was popularized in 2003 by the Clinton Health Access Initiative, which facilitated negotiations between firms to make fair price ceilings that still gave originator firms a sustainable profit margin [28].

Licensing agreements allow the originator firms to charge high prices in industrialized markets while licensing production to multiple firms in low-income markets. The practice of non-exclusive licensing creates competition driving prices down supplying markets with cheap, generic ARVs. Generic producers are able to legally produce ARVs at a fraction of the price improving on accessibility and adherence. As voluntary licensing became popular, the UN-backed organization Medicines Patent Pool emerged connecting transparently drug originators with licensees.

The Global Fund to Fight AIDS, Tuberculosis, and Malaria (GFATM), an NGO mainly supported by government donors, created a voluntary pooled procurement scheme in 2009 to encourage country-level participation. Pooled procurement was found to have reduced the ex-works price of the generic drug Efavirenz by 16.2% from 2009 to 2013 [29].

Challenges Beyond Drug Pricing

While advances have been made in reducing price and increasing the share of HIV positive people on ARVs, not only do challenges remain relating to price but they also extend beyond the economic dimension.

Patenting, Drug Resistance, and Adherence

While HIV treatment started out as a single drug therapy, it was discovered that combining multiple drugs in combination cocktails is highly effective. The newest HIV drug to date is a quadruple therapy, intended for severe cases of HIV. The drugs involved in a combination therapy may all be beyond patent, but a firm may still patent their combination as a novel therapy often delivered as one pill per day. Patents prevent generic producers from being able to legally combine these drugs into one pill forcing them to produce multiple pills. While complex dosing schedules (multiple pills per day) increase non-compliance, reduced pill burden (e.g., one pill a day) is an effective way to improve medication adherence and reach viral suppression [30]. Furthermore, patients may develop drug resistance requiring the use of a new drug class or may suffer from side effects like nausea, anxiety, confusion, vision problems, anorexia, insomnia, taste perversion, and abnormal fat distribution, increasing the risk of patient non-adherence [31, 32]. Patent protection may reduce the ability of generic manufactures to produce convenient delivery forms (one pill a day), or drugs avoiding side effects, influencing treatment adherence, or make affordable new ARV classes unobtainable.

Social Stigma, Prevention, and Diagnosis

With currently 1.8 million new annual infections the UN is trailing in reaching its goal of annually 500,000 new infections by 2020. Unsuccessful prevention increases the number of people who need ARVs which reinforces resource scarcity and creates challenges in meeting ARV demand, which subsequently reinforces problems related to drug resistance and distribution.

Prevention is a multidimensional problem directly associated with human behavior change. To successfully limit HIV, prevention needs to be integrated with treatment, medical approaches to prevention, and institutional prevention approaches. To achieve changes in behavior, Coates et al. suggest that education, stigma reduction, use and access to contraceptives, and changes in sexual behavior need to be addressed [33].

Stigma, as a result of a lack of knowledge, is a barrier to testing and thus taking effective steps to stop the spread of HIV. The tendency to blame infected, and the perception that HIV is associated with death leads to people expressing shame, guilt, and social disapproval to people living with HIV. People who associate a larger stigma with HIV express strong negative attitudes towards testing which may imply that they are less likely to be tested [34]. People fearing to be stigmatized by their partner show much lower adherence rates hiding medication from discovery [35].

Distribution of Medicine

Reaching low income patients is challenging. Access to health care for low income population is limited by affordability but also physical access to health care, a lack of financing/insurance options, unawareness, and a lack of access to information [36]. Successful approaches to reach low income population need to rely on distribution, information, and marketing channels adequate for the intended audience and their realities. For example, long travel to and wait times at hospitals or pharmacies require patients to take a day off to pick up their drugs. This decreases treatment adherence as it is inconvenient and results in loss of income for a household reducing its financial capabilities further. Decentralized distribution system may be solution to reduce travel time and wait time. South Africa for example experiments with automated ATM-like distribution systems providing patients via ID card access to their prescription reducing wait times [37].

Conclusions

In sum, research in HIV treatment has long focused on the pharmacological technologies that can be applied to improve patient health. Business innovation may prove to be a major player in ending the HIV and AIDS epidemic. Without taking a closer look at the strategies at play and the results they yield, we will consistently face the same issues of cost creating a barrier to positive health outcomes. In addition, future trade agreements and business models must take into account the way ARV demand may increase with the adoption of the test and treat paradigm aiming to supply patients directly after diagnosis with ARVs and achieve the 90-90-90 goal of UNAIDS. While available in developed countries like the USA, the use of ARV as pre-exposure prophylaxis (PrEP, a form of preventive healthcare) for people who are HIV negative but with ongoing risk of getting infected may be an important element in reducing the number of new HIV infections in developing countries and protect groups at risk of infection. This would require broad availability and affordability of suitable ARVs in developing countries. Finally, as newer, more effective, more tolerable therapies emerge, trade regulators must reconsider the existence of other ARV options as justification for monopolies on the improved drugs. It is also recommended that more research look into the potential positive and negative externalities of the current business interventions being employed in ARV delivery. There is room for significant growth in the quality and quantity of HIV care in least-developed countries. The response to HIV focuses on questions of price and technology but drug adherence, stigma, education, and access are dimensions important in dealing with HIV. The response needs to go beyond technological innovation related to drugs and price and has to include new forms of partnerships, organizational forms, business models, and education services, by companies and organizations, while being responsive to social and cultural context. Adopting to the local context combining prevention and treatment efforts while relying on local stakeholders may increase drug penetration, treatment adherence, and the effectiveness of prevention while at the same time promoting business opportunities for low income population.

References

1. UNAIDS. UNAIDS DATA 2017. 2017. http://www.unaids.org/sites/default/files/media_asset/20170720_Data_book_2017_en.pdf. Accessed 27 Apr 2018.
2. US Department of Health and Human Services. The stages of HIV infection understanding HIV/AIDS. n.d. https://aidsinfo.nih.gov/understanding-hiv-aids/fact-sheets/19/46/the-stages-of-hiv-infection. Accessed 27 Apr 2018.
3. Rodger A, Bruun T, Cambiano V, Vernazza P, Strada V, Van Lunzen J. 153LB: HIV transmission risk through condomless sex if HIV+ partner on suppressive ART: PARTNER study. In 21st Conference on Retroviruses and Opportunistic Infections. p. 3–6.
4. United Nations, Department of Economic and Social Affairs, Population Division. The impact of AIDS. New York: United Nations; 2004.
5. Dixon S, McDonald S, Roberts J. AIDS and economic growth in Africa: a panel data analysis. Journal of International Development. 2001;13(4):411–26.
6. Kambou G, Devarajan S, Over M. The economic impact of AIDS in an African country: simulations with a computable general equilibrium model of Cameroon. J Afr Econ. 1992;1(1):109–30.
7. World Health Organization. The global burden of disease: 2004. Part 4. Burden of disease: DALYs. 2004. WHO. www.who.int/healthinfo/global_burden_disease/2004_report_update/en/. Accessed 24 Apr 2018.
8. Ortblad KF, Lozano R, Murray CJL. The burden of HIV: insights from the Global Burden of Disease Study 2010. AIDS. 2013;27(13):2003.
9. World Health Organization. Global health estimates 2015: deaths by cause, age, sex, by country and by region, 2000–2015. Geneva: World Health Organization; 2016.
10. The U.S. President's Emergency Plan for AIDS Relief. Fact sheet DREAMS Partnership. 2017. https://www.pepfar.gov/documents/organization/252380.pdf. Accessed 27 Apr 2018.
11. World Health Organization. Global price reporting mechanism for HIV, tuberculosis and malaria. n.d. www.who.int/hiv/amds/gprm/en/. Accessed 27 Apr 2018.
12. Global Health Observatory. Global Health Observatory data HIV/AIDS. n.d. www.who.int/gho/hiv/en/. Accessed 27 Apr 2018.
13. BBC. Indian firm offers cheap AIDS drugs. BBC. 2001. http://news.bbc.co.uk/2/hi/health/1158637.stm. Accessed 27 Apr 2018.
14. Merson MH, Curran JW, Griffith CH, Ragunanthan B. The President's Emergency Plan for AIDS Relief: from successes of the emergency response to challenges of sustainable action. Health Aff. 2012;31(7):1380–8.
15. Mothers2mothers. mothers2mothers: a simple solution to a complex problem. Frequently asked questions. 2015. https://www.m2m.org/wp-content/uploads/2015/08/Frequently-Asked-Questions-Aug-2015.pdf. Accessed 27 Apr 2018.
16. Medecins Sans Frontieres. AIDS triple therapy for less than $1 per day. 2001. http://www.msf.org/en/article/aids-triple-therapy-less-1-day. Accessed 27 Apr 2018.
17. World Trade Organization. TRIPS and public health. n.d. https://www.wto.org/english/tratop_e/trips_e/pharmpatent_e.htm. Accessed 27 Apr 2018.

18. Mothers2mothers. External evaluation and cost-benefit analysis of mothers2mothers' mentor mother programme in Uganda. 2015. https://www.m2m.org/wpcontent/uploads/2015/01/Uganda_CostBenefitAnalysis.pdf. Accessed 27 Apr 2018.
19. Mothers2mothers. 2016 impact report. 2017. https://www.m2m.org/wp-content/uploads/2017/10/m2m-2016-AnnualEvaluation.pdf. Accessed 27 Apr 2018.
20. Mothers2mothers. Mothers2mothers. n.d. https://www.m2m.org/. Accessed 27 Apr 2018.
21. The U.S. President's Emergency Plan for AIDS Relief. Dreams innovation challenge. n.d. https://www.pepfar.gov/documents/organization/247602.pdf. Accessed 27 Apr 2018.
22. The U.S. President's Emergency Plan for AIDS Relief. The United States President's emergency plan for AIDS relief. n.d. https://www.pepfar.gov/. Accessed 27 Apr 2018.
23. PEPFAR Solutions. Improving access to HIV treatment services through community ART distribution points in Uganda. 2018. https://www.pepfarsolutions.org/women/2018/1/13/improving-access-to-hiv-treatment-services-through-community-art-distribution-points-in-uganda. Accessed 27 Apr 2018.
24. Gebo KA, Fleishman JA, Conviser R, Hellinger J, Hellinger FJ, Josephs JS, Keiser P, Gaist P, Moore RD. Contemporary costs of HIV health care in the HAART era. AIDS. 2010;24(17):2705–15.
25. World Health Organization and UNAIDS. Accelerating access initiative – widening access to care and support for people living with HIV/AIDS - progress report, June 2002. 2002. http://apps.who.int/iris/bitstream/10665/42550/1/9241210125.pdf. Accessed 27 Apr 2018.
26. UNAIDS. Technical brief. Implementation of TRIPS and access to medicines for HIV after January 2016: strategies and options for least developed countries. UNAIDS, 2011. 2016. http://files.unaids.org/en/media/unaids/contentassets/documents/unaidspublication/2011/JC2258_techbrief_TRIPS-access-medicines-LDC_en.pdf. Accessed 27 Apr 2018.
27. Committee on Trade and Development. Special and differential treatment provisions in WTO agreements and decisions WT/COMTD/W/219 (22 September, 2016). 2016. p. 135–6.
28. Waning B, Kaplan W, King AC, Lawrence DA, Leufkens HG, Fox MP. Global strategies to reduce the price of antiretroviral medicines: evidence from transactional databases. Bull World Health Organ. 2009;87:520–8.
29. Kim SW, Skordis-Worrall J. Can voluntary pooled procurement reduce the price of antiretroviral drugs? A case study of Efavirenz. Health Policy Plan. 2017;32(4):516–26.
30. Hanna DB, Hessol NA, Golub ET, Cocohoba JM, Cohen MH, Levine AM, Wilson TE, Young M, Anastos K, Kaplan RC. Increase in single-tablet regimen use and associated improvements in adherence-related outcomes in HIV-infected women. J Acquir Immune Defic Syndr. 2014;65(5):587–96.
31. Ammassari A, Murri R, Pezzotti P, Trotta MP, Ravasio L, De Longis P, Lo Caputo S, et al. Self-reported symptoms and medication side effects influence adherence to highly active antiretroviral therapy in persons with HIV infection. J Acquir Immune Defic Syndr. 2001;28(5):445–9.
32. Clavel F, Hance AJ. HIV drug resistance. N Engl J Med. 2004;350(10):1023–35.
33. Coates TJ, Richter L, Caceres C. Behavioural strategies to reduce HIV transmission: how to make them work better. Lancet. 2008;372(9639):669–84.
34. Kalichman SC, Simbayi LC. HIV testing attitudes, AIDS stigma, and voluntary HIV counselling and testing in a black township in Cape Town, South Africa. Sex Transm Infect. 2003;79(6):442–7.
35. Nachega JB, Stein DM, Lehman DA, Hlatshwayo D, Mothopeng R, Chaisson RE, Karstaedt AS. Adherence to antiretroviral therapy in HIV-infected adults in Soweto, South Africa. AIDS Res Hum Retroviruses. 2004;20(10):1053–6.
36. Bagchi-Sen S, Schunder T, Bourelos E. Foreign direct investment in R&D and the base of the pyramid: is a new space of innovation emerging in India? In: Innovation spaces in Asia: entrepreneurs, multinational enterprises and policy. Cheltenham: Edgar Elgar; 2015. p. 256–78.
37. McVeigh T. South Africa's latest weapon against HIV: street dispensers for antiretrovirals. The Guardian. 2016. https://www.theguardian.com/world/2016/jul/17/atms-dispense-antiretrovirals-south-africa-hiv. Accessed 27 Apr 2018.

Chapter 17
Building Policies, Plans, and Cities to Manage Extreme Weather Events: Perspectives from Urban Planning and Landscape Architecture

Zoé Hamstead and Paul Coseo

Urban Extreme Events

By the time the heat wave subsided in late August of 2003, France had lost an estimated 15,000 people to heat-related illnesses [1, 2]. On August 12, temperatures in Paris approached 40 °C (104 °F) and the low that night (24 °C; 75 °F) exceeded the normal daytime high (22 °C; 72 °F) [3]. Since air conditioning is uncommon given the typically mild Parisian summers, residents had little relief from the scorching temperatures. If you were poor, elderly, or had other complicating health factors, you were at greater risk. Socially isolated elderly residents who lived in older buildings without private toilets and in proximity to urban heat islands were particularly vulnerable [2].

During the summer of 2003, atmospheric patterns shifted north, leading to a series of high pressure systems that dried out the air mass and displaced weather patterns over central Europe. That combined with the storm track further south than normal resulted in clear, stagnant, dry conditions. Those conditions led to a feedback loop through which warming led to drying and drying led to additional warming [4]. The drought intensified as part of a reinforcing drying-warming cycle akin to the North American Dust Bowl of the 1930s.

The cities where most Europeans reside were already drier than the countryside due to their very design. Monumental, masonry buildings and impervious roads of Paris soaked in the abundant sunshine. Those materials absorbed the sun's energy and emitted it into the stagnant atmosphere, heating the city several degrees warmer than the surrounding French countryside. The heat wave that killed more than

Z. Hamstead (✉)
University at Buffalo, State University of New York (SUNY), Buffalo, NY, USA
e-mail: zoehamst@buffalo.edu

P. Coseo
Arizona State University, Tempe, AZ, USA

© Springer Nature Switzerland AG 2020   261
K. H. Smith, P. K. Ram (eds.), *Transforming Global Health*,
https://doi.org/10.1007/978-3-030-32112-3_17

70,000 people across Europe was driven by not only atmospheric conditions, but also urbanization and socio-ecological dynamics [5]. Extreme heat events quietly extinguish lives and livelihoods to reveal hidden inequities. As weather-related disasters like the 2003 European heat wave become more common and intense, extreme weather events are an urgent focus for our twenty-first century urban planning agenda.

Since the middle of the twentieth century, scientists have observed increases in warm temperature extremes, high sea level extremes, and heavy precipitation events. Globally, weather-related disasters have more than tripled since the 1960s [6]. In some parts of the world, it is estimated that human-induced climate change has more than doubled the probability of heat waves. A large proportion of climate change will not be reversible within the next 100–1000 years [7]. Even if humans were to completely abate greenhouse gas emissions, warming and sea level rise will continue. These warming trends will increase the frequency, intensity, and duration of heat waves and extreme precipitation events while continuing to elevate sea levels, warming and acidifying oceans. Climate-induced heat waves, droughts, floods, cyclones, and wildfires reveal vulnerabilities in ecosystems, infrastructure, economic systems, and human health. It is estimated that climate change-mediated heat exposure, diarrhea, malaria, and childhood undernutrition will cause approximately 250,000 additional deaths per year between 2030 and 2050 [8]. Warming temperatures can increase the spread of disease as reproductive rates of organisms that transmit infectious diseases are affected by climate variation. Malaria transmission has been associated with increased temperatures in parts of Kenya and Ethiopia and waterborne diseases have been linked to the 1997–1998 El Nino event in Peru during which daily admissions for diarrhea more than doubled [9]. Extreme weather poses risk to critical food, water, energy, transportation, and healthcare infrastructure. Disruptions caused by extreme weather may exacerbate political violence and trigger socio-economic instability through processes such as climate migration. The United Nations High Commissioner for Refugees estimates that 22.5 million people were displaced by climate or weather-related events between 2008 and 2015, equivalent to 62,000 people each day [10]. From a national security perspective, countries like the USA are concerned with the high potential for refugee flow due to humanitarian disasters in neighboring Caribbean countries, damage to national and international military bases in coastal communities, and politically unstable regimes that could be further destabilized by climate change [11].

Human civilization relies upon an overlap between the rise of human intelligence alongside favorable and somewhat reliable climate conditions. Yet, extreme events across all regions are destabilizing these favorable conditions and unevenly distributing risks by placing a disproportionate burden on the poor, people of color, people who live in the global south, and other marginalized groups. To consider how policies, plans, and cities should manage extreme events, we need to understand how urban systems make extreme events worse, and how those extreme events in turn disrupt urban systems. We also must understand the professional systems of engineers, architects, urban planners, public works professionals, meteorologists, emergency managers, public health practitioners, and others who shape planning and design responses to extreme weather events such as heat waves [12].

Global Impacts

The Global Climate Risk Index developed by Germanwatch quantifies impacts of extreme events across countries. Between 1997 and 2016, 11,000 extreme events directly led to more than 524,000 deaths and losses of US$ 3.16 trillion. However, indirect mortality rates such as those associated with heat and cold extremes are more difficult to capture. Since hot and cold temperatures interact with co-morbidities such as cardiovascular disease, hospital admissions codes generally underestimate temperature-related death and illness. Heat waves can also lead to indirect impacts via droughts and food scarcity, particularly in African countries. Of the ten countries most affected by extreme events in 2016, nine were developing, low- or middle-income countries. Haiti, Zimbabwe, and Fiji were most affected by extreme events, followed by Sri Lanka, Vietnam, and India. In September of 2016, Haiti was struck by Hurricanes Matthew and Nicole. Hurricane Matthew killed over 500 people, threatened food security and exacerbated cholera outbreaks. In 2016, Zimbabwe experienced extreme droughts and heat waves that caused agricultural losses, followed by a massive precipitation event that led to flooding which killed 250 people, left several thousand people homeless, and destroyed public infrastructure such as dams and bridges. Countries most affected by extreme events between 1997 and 2016 include Honduras, Haiti, and Myanmar (Fig. 17.1) [13].

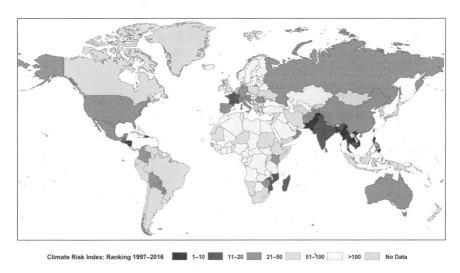

Climate Risk Index: Ranking 1997–2016   ■ 1–10   ■ 11–20   ■ 21–50   □ 51–100   □ >100   □ No Data

Fig. 17.1 World map of the Global Climate Risk Index for 1997 and 2016. (Source: reproduced with permission from Eckstein et al. [13], Germanwatch and Munich, Re NatCatSERVICE)

Since most people live in cities, extreme events occurring in urban systems substantially impact large populations, infrastructure, and ecosystems. By 2050, the proportion of people living in cities is expected to double to 70% [14]. In addition to this transition from rural to urban environments, also shifting is the geography of urbanization. Whereas current urbanization rates in the Americas and Europe are between 75% and 80%, less than 40% of African and Asian populations are urban. Urban populations in Africa and Asia are expected to increase more than 50% by just 2025 [15]. Following extreme events, many urban communities experience slow or incomplete recovery. Given urban warming trends and increased demands placed on infrastructure that is already insufficient to support population increases, urbanizing populations will be particularly vulnerable to extreme events. In addition to these direct impacts, urban areas will also be impacted by climate migration patterns. One of the most poignant ways in which people are impacted by warming trends and extreme events is through displacement. In the Arctic, temperatures are rising at more than twice the rate of the global average, leading to violent ocean storms, flooding, and erosion beneath peoples' homes [16]. The community of Kivalina, Alaska, USA is a barrier island that is shrinking due to sea level rise and storm surges. According to the United States Army Corps of Engineers, the island with a population of 400 will no longer be habitable in the next decade [17]. As the people of Kivalina and other communities compromised by climate change must relocate, it will largely fall on cities to host these climate refugees and manage their transition to new ways of life.

Breaking Records: "Natural" Fluctuations Across Scales

Extreme weather and climate phenomena occur on the outer boundaries of normal temperature, precipitation, and other meteorological distributions [18]. Since extremes are observed in relation to all weather and climate events in a particular geographic context, an extreme in one region may be considered normal in another. Societal risks associated with extreme events are driven by weather and climate, as well as the biophysical urban environment which can exacerbate or ameliorate climate and weather and determine spatial patterns of risk (Fig. 17.2).

Urban planners, designers, and developers have constructed the urban built environment based on the assumption of an unchanging atmosphere and climate system. The term stationarity describes this static, unchanging view of atmospheric processes. Increasingly, scientists recognize that atmospheric processes are nonstationary, describing "…a shifting mean value or a changing shape of the probability distribution with time." [19] The known boundaries of our climate systems are based on empirical data collected only over the last 150 years, and that record is geographically and temporally incomplete [20]. Although incomplete, the record clearly indicates that the climate boundaries within which humans built industrial cities are rapidly shifting due to climate change.

Fig. 17.2 The same meteorological event can result in uneven impacts for residents within adjacent neighborhoods because of differences in risk and vulnerabilities such as the quality of buildings and infrastructure. In urban systems, the impacts of extreme events are driven by how weather and climate interact with infrastructure, ecosystems and humans. (Source: Paul Coseo, YY, Brazil, 2012, https://yyinbrazil.wordpress.com/2012/12/01/yy-in-brazil-rio-de-janeiro-favela-rocinha-day-4/)

Extreme atmospheric events (weather and climate) are bounded and influenced by the limits of our earth systems. We use the term extreme weather events to describe short-term extreme changes in our atmosphere measured in hours and days, while extreme climate events relate to longer-term extreme atmospheric conditions that last over months to years. Climate trends over large spatial scales also function on longer time scales and partially control weather phenomena at smaller scales. The weather you experience in your backyard is a manifestation of global wind patterns, regional macro dynamics, local climate influences, and the biophysical conditions of your backyard (Fig. 17.3).

Biophysical Urban Infrastructure

When we build cities, we replace agricultural and rural landscapes with mineral-based materials to construct infrastructure such as buildings and roads, and we plant novel, non-native vegetation. Those altered urban biological and physical environments play a key role in how extreme events impact our lives. How we build, where we build, and for whom we build all factor into the degree, intensity and burden of impacts. We choose many non-native plants for our yards and parks that modify

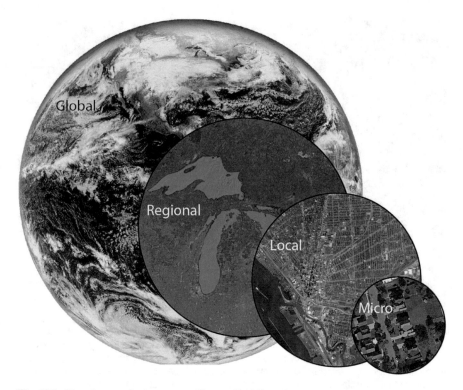

Fig. 17.3 Nested atmospheric systems. (Source: Paul Coseo)

native plant communities, change biological compositions and in some cases, use more water and nutrients. We construct buildings and roads composed of mineral-based engineered materials that seal soils and warm the environment. Changing both biology and physical environments alters the hydrological and heat energy budgets that influence the quality and quantity of water, heat, and wind.

The term biophysical environment describes the biological and physical compositions of urban systems. Three aspects of cities' biophysical environment exacerbate extreme atmospheric events. First, impervious surfaces seal soil, which causes more stormwater to run-off. Materials that are 100% impervious, such as concrete and asphalt, completely restrict water or air from moving through the material and reduce the land's natural ability to absorb stormwater and cool itself. Second, engineered materials alter the albedo or reflectivity of the land surface. Cities absorb 80–85% of incoming solar radiation, making them hotter than non-urban locations [21]. Finally, the design of a city's biophysical environment is a social and political endeavor. Existing social inequities in decision-making, resource allocation, and services are manifest in the biophysical environment of our cities. These inequities result in uneven impacts from extreme events.

Historic and Contemporary Developments

Trend 1: The Carbon Metropolis

During industrialization of the 1800s, society adopted carbon as the main source of energy on which to build cities (Fig. 17.4). Human energy systems use Earth's resources for survival and quality of life [22]. These energy systems have three main components: natural energy sources, conversions of those sources, and uses of energy flows. Humans transitioned from plant and animal sustenance to wood to water and more complex carbon-based resources (e.g., coal) for societal survival and quality of life [23]. At the beginning of the industrial revolution and continuing through modernization of cities, a variety of renewable and non-renewable energy sources have all vied for dominance to fuel urban and economic growth. George Perkins Marsh, in Man's Responsibility for the Land, recognized the power of natural processes to fuel society.

> Mechanical philosophers have suggested the possibility of accumulating and treasuring up for human use some of the greater natural forces, which the action of the elements puts forth with such astonishing energy. [24]

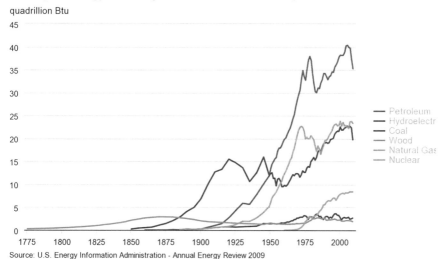

History of energy consumption in the United States, 1775-2009

Source: U.S. Energy Information Administration - Annual Energy Review 2009

Fig. 17.4 Energy consumption in the U.S. since 1775 in quadrillion Btu. Led by coal and then petroleum, carbon-based energy became an increasingly large share of the U.S. and world's energy system starting in the late 1800s. (Source: U.S. Energy Information Administration, 2009. https://www.eia.gov/totalenergy/data/annual/archive/038409.pdf the 2009 Annual Review report from the U.S. Energy Information Administration. Direct Link to Data: https://www.eia.gov/totalenergy/data/annual/index.php)

Although a mix of energy sources fuels cities today (Fig. 17.4), the majority of the built environment is a direct product of, and reliant upon carbon-intensive modes of production. Wuebbles et al. [25] report that under higher emissions scenarios, annual global carbon emissions are projected to approach 30 gigatons of carbon per year by 2100, while projected global temperatures may range from 5.0° to 10.2 °F above 1901–1960 averages (Fig. 17.5). Although a warming world will likely lead to more extreme events, scientists are not able to attribute a particular extreme event at a specific location (e.g., Hurricane Katrina) to human-induced global climate change from natural variability of extreme events [26]. This is because we are only beginning to understand natural variability in higher-level (e.g., global) atmospheric patterns that last from several weeks to months—such as El Niño-Southern Oscillation (ENSO), the North Atlantic Oscillation (NAO), the Pacific Decadal Oscillation (PDO), and the Artic Oscillation (AO)—within which local-scale extreme events are nested. For instance, in 2014 and 2019, polar vortices with extremely cold temperatures over mid-latitudes in North America were driven by higher-level atmospheric patterns (a negative phase of the AO) coupled with climate change (unusual arctic warming). Trenberth et al. [26] suggest that rather than asking whether a particular weather event is attributable to climate change, one should instead ask, "… given the change in atmospheric circulation (natural global variability) that brought about the event, how did climate change alter its impacts?" Scientists must be able to articulate climate trends with both accuracy and urgency. A tendency to describe climate change in conservative, non-alarmist terms makes room for denial by special interests whose power and wealth depend on a carbon-based economy.

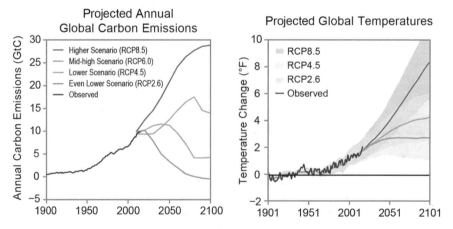

Fig. 17.5 Projected annual global carbon emissions and projected global temperatures. (Source: Wuebbles et al. [25])

Trend 2: Industrial Material Cities

The carbon-based city caused two key changes to the frequency and intensity of extreme events. First, cheap carbon-based energy fueled urban-induced local climate change through conversion of unbuilt land to urban industrial produced materials, configurations, and uses. Local climate change—such as the formation of urban heat islands—tends to be worse in areas with high percentages of impervious pavements and buildings with little to no vegetation (explained more below) and exacerbates extreme events such as: (1) hot temperatures; (2) heavy rainfall; and (3) stagnant, polluted air conditions in concentrated areas. Our carbon-based economy compounds urban-induced warming by enabling sprawling expanses of mineral-based urban landscapes that exacerbate waste emissions. At the same time, carbon emissions cause global climate change from the same consumptive, decentralized auto-oriented settlement patterns of buildings and uses. Although these global forces will result in higher than average global temperatures, they also create increased variability in bioclimatic regions, warming some bioclimatic regions, while cooling others [27, 28]. Local and global climate change interact in complicated and dynamic ways that scientists are only beginning to understand. Moreover, many global temperature projections are derived at a scale that does not incorporate the effects of urban-induced warming. Stone [29] compared urban to rural air temperature trends over 50 years for the 50 largest American cities finding that, on average, UHIs make these urban areas 1.5 °F warmer than rural areas and this average difference is increasing at a rate of 0.14 °F per decade. Thus, UHIs in the U.S. are intensifying over time. The global scale at which climate change projections are often provided limits their utility by ignoring important local factors and underestimating warming impacts in many urban areas. However, even as we recognize the limitations of global climate projections, decision-makers need to make informed decisions about how to mitigate weather impacts.

Trend 3: Fail-Safe Management

Institutions that plan for and respond to extreme events often use standard operating procedures—rules, norms, or routines that promote efficiency. With standard operating procedures in place, institutional behavior is largely consistent and change tends to occur incrementally [30]. This is problematic given that extreme events are occurring with more frequency and unanticipated intensities. Events such as the heat wave that struck Europe in 2003 are sobering reminders that we underestimate risks posed by extreme events. As many physicians and government agents were on holiday during that summer and many vulnerable elderly people who lacked constant medical care did not have support systems in place, institutions were unable to respond to people in need.

Environmental policies have traditionally been constructed under the command-and-control approach, which bases regulation on current conditions and technology currently available to manage those conditions. This approach assumes that we can predict how natural and cultural systems will behave and therefore, managers can use science and technology to reduce environmental pollution and meet human needs. Yet, in reality we find that natural and cultural systems are unpredictable, and that technology (such as emissions control) can fail during surprise events [31]. When Hurricane Harvey hit Texas, USA in the fall of 2017, unusually warm waters in the Gulf of Mexico led to unusually heavy rainfall; on the order of 24.5 trillion gallons of water fell on southeast Texas and southern Louisiana [32]. Floods caused by heavy rainfall were worsened by impenetrable concrete land cover and lack of soil absorbency in the low-lying coastal plain on which Houston, Texas is built. The interaction between ocean warming and impervious surfaces not only damaged property and directly claimed lives, but also further endangered public health by triggering chemical discharges. As power outages, backup failures and other storm damage occurred at chemical plants and oil refineries, fires that released air toxins broke out. Oil, refined fuel and chemicals, natural gas, and hundreds of tons of other toxic substances were spilled across Texas [33]. Regulators and company executives did not anticipate the magnitude of emissions releases. State officials shut off air quality monitors to protect them from storm-related damage, increasing uncertainty about public health impacts [34]. Lacking recognition of dynamic relationships across our climate, landscapes, and industrial infrastructure, managers are blindsided by the cascading spin-off effects of extreme events.

The risk of future climate change impacts rises with temperature. Our tendency to design policies, plans, and infrastructure based on average conditions is increasingly ineffective in the context of climate change. Countries at all levels of development lack preparedness for extreme events that are altering ecosystems, disrupting food production and water supply, damaging infrastructure and settlements, causing human morbidity and mortality, and impacting human well-being [7].

Trend 4: Vulnerable Communities

Due to ways in which institutions directly and indirectly disenfranchise minority social groups, some communities and households have less capacity to cope with extreme events than others. For instance, throughout the twentieth century in the USA, federal housing policy created an enormous wealth gap between black and white households. In 1935, the Home Owners Loan Corporation (HOLC) was established to insure home mortgages following the banking crisis of the 1930s that had decreased home ownership and resulted in many foreclosures. However, this insurance was unavailable to racial minorities, ethnic minorities, immigrants and urban communities. It was the HOLC's policy to ensure that "incompatible racial groups should not be permitted to live in the same communities." [35] This policy was enforced through redlining maps, which delineated red boundaries around communities with African Americans, immigrants, and that were located near central business districts. The HOLC also recommended that highway construction be used to separate "incompatible races," further isolating urban black neighborhoods. Since the middle of the last century in the U.S., home ownership has been a primarily means through which middle class families generate wealth and pass on wealth to their children in the form of higher education investments and direct bequeathment. In addition, residential property tax forms an important means through which communities collectively invest in local infrastructure. Housing policy not only led to wealth disparities (African American wealth in the U.S. is only 5% that of white wealth), but this wealth disparity was also compounded at the neighborhood-level by uneven community investments through which white children could access better opportunities and living conditions through higher quality education and other public investments. Household and neighborhood-level resources can be crucial for resilience to extreme events. Access to air conditioning on hot days, or flood insurance during storms is largely a matter of income and wealth.

Just as institutions tend to prepare for average rather than extreme conditions, there is also a tendency to neglect human outliers, and further marginalize people who are already disenfranchised members of society. Disasters are worse for low income communities of color. During storm events, low income communities sustain more damage because they tend to be built on cheaper under-regulated land (Fig. 17.2) that is more flood-prone, and are more likely to lack personal transportation and other resources needed for evacuation. Adding insult to injury, marginalized groups are often stigmatized during disaster relief processes. During Hurricane Katrina in New Orleans, Louisiana, USA, poor and African American victims were labeled as looters while trying to procure necessities through community support systems. Similar efforts in whiter, wealthier areas were characterized as "finding." [36] Since recovery tends to be more difficult for low-income communities of color, extreme events will continue to exacerbate existing inequalities unless sensitivities are explicitly addressed through a social justice lens (Fig. 17.6).

Barriers to Transformation

Scientific Knowledge

Increasing intensities, frequencies, durations, and impacts of extreme events in cities are driven by not only greenhouse gas-related climate change, but also interactions across ecosystems that govern the energy and water balance and human-driven processes of urbanization (Fig. 17.6). Urbanization processes determine both the composition of built materials that can stymie climate moderation, as well as which human populations and infrastructure are in the path of these events. Although human and ecological systems interact with one another, they have traditionally been studied and managed separately. Traditional engineering practices set

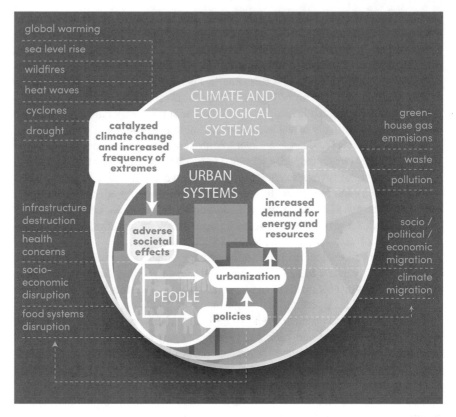

Fig. 17.6 Climate and ecological systems. (Source: Nicole C. Little, at the request of the Editors)

acceptable levels of risk for infrastructure to withstand environmental conditions such as rainfall, which depend on the assumption of a stationary climate [37]. These assumptions of predictability coupled with lack of recognition for human and climate drivers across scales has limited the extent to which scientific knowledge could inform environmental planning and policy.

Politics and Legislation

In the early 1990s—just as global leaders were negotiating how to reduce greenhouse gas emissions through the United Nations Framework Convention on Climate Change (UNFCCC)—the World Trade Organization (WTO) adopted a set of trade policies that undermined the UNFCCC. Enlarging distances over which goods travel, these trade policies increased global greenhouse gas emissions. Aggressive technology patents have inhibited the transfer of renewable energy technology, and governments have been reluctant to adopt anti-pollution policy for fear of being sued by private industry [38]. Local governments' ability to address climate change through legislation that would reduce greenhouse gas emissions is undercut by trade policies that favor liberal global trade.

Since the early 1990s, party activists in the USA accelerated political conflict on a range of issues, which more distinctly sorted individuals along party lines [39]. Climate change is one of the issues that has become highly politicized. A study of Gallup Polls between 2001 and 2010 found that liberals and Democrats are more likely to believe the scientific consensus on global warming than are conservatives and Republicans [40]. Thus, national-level support for climate legislation may depend on which party is in leadership.

Recognition and Procedural Injustice

Increasing intensity and duration of weather events will fundamentally change human society as a whole. But the poor, people of color, and people living in the global south will be disproportionately impacted. In modern history, institutions have oppressed racial and ethnic minorities, women, the poor, people who do not identify with gender binaries or heteronormative roles, and people living in the global south. Ecofeminist scholars argue that constructs such as nature, culture, reason, and humanity are the very vehicles of oppression [41]. For instance, Western culture conflated women/nature in opposition to men/reason. Casting nature and reason as opposites creates cultural opposition between men and women. Critical environmental justice scholars argue that oppressed people are treated as expendable to our collective future [42]. White supremacy regimes attempt to marginalize non-White populations, as human dominionism claims human supremacy over other species, ecosystems, and Earth systems. The divisive nature of identity-based associations affects norms and policies that govern our relationship with nature and with each other.

Climate-driven extreme events are threat multipliers that compound existing injustices and power imbalances. Uneven impacts of extreme events are as much about environmental injustice as they are about unfair institutions of housing, transportation systems, and economic development. In Gulf region states of the USA affected by Hurricane Katrina, women and women of color were the most likely to lack access to health insurance, experience disproportionately high rates of poverty and engage in low-wage work despite high work participation rates. Jones and Hartman [43] argue that women of color and elderly women were "abandoned before the storms" and thus experienced particularly great difficulty relocating and recovering in their wake. Institutions of governance and industry both play key roles in exacerbating existing inequalities at all geographic scales—from global to national and local levels. Although democratic decision-making processes offer opportunities to identify solutions for protecting people who are most at risk, all existing democratic nations engage in various forms of representative democracy in which elected representatives—rather than all citizens—make decisions. Thus, decision-making processes tend to be democratically weak at best and authoritative at worst.

Integrative and Ethical Approaches

Urban Resilience

Scientific efforts over the last decades have begun to address the complex nature of extreme events. Urban ecologists strive to understand how both human and ecological processes work together to create more or less resilient urban systems [44–46]. This systems perspective provides a way to consider how humans, infrastructure, and ecosystems all interact with each other at multiple scales and in often complex and surprising ways. The resilience of urban systems—their ability to maintain essential functions—is tested during times of social or ecological disturbance. Resilience defines how much a system can change before losing its ability to recover. Adaptability is the capacity of actors to influence a system's resilience, by enabling a system to recover more successfully or by preventing a level of change under which it would not be able to recover [47]. *Adaptive governance* is an approach by which resource management strategies are seen as experiments that test how managers can best deliver services in the face of imperfect information about how our social, economic, infrastructure, and ecological systems will respond during disturbances such as extreme events [48]. Command-and-control strategies—which have dominated contemporary natural resource management—aim to control the variability of resources such as coastal fishery or agricultural yields [49]. These approaches do not work well in times of crisis because they are optimized for everyday occurrences rather than extreme events. For instance, engineered stormwater systems in which streams have been buried in underground tunnels and which largely rely on grey infrastructure to control flooding may function well in the context of a stationary climate, as they are designed to handle rain events with particular occurrence probabilities and levels. However, when the return period becomes more frequent and rain event levels increase under non-stationary conditions, the system may fail. Adaptive governance recognizes that the behavior of social-ecological systems is not entirely predictable, and therefore one-size-fits all approaches which limit redundancy are less resilient to change [50]. It centers around monitoring as a way for managers to incorporate system feedback into maintenance and to inform future design. For instance, the performance of a variety of green infrastructure projects such as raingardens, bioswales, and rooftop gardens could be tested under conditions of different rain events through a quasi-scientific process [51].

Transportation and critical infrastructure are often developed as "fail-safe" systems designed to withstand known threats. In an world of increasingly unknown threats, Ahern suggests that we instead design systems using a "safe-to-fail" approach, according to principles of redundancy, diversity, connectivity, and adaptive planning and design [31]. Redundancy and diversity are ways of providing "backup" systems that perform the same function, but are not reliant on the same support structure and thus spread risks across time and space. Connectivity supports functionality at multiple scales. Experimental approaches that pilot responsible safe-to-fail design experiments enable managers to "learn-by-doing," and adapt management strategies with changing conditions over time.

Earth Ethics and Social Movements

The largest climate march in history occurred in New York City in 2014, galvanizing over 1500 environmental justice organizations. An estimated 311,000 people marched in New York, and spurred thousands more to march worldwide. The staging area, located in a blocked off area of Central Park West between 59th and 86th streets, was divided into six sections: (1) Frontlines of Crisis, Forefront of Change; (2) We can Build the Future; (3) We Have Solutions; (4) We Know Who is Responsible; (5) The Debate is Over; and (6) To Change Everything, We Need Everyone. The title of the sixth group was part of a message in The Guardian earlier that month by a group of non-governmental organizations that included Avaaz, 38 Degrees, Greenpeace UK, and Oxfam. "From New York and London to Paris, Berlin, Delhi and Melbourne we'll demonstrate demand for an economy that works for people and the planet; a world safe from the ravages of climate change; a world with good jobs, clean air and water, and healthy communities. There is only one ingredient that is required: to change everything, we need everyone." [52]

Climate scholars and activists argue that although many of our legal and governance structures treat future generations, marginalized peoples, and natural systems as expendable, the well-being of all people, species, and ecosystems are actually critical to our collective existence [42]. Injustice to nature comes about through disrespect and status injuries; through extractive industries, humans dominate nature, render it politically invisible and disparage the natural world [53]. Our Children's Trust is a legal team representing a group of youth who filed suit against the United States federal government for intergenerational injustice. The group claims that the federal government has not taken sufficient action to protect the nation's youth and future generations against climate change. In addition to legal action, climate protests are another means by which activists aim to motivate climate action. Drawing on civil rights movement-era civil disobedience strategies, notable activist-scholars including Bill McKibben (co-founder of 350.org) and climatologist James Hansen (former director of the NASA Goddard Space Institute) have been arrested during protests such as that of the Keystone-XL pipeline. Just as the civil rights movement of the 1960s brought national attention to racial inequalities, climate justice legal battles and political protests are drawing attention to intergenerational and ecological injustice on the world stage.

Collaborative Management Across Disciplines

In addition to addressing climate change and adaptation at global and national levels through social movements and legal means, vision-oriented planning processes at the local level can directly engage those who are most impacted. Collaborative climate planning approaches bring together a range of citizens, scientists, and professionals to integrate knowledge about environmental planning and design (green infrastructure, energy policy); health (determinants of health-related outcomes); emergency management (protocols for reaching most vulnerable); ecological sciences and engineered infrastructure managers. Climatologists and other scientists contribute knowledge for setting benchmarks for pollution reduction. Engineers evaluate and design strategies for guarding against impacts on critical infrastructure. Planners and designers can address ways in which our built environment mitigates climate change and urban warming. Emergency responders develop protocols for responding to critical human health needs when other systems fail.

Given that urban areas have significant impact on extreme events at local scales, and that urban populations are most impacted by these events, collaborative planning processes hold important potential for mitigating and adapting to climate extremes at the urban scale. By integrating knowledge from multiple scientific domains (Table 17.1) and recognizing critical knowledge and experiences of those who are most vulnerable to extreme events, local-level planning can enable urban communities to transition to a more resilient and equitable future.

Table 17.1 Professional knowledge domains for management of extreme events in the USA

Area of expertise	Primary jurisdiction level	Profession	Arena
Science	Federal, state	Climatologists	Weather forecast and prediction
	Federal, state	Ecologists	Ecosystem health and restoration
	Federal, state	Biologists	Wildlife management
Planning and design	City	Architect	Building infrastructure planning, design, and management
	City	Landscape architect	Green infrastructure planning, design, and management
	City	Urban planner	Urban infrastructure planning (includes ecosystems, technological infrastructure, and social institutions)
	Federal, state, county, city	Engineer	Technological infrastructure planning, design, and management

(continued)

Table 17.1 (continued)

Area of expertise	Primary jurisdiction level	Profession	Arena
Business and finance	City	Real estate developer	Neighborhood economic development
	State, city	Land use lawyers	Land use regulation navigation and creation
	Global, federal, state, city	Financier	Urban infrastructure investment
	Global, federal, state, city	Insurance agents	Risk assessment and management
Advocacy	Global, federal, state, county, city	Conservancies	Green infrastructure conservation, care, and advocacy
	Global, federal, state, county, City	Non-profits	Varies by focus---protection, care, and advocacy for interests of the organization
Governance	City	City councils	Urban policy creation and accountability
	City	Mayor/city manager	Urban policy creation and accountability
	Metropolitan region, counties	Regional planning associations	Regional policy creation and management
	Federal, state, county, city	Representatives	Urban policy creation and accountability
	Neighborhood	Home owners' association (HOA) boards	Neighborhood policy creation and accountability
Emergency response	City	Police and fire	Human protection, health, and safety
	Global, federal, state, county, city	Public health	Human health, safety, and disease prevention
	County, city	Health care systems	Human health, safety, and disease treatment
Social services	County, city	Social workers	Human health, dignity, safety, and Well-being
	County, city, neighborhood	Religious organizations	Human spirituality and Well-being

Voice from the Field: Kizzy Charles-Guzman

1. Describe your position at the New York City Mayor's Office of Recovery and Resiliency, and one or more initiatives that you have undertaken to protect city residents from extreme climate and weather events.

I am a Deputy Director, leading the social resiliency team at the NYC Mayor's Office of Resiliency. We develop policies and programs that protect New York City communities against climate change impacts. These initiatives are rooted in our philosophy that residents, community-based organizations, and small businesses are the fabric of our neighborhoods. My team supports and brings capacity to these institutions through multi-agency initiatives that improve the physical environment, foster social cohesion, and strengthen their services and ability to conduct neighborhood-level emergency and climate preparedness planning. For example, we promote climate-smart adaptation measures for community organizations and small businesses that minimize the risk of lasting service disruptions in our communities. Lastly, we developed and are working to implement Cool Neighborhoods NYC, the City's first heat resiliency strategy. This plan is informed by health and climate data to achieve climate equity goals, focusing on historically underserved communities and at-risk populations.

2. In your view, what are the major barriers that New York City faces in managing extreme heat and other climate-related threats?

Heat is often referred to as a silent killer because heat-related deaths and illnesses occur behind closed doors and often don't receive much media attention. Our work also shows that even when the media reports on heat waves and extreme heat days, the focus of their messaging is misplaced. Most images that are used on TV and other media show people cooling down at beaches, and emphasize risks from outdoor activity, like jogging or working outdoors, instead of the much greater risk that older adults confined in unair-conditioned homes endure. So, people who are at risk and who should take action to protect themselves or their loved ones might see the news, not see themselves represented, and then take no protective actions to stay cool. We are working to educate health and weather reporters on the types of messages and images that can help at-risk New Yorkers to stay safe.

Another challenge in NYC is that our communities include people from many countries and cities that are much warmer. Many people don't like the feel of air conditioning or have concerns about the cost of energy bills. We emphasize that getting to an air-conditioned place such as a library, a cooling center, or someone else's home can save lives. We also emphasize that an AC unit can be set at 78 degrees, which is a way of staying safe while not running up the energy bill. Finally, many people don't perceive themselves to be at risk if they don't *also* know that certain conditions may lead to heat illness and heat death, such as taking certain medications, abusing drugs or alcohol, or having diabetes, cardiovascular, or respiratory disease. So, we have to work hard to ensure that we're reaching people in their languages and that we are disseminating information inclusively via their trusted messengers and the media channels they access day-to-day.

3. What lessons have you gleaned from your experience that could inform other communities' efforts to protect people against the impacts of climate-related threats?

Resiliency starts with preparation and the literature is clear that communities with stronger "social capital" and "social cohesion" recover faster from disasters. So, it is critical for communities to work on increasing civic engagement and volunteerism and for their leaders to support and build the capacity of community-based organizations. Residents and organizations are usually first on the scene in a disaster, remain long after official services have ended, and often play vital roles in helping those affected to respond and recover. This idea is to build relationships with neighbors, feel invested in your community, receive proper training during "blue skies" and not during times of emergency, and having adequate, if not excellent, neighborhood gathering and resource centers.

Second, it is important to expand the participation of "trusted messengers," including health professionals, weather reporters, health services providers, home health aides, and faith-based groups in emergency preparedness planning and climate resiliency efforts. This is particularly important in communities that have low trust in government and who will rely on their existing networks and local leaders to endure climate-related threats.

Finally, the main lesson is to convey that climate-related threats aren't expected to happen in *the future*. Instead, we're all facing those risks now. Local-level impact information is critical to encouraging people to be prepared and a key action to protect people is to ensure that we're checking on each other during warm weather. Practicing safety checks and building community during heat waves—which occur consistently every summer—are habit forming! If a community then faces a different disaster, many can already have a habit of checking on or helping at risk, or isolated neighbors.

4. What educational and training experiences best prepared you for your job? What do you wish you had learned during your training that would have prepared you to better address extreme climate events? What advice would you give to students who are eager to make a contribution to this field?

I have spent well over a decade working to ensure that social, public health, and environmental justice priorities are integrated into climate adaptation plans and environmental policies in NYC and I intentionally work at the intersection of my passion and abilities. I have a BA in Geology from Carleton College, and a Master of Science in Natural Resources Policy and Management from the University of Michigan at Ann Arbor. In both instances, I sought out minors, concentrations, or tracks that focused on environmental health and racial disparities in the distribution of environmental benefits and burdens. My calling is to enhance environmental quality while redressing those disparities. This also meant being OK with seeking elective courses outside of my departments and graduating with more credits than required for my degrees. Not only is this a good use of your time in academia, but it is also critical to developing a fuller picture of environmental and climate issues, which are inherently inter-disciplinary, multi-faceted, and cross-sectoral. I also completed several leadership development, negotiation skills, management training, and environmental fellowship programs which also built my capacity as a sustainability leader and as a manager.

My advice to students who are eager to make a contribution to this field is to focus on learning rather than on grades. I have never hired someone based on their GPA. Read. Go to community meetings. Lend your skills. Don't do science for science's sake. Don't get indebted over a professional degree that doesn't stir your passion. Don't spend your entire career behind a computer *planning for* communities. Do place-based work.

We need people who can think clearly, take data-driven approaches, incorporate health and equity into environmental and climate policy, and develop science for the public's interest. Exposure to environmental pollution, housing conditions, rising temperatures, and other climate risks disproportionately affects people that also face economic inequality, housing pressures, and limited access to neighborhood resources such as open space or healthy food options. People from any discipline can, and must, become activists to address the rapid deterioration of our natural resources and improve global health.

References

1. Kovats RS, Hajat S. Heat stress and public health: a critical review. Annu Rev Public Health. 2008;29:41–55. https://doi.org/10.1146/annurev.publhealth.29.020907.090843.
2. Vandentorren S, Bretin P, Zeghnoun A, Mandereau-Bruno L, Croisier A, Cochet C, Ribéron J, Siberan I, Declercq B, Ledrans M. August 2003 heat wave in France: risk factors for death of elderly people living at home. Eur J Pub Health. 2006;16:583–91. https://doi.org/10.1093/eurpub/ckl063.
3. Weather Underground. Weather records for Charles de Gaulle, France. 2003.
4. Black E, Blackburn M, Harrison RG, Hoskins BJ, Methven J. Factors contributing to the summer 2003 European heat wave. Weather. 2004;59:217–23.
5. Robine JM, Cheung SLK, Le Roy S, Van Oyen H, Griffiths C, Michel JP, Herrmann FR. Death toll exceeded 70,000 in Europe during the summer of 2003. C R Biol. 2008;331:171–8. https://doi.org/10.1016/j.crvi.2007.12.001.

6. World Health Organization. Climate change and health. Geneva: WHO; 2018.
7. IPCC. Climate change 2014: synthesis report. In: Core Writing Team, Pachauri RK, Meyer LA, editors. Contribution of Working Groups I, II and III to the Fifth Assessment Report of the Intergovernmental Panel on Climate Change. Geneva: IPCC; 2014. https://doi.org/10.1017/CBO9781107415324.
8. Hales S, Kovats S, Lloyd S, Campbell-Lendrum D. Quantitative risk assessment of the effects of climate change on selected causes of death, 2030s and 2050s. Geneva: World Health Organization; 2014.
9. Patz JA, Campbell-Lendrum D, Holloway T, Foley JA. Impact of regional climate change on human health. Nature. 2005;438:310–7. https://doi.org/10.1038/nature04188.
10. Yonetani M, Albuja S, Bilak A, Ginnetti J, Glatz A-K, Howard C, Kok F, et al. Global estimates 2015: people displaced by disasters. Geneva: IDMC; 2015.
11. Busby JW. Climate change and national security: an agenda for action. New York: Council on Foreign Relations; 2007. CSR No.32.
12. Rajkovich NB. A system of professions approach to reducing heat exposure in Cuyahoga County, Ohio. Mich J Sustain. 2016;4:81–101. https://doi.org/10.3998/mjs.12333712.0004.007.
13. Eckstein D, Künzel V, Schäfer L. Global climate risk index 2016: who suffers most from extreme weather events? Weather-related loss events in 2014 and 1995 to 2014. Bonn: Germanwatch; 2017. ISBN: 978-3-943704-04-4.
14. United Nations, Department of Economic and Social Affairs, and Population Division. World urbanization prospects: the 2014 revision, highlights (ST/ESA/SER.A/352). New York: United Nations; 2014.
15. Seto KC, Parnell S, Elmqvist T. A global outlook on urbanization. In: Elmqvist T, Fragkias M, Goodness J, Güneralp B, Marcotullio PJ, McDonald RI, Parnell S, Schewenius M, Sendstad M, Seto KC, Wilkinson C, editors. Urbanization, biodiversity and ecosystem services: challenges and opportunities. Dordrecht: Springer; 2013. p. 1–12. https://doi.org/10.1007/978-94-007-7088-1.
16. National Oceanic and Atmospheric Administration. Relocating Kivalina. U.S. Climate Resilience Toolkit. 2017.
17. Robinson M. This remote Alaskan village could disappear under water within 10 years — here's what life is like there. Business Insider 2017 Sept 27.
18. Task Team on the Definition of Extreme Weather and Climate Events. Guidelines on the definition and monitoring of extreme weather and climate events. Geneva: World Meteorological Organization; 2016. p. 62.
19. Rahmstorf S, Coumou D, Rahmstorf S, Coumou D. Increase of extreme events in a warming world. Proc Natl Acad Sci. 2016;108:17905–9.
20. Easterling DR, Meehl GA, Parmesan C, Changnon SA, Karl TR, Mearns LO. Climate extremes: observations, modeling, and impacts. Science. 2000;289:2068–75. https://doi.org/10.1126/science.289.5487.2068.
21. Taha H. Modeling the impacts of large-scale albedo changes on ozone air quality in the south coast Air Basin. Atmos Environ. 1997;31:1667–76. https://doi.org/10.1016/S1352-2310(96)00336-6.
22. Smil V. Energy transitions: history, requirements, prospects. Santa Barbara: ABC-CLIO; 2010.
23. Bridge G, Bouzarovski S, Bradshaw M, Eyre N. Geographies of energy transition: space, place and the low-carbon economy. Energy Policy. 2013;53:331–40. https://doi.org/10.1016/j.enpol.2012.10.066.
24. Marsh GP. Man's responsibility for the land. In: Nash R, editor. The American environment: readings in the history of conservation. Reading: Addison-Wesley Publishing Company; 1864. https://doi.org/10.1080/19397030902947041.
25. Wuebbles D, et al. Our Globally Changing Climate. In: Climate science special report: fourth national climate assessment, vol. I. Washington, DC: U.S. Global Change Research Program; 2017. p. 12–34. https://doi.org/10.7930/J0DJ5CTG.
26. Trenberth KE, Fasullo JT, Shepherd TG. Attribution of extreme events. Nat Clim Chang. 2015;5:725.

27. Kennedy C. Does "global warming" mean it's warming everywhere? ClimateWatch Magazine. 2014.
28. Rogers JC. The 20th century cooling trend over the southeastern United States. Clim Dyn. 2013;40:341–52.
29. Stone B Jr. The city and the coming climate: climate change in the places we live. New York: Cambridge University Press; 2012.
30. Allison G, Zelikow P. Essence of decision: explaining the Cuban missile crisis. 2nd ed. New York: Addison-Wesley Education Publishers Inc.; 1999.
31. Ahern J. From fail-safe to safe-to-fail: sustainability and resilience in the new urban world. Landsc Urban Plan. 2011;100:341–3. https://doi.org/10.1016/j.landurbplan.2011.02.021.
32. Fritz A, Samenow J. Harvey has unloaded 24.5 trillion gallons of water on Texas and Louisiana. Washington Post 2017 Sept.
33. Flitter E, Valdmanis R. Oil and chemical spills from Hurricane Harvey big, but dwarfed by Katrina. Reuters 2017 Sept 15.
34. Olsen L. After Harvey, a "second storm" of air pollution, state reports show. Houston Chronicle 2018 Mar 30.
35. Federal Housing Administration. Underwriting manual: underwriting and valuation procedure under title II of the national housing act. Washington, D.C.: United States Government Printing Office; 1936.
36. Shalby C. What's the difference between "looting" and "finding"? 12 years after Katrina, Harvey sparks a new debate. El Segundo: Los Angeles Times; 2017.
37. Mcphillips LE, Chang H, Chester MV, Depietri Y, Friedman E, Grimm NB, Kominoski JS, et al. Defining extreme events: a cross-disciplinary review. AGU Earth's Futur. 2018;6(3):1–15. https://doi.org/10.1002/2017EF000686.
38. Klein N. Hot money: how free market fundamentalism helped overheat the planet. In: This changes everything: capitalism vs the climate. New York: Simon & Schuster; 2014. p. 64–95.
39. McCarty N, Poole K, Rosenthal H. Polarized America: the dance of ideology and unequal riches. Cambridge: MIT Press; 2006.
40. McCright AM, Dunlap RE. The politicization of climate change and polarization in the American public's views of global Warming. Sociol Q. 2011;52:155–94.
41. Gaard G. Toward a queer ecofeminism. In: New perspectives on environmental justice, vol. 12. New Brunswick: Rutgers University Press; 1997. p. 114–37. https://doi.org/10.1111/j.1527-2001.1997.tb00174.x.
42. Pellow DN. Toward a critical environmental justice studies: black lives matter as an environmental justice challenge. Du Bois Rev. 2016;13:221–36. https://doi.org/10.1017/S1742058X1600014X.
43. Jones-DeWeever AA, Hartman H Abandoned Before the Storms: The Glaring Disaster of Gender, Race and Class Disparities in the Gulf. There Is No Such Thing as a Natural Disaster: Race, Class and Hurricane Katrina, eds Squires G, Hartman C (Routledge, New York, NY). 2006, pp 85–102.
44. Alberti M. Advances in urban ecology: integrating humans and ecological processes in urban ecosystems. New York: Springer Science + Business Media; 2008.
45. Liu J, Dietz T, Carpenter SR, Alberti M, Folke C, Moran E, Pell AN, et al. Complexity of coupled human and natural systems. Science. 2007;317:1513–6. https://doi.org/10.1126/science.1144004.
46. Holling CS. Understanding the complexity of economic, ecological, and social systems. Ecosystems. 2001;4:390–405. https://doi.org/10.1007/s10021-00.
47. Walker B, Hollin CS, Carpenter SR, Kinzig A. Resilience, adaptability and transformability in social-ecological systems. Ecol Soc. 2004;9:9.
48. Gunderson LH. Adaptive dancing: interactions between social resilience and ecological crises. In: Berkes F, Colding J, Folke C, editors. Navigating social-ecological systems: building resilience for complexity and change. Cambridge: Cambridge University Press; 2003. p. 33–52.
49. Folke C. Resilience: the emergence of a perspective for social–ecological systems analyses. Glob Environ Chang. 2006;16:253–67. https://doi.org/10.1016/j.gloenvcha.2006.04.002.

50. Low B, Ostrom E, Simon C, Wilson J. Redundancy and diversity: do they influence optimal management? In: Navigating social-ecological systems: building resilience for complexity and change. Cambridge: Cambridge University Press; 2003. p. 83–108.
51. Schreiber E, Sabine G, Bearlin AR, Nicol SJ, Todd CR. Adaptive management: a synthesis of current understanding and effective application. Ecol Manag Restor. 2004;5:177–82. https://doi.org/10.1111/j.1442-8903.2004.00206.x.
52. Patel R, Babbs D, Sauven J, Frost M, Goldring M, Forsyth J, Nussbaum D, et al. Climate change back on the agenda. The Guardian. 2014 Sept 7.
53. Schlosberg D. Defining environmental justice: theories, movements and nature. New York: Oxford University Press; 2007.

Chapter 18
Alternative Housing Delivery for Slum Upgrades: Perspectives from Architecture, Engineering, and Non-governmental Organizations

Sheela Patel, Young Seop Lee, Jin Young Song, and Greg Sloditskie

Introduction

Influencing physical, mental, and social well-being, our home's physical character is one of the most critical elements of our lives. Extensive research has proven the dramatic impact of poor living environments on health [1–3]. Informal settlements, therefore, must be one of the most extreme examples exposing the critical influence of architecture on health. Such poor living conditions are increasing worldwide, despite record levels of global wealth, advances in architecture and construction technologies, and urban development. According to UN Habitat, one in eight people (about 1 billion) currently live in slums [4]. There are significant health and safety problems in these living environments due to lack of safe housing and proper public services. For example, in Dharavi, Mumbai, most of the 100 sq. feet slum houses do not have sanitation and water facilities. Many community toilets are suffering with lack of water supply, electricity, and basic safety such as proper doors.

S. Patel
SPARC (Society for the Promotion of Area Resource Centers),
Mumbai, India

Y. S. Lee
Gensler, Los Angeles, CA, USA

J. Y. Song (✉)
University at Buffalo, State University of New York (SUNY), Buffalo, NY, USA
e-mail: jsong11@buffalo.edu

G. Sloditskie
MBS Consulting, Walnut Creek, CA, USA

© Springer Nature Switzerland AG 2020
K. H. Smith, P. K. Ram (eds.), *Transforming Global Health*,
https://doi.org/10.1007/978-3-030-32112-3_18

Dharavi is one of the largest slums in Asia and suffers from inadequate access to safe water, sanitation and other infrastructures, insecure structure of housing, and over-crowding. Efforts to improve these conditions have been inefficient due to conventional top down development practices which rely on commercial development with tall housing without substantial demographic, socio-economic data, ownership, and physical surveys [5]. Lack of proper local information results in inadequate consideration of rehousing and resettlement of current slum residents. In many ongoing cases, redevelopment of informal settlements bears significant socio-economic costs during multiple years of construction processes due to mass evictions, relocation, and effectively, the erasure of communities and micro-cultures. However, we should note that slums develop over long periods of time with diverse communities and cultures rather than being a result of a quickly formed temporary settlement. Top down development without considering the dwellers' ongoing lives not only counters fundamental human rights, but also negatively impacts the rest of the city which is organically connected by industry, employment, and culture. However, a bottom up strategy, focusing on individual projects and gradual improvement based on slum dwellers' needs, is outpaced by rapidly increasing slum populations. Lack of government funding and private investment exacerbate this problem.

Responding to this challenge, we suggest that architects, engineers, and urban planners consider a slum upgrade with an incremental prefabricated delivery model to upgrade informal settlements (Fig. 18.1). Aiming for the wellness of the inhabitants, architecture can work within existing urban frameworks where slum residents can maintain their dynamic urban activities while promoting larger economic development. Due to advances in digital tools, manufacturing technology, and the standardization of components, prefabricated modules and components have become common in many countries. This chapter explores the design of a prefabricated building system that could be easily constructed by low-skilled workers in a factory located close to an informal settlement. This approach allows for participatory urban upgrades with reduced costs, construction time, and disruption. It also provides new jobs and training for the residents of the informal settlements.

Sheela Patel discusses the nature of problems regarding the housing in Mumbai based on her valuable experiences in Dharavi and elsewhere. Youngseop Lee describes the core benefit of prefabrication in architecture and how it can change the construction industry by making low cost, well designed buildings possible. Jin Young Song and Greg Sloditskie will describe a theoretical supply chain that leverages a Mumbai factory to radically change the upgrade process.

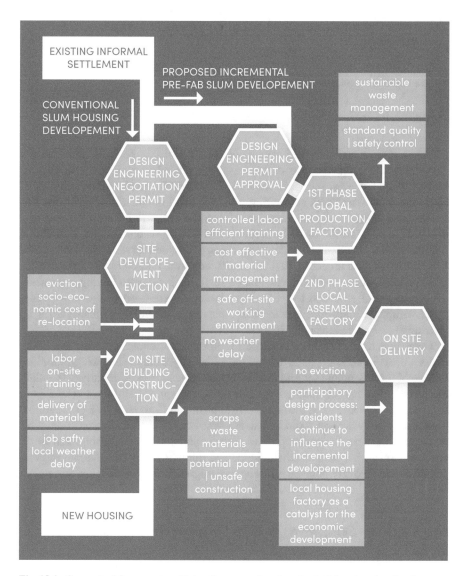

Fig. 18.1 Conceptual framework outlining the conventional method of slum housing development compared to the proposed incremental prefabrication method of slim development. (Source: Jin Young Song)

Toward a New Paradigm of Slum Upgrade in India and Elsewhere

Sheela Patel

Dharavi, Mumbai, India

Mumbai, India is a city created from a colonizer's global trade aspiration. The city developed around a natural harbor that was much needed by the British colonists for trading. A city grew around it in the early 1880s, and a municipality was created along the same framework set by English cities like London and Manchester. The land was a group of seven marshy islands inhabited by native fisher folk. Labor was drawn to the city from nearby hinterlands and a business community from Surat, now in Gujarat, who built ships, traded various goods, and served the colonists. The labor crews worked on docks, in textile mills, and were housed in tenements. Many more people came to work informal labor, servicing the city and living informally. Whenever the city needed to grow, these laborers were pushed outward—sometimes without explanation, and at other times with the colonists justifying it by "wanting to beautify the public spaces."

Over time, as the city grew northward, informal households working and living in dock areas in various places around the island city were pushed to communities outside of the city boundary, and marshy lands were claimed informally. In parallel, the formal city was reclaiming land and filling the spaces between the islands to form what is now known as the island city of Mumbai. Currently, 70% of the formal houses in the city are one room tenements—the houses historically constructed for textile mill workers. Since these homes are not enough to house the migrated workers, large townships such as Dharavi emerged through either the eviction, pushing people outside the city boundary, or ongoing migration through kinship groups.

Origins of the Settlements and Its Location Over the Years in the City

Until 1956, Dharavi was outside the city limits and comprised of self-built, self-regulated informal "nagars" or neighborhoods which the city ignored. These neighborhoods had no toilets, amenities, or drinking water—even though the main water supply from lakes went directly through the township. One hardly needs statistics to tell the story of health concerns of the informal habitats in the city. This was further exacerbated by the fact that, in a period of prohibition in the 1960s, illegal production of liquor and the curing of leather permeated dangerous chemical into soil, which then contaminated drinking water in the bored wells dug by residents.

It was only in the late 1970s that the state government of Maharashtra and the city of Mumbai focused on Dharavi albeit for the wrong reasons. In the absence of any formal governance in dealing with informal areas, informal governance managed all businesses in Dharavi. Attempting to clamp down on the sale of illicit liquor and other illegal activities was impossible, as police could never enter and locate the miscreants. The huts, densely built with narrow pathways, and a community fearful of police meant that the law and order situation just kept getting worse.

Apart from this challenge, the city had grown around Dharavi to over 500 hectares of land. Interestingly, it was only through Dharavi that the eastern and western extension of the city could be accessed via roads and railways. So a road was built through Dharavi, demolishing a large number of houses and cutting Dharavi for the city traffic to pass though the neighborhoods.

Dharavi Redevelopment Strategies by the State Fail Time and Time Again

In India as elsewhere in the world, informal settlements grow out of an inability for the state to effectively use planning instruments to embrace the migration of people into cities. Poor, unskilled, and/or asset-less people often come to find work in the city, but cannot find neighborhoods to stay in. Therefore, they build their own homes incrementally—collectively, neighborhoods were produced. City planning frameworks set up by the British colonists in the late 1800s, often modeled after their own cities, remain exclusionary even after many years of independence. State interventions in these neighborhoods come with vertical, top down developed solutions that come too late—solutions whose frameworks are clearly unviable—all in the name to improve the lives, health and well-being of communities and fail time and time again without even understanding the reality on the ground of increased densification and worsening habitat indicators. Even more challenging is the fact that most of the attempts would gentrify the neighborhoods and expel residents.

Dharavi has had several such interventions: In each instance, the policy was announced and plans made by the state with huge publicity only to fritter away the opportunity to improve the well-being of residents due to an inability of the state to involve residents' associations and listen to what people wanted.

The PMGP (The Prime Minister's Grant Project) Project 1985

In 1985 the prime minister of India gave 100 crores, a substantial investment at that time, to Mumbai for its improvement to celebrate the 100 years of the Congress Party's inception in the city, which had won India its independence. He sought to give a portion of the award, 35 crores, to Dharavi to demonstrate the party's commitment to the urban poor. The state government set up a department in its housing authority to provide special assistance to Dharavi and other neighborhoods. The project proposed, assumed that there were 50,000 huts or structures of which 15,000 would have to be removed to make Dharavi habitable. The residents of Dharavi created two of their organizations with the help of NSDF (National Slum Dwellers Federation) and undertook surveys to demonstrate who lives where and seeks identity and voice for their demands. Their survey identified over 80,000 structures which housed over 120,000 households and innumerable informal businesses in contrast to the state surveys. Out of that activism arose two organizations representing this challenge. One was Dharavi Vikas Sammittee, a residents' network, and the other was the committee of business in Dharavi which provided employment to more than two thirds of the residents and were linked to the city's food, clothing, and manufacturing industries.

While some critical infrastructure was provided—roads were widened and some housing stock developed—the plan underestimated the number of households who lived there, did not take into consideration the huge informal businesses that were based in Dharavi, and despite taking the bold step to treat Dharavi as a special planning area, did little to explore ways to develop planning norms that worked for the residents. Its main claim to fame was that it removed the leather curing activities and relocated this outside the city, thus reducing the toxins from chemicals that impacted the ground water.

The Dharavi Redevelopment Plan 2004–2014

In the next intervention, the government of Maharashtra was recommended by a consultant to redevelop Dharavi through invitation of international real estate agencies who would pay the state for value of land and build free tenements (250 sq. feet per residents), which would be multi-storied and create additional housing and commercial hubs to cover their developmental costs. This has been a long and contested activity which the residents of Dharavi opposed. In 2014, the city concluded on this possibility, continuing to tweak it in different ways without making much headway. In the meanwhile, Dharavi continues to densify, and the possibilities of solutions become even more improbable.

Searching for Sustainable Solutions for the Informal City

Today, Dharavi is no longer the largest slum in Mumbai, or India, or even Asia for that matter. Census data indicates that over 60% of the over 12 million people living in Mumbai's slums, the informal habitats of the island city, have been reduced. This was done through evictions initially, and through relocation more recently—although many pockets of informal dwellers still cling to their locations. However, most of the informal habitats are in both eastern and western suburbs of the city, and this majority of population inhabits only 8% of the land in the city. Access to water sanitation and electricity is still very low, and the quality of life in informality remains very poor in general.

The Mumbai metropolitan region now has 12 municipalities that have very large informal settlements facing various challenges. Due to the delays in redevelopment and intensified growth in the city, the densification went over the limit of horizontal expansion. People began to grow vertically, building two or three floors on top of the existing structures. So what can be done in a situation where ongoing delays allow for further densification, and growing populations have no other choice but to grow vertically? How can the production of these unsafe structures, built precariously on existing habitats, be avoided?

In Dharavi, there has been a lack of meaningful or effective interventions to provide even basic infrastructure if nothing else, to these households. This is not unlike many towns within cities that grow organically (and often illegally) right under the eyes of the municipal and provincial administration. The provision of basic amenities remains the most cost effective investment. Paradoxically, when the neighborhoods become recognized as vote banks, and some basic infrastructure is attempted to appease the settlements, it is the very dense habitat itself which makes the investment of laying out water pipes, sewerage, and drainage systems impossible without demolishing many of the existing houses. And so, the viscous cycle continues with no real solution in sight.

How can such retrofitting of basic services be done? Can the mistakes of delaying investments inform to produce a better response from the city, to serve more recently emerging informal settlements? Given that almost all housing stock is built informally by the residents, how can we produce a response from the construction industry that can provide better incremental upgrading possibilities that are safer and cost effective for the residents?

Prefab Now/Prefab Future

Young Seop Lee

A building is a unique type of man-made product. Due to its size, transporting a building as a completed product imposes many challenges, and often times, is not feasible. Therefore, we tend to construct buildings on a site. This very nature of constructing on a site causes many inefficiencies in cost, time, and materials that are not common in other manufacturing industries. For example, all building materials and personnels have to be transported to a construction site which can be in a remote area or a dense urban area which causes many logistic problems. It also creates high risks such as hundreds of construction workers having to access and work on non-enclosed high-rise floors. These are some of the reasons that the construction industry is struggling to match the growth of other manufacturing industries like auto and aerospace with their highly automated efficiency. However, the recent development of prefab construction is opening an alternative way to improve this centuries-old building construction method by merging construction and benefits of manufacturing. Prefab construction is an off-site construction that manufactures a building in a factory by building various elements or an entire building divided in deliverable modules, which can then be transported to a construction site to install. Although this type of construction has been practiced to some degree historically, recent technical developments and innovations are fueling a global rise of prefab construction as a prominent alternative construction method with promise of faster, better, cheaper, and greener solution to tackle major challenges our society faces today like shortage of affordable housings in many fast urbanizing cities around the world.

Prefab construction creates its own unique challenges that limit its applicability project by project and per building types. This led to the development of various types of prefab construction with varying degrees of prefabrication. Currently, there are three types of widely used prefab construction. The first is called modular construction, which constructs the entire building in deliverable modules that are transported and assembled on a construction site. The modules can be delivered at a 60–90% completed stage with mechanical, electrical, and plumbing, and even furniture, fixtures and interior finish installed in the factory. The second type is called pod construction which has only labor and cost intensive portions of a building like a kitchen, restroom, and hospital room headwalls built in a factory. Delivered pods will be crane-lifted and anchored on an already site-built building floor slabs. The last type is called panelized construction which mass produces building components in a factory that will be shipped and assembled on the site. For an efficient assembly on-site, each panelized system is fully engineered with all joints and connections to minimize installation time and labor and to shorten construction sequence and time. However, compared to modular or pod construction, panelized construction requires more on-site labor and time for assembling smaller panelized pieces and defeats some of prefab construction's advantages.

To justify the threshold for prefab construction cost to be competitive compared to the on-site construction, a general rule of thumb is to construct a building type and program with a large number of repetitive modules like residential or educational buildings rather than arts and cultural buildings that are more custom-designed and has less repetitive modules. Therefore, prefab construction is becoming a suitable solution to mass residential developments in many urbanized cities in the world with huge housing shortages due to mass migrations and skyrocketing housing prices. The prefab construction industry is expanding and growing fast due to this global mass housing needs, primarily because of its fast construction schedule and more economic construction costs compared to the traditional site-build construction. Prefab construction reduces overall construction time as its simultaneous multi-lateral construction progresses in multi-locations while the site-build construction is limited with its single linear construction progress. Up to 40–50% time savings are possible by manufacturing modules off-site while site-development and foundation work are undergoing at a site. This time saving is one of the key advantages prefab construction offers. Fast project completion provides a faster return on investment and lowers hard and soft development costs like construction overheads and insurance. Another advantage of prefab construction is that it produces higher quality and more controlled craft than on-site construction by constructing in a factory, which offers optimally designed environment and a system like lean manufacturing. For example, air and waterproofing of a building are two of the most challenging areas in construction that can cause huge issues after the completion. Those issues can be minimized in higher quality control and quality assurance in prefab construction with tighter tolerance in a controlled factory environment. Job site accidents are also reduced in a factory as it is designed and equipped with a safer environment and system.

In addition, prefab construction offers significant environmental benefits. Even though prefab construction might be seen as adding more energy for transportation from factory to the site, prefab construction is generally much more environmentally friendly and energy saving than on-site construction. There are several researches done on the prefab construction life cycle analysis. Most research has found that prefab construction uses much less energy and generates much fewer pollutants. One key reason is that construction time is much shorter than on-site construction. It reduces energy uses and environmental impacts in all categories. Moreover, constructing most portion of the building in a factory minimizes material waste, site disturbance, noise to neighbors, traffic, and staging that can cause huge challenges and cost if a site is in a dense urban neighborhood or a specially challenged site like hillside sites. Many prefab constructed projects earn high LEED rating, which is a U.S. Government Building Committee's internationally used environmental protecting and energy efficiency rating. Prefab buildings usually gain better ratings in sustainable sites, water efficiency, and innovation in design categories.

Innovative use of prefab construction has also been developed due to prefab construction's restrictions and limitations like transportable size. Previously, it was not feasible to use prefab construction for some types of building programs like an auditorium or a banquet hall, which require a long-span column-free space or high ceiling heights. A "hybrid" of prefab/off-site and on-site constructions has been developed which builds a custom portion of a building on-site and completes with installing repetitive modules later. It is necessary that we consider prefab construction not as a finished solution but consider to use with creativity and innovations to maximize its potential for each project. It is also recommended to move away from the conventional linear building-design and construction process to a more integrated process between architect, manufacturer, owner and contractor. Recent development in Building Information Modeling (BIM) designing, clouding, and other technologies accelerate this paradigm shift by linking and integrating design, procurement, manufacturing, and assembling.

Today, we are seeing innovative prefab construction widely used against global challenges our society faces. The New York City is promoting prefab construction as a solution to the city's housing shortage by hosting many prefab programs and constructions. Many school districts in the US are investigating to improve current cheap and utilitarian mobile classrooms by building prefab constructed schools as they cannot provide permanent schools to match fast growing and changing demography. Some global hotel chains have been developing their hotel designs around the world on a modular design for their projects around the world. Globally manufactured and shipped, quality and design assured hotels can be assembled any-where in the world. Prefab provides more controlled quality and branded hotel experiences to its hotel guests throughout the world, including remote/ under-devel-oped countries where no-trained vendors are available. Canadian Olympic Committee has also implemented prefab construction to reuse its modular buildings for the 2010 Vancouver Winter Olympic Village for staffs and volunteers. After the Vancouver Winter Olympic, the modular buildings were disassembled and shipped to six different cities in the British Columbia and re-configured into new communal and residential buildings. It was an innovative solution to the growing criticism for mega sports events like Olympic or FIFA world-cup, for which hosting cities had developments that wasted billions of dollars after the games.

Briefly reviewing global market development, it is considered that European countries like England, Germany and Japan have the most developed prefab con-struction market. For example, the prefab construction industry in Japan is very mature. Its mass-producing factories are served with robotic and automated manu-facturing, and about 12.9% of the entire single-family houses in Japan in 2013 were constructed by prefab construction. East Asia including China have fast-growing prefab construction markets recently due to its shortage of housing due to exploding growth and urbanizations in their mega cities. Malaysia, Singapore, and South Korea have government supported or even government led prefab construction markets. In China, for example, the Broad Group has developed and is leading the panelized type of prefab construction. They are known for completing a high-rise hotel tower in just 15 days. Also, CIMC in China, the largest shipping container manufacturer in the world, is expanding its manufacturing and logistic capabilities into a modular type of prefab construction and serving global markets by exporting custom design-able modules in shipping container sizes from its factories in Southern China.

Prefab construction is bringing a paradigm shift toward the future of construc-tion. Prefab construction is still in the developing stage, and it is rather an open platform where many new ideas and innovations can be tested and used creatively. It will be a powerful tool to the construction industry to better serve our fast chang-ing society's needs.

Grand Upgrade Without Mass Demolition

Jin Young SongGreg Sloditskie

Improving the lives and health of the residents in informal settlements should be the main focus of this redevelopment strategy. In the majority of cases where a process of eviction is followed by the construction of taller public housing, community par-ticipation is not seriously considered and the socio-economic cost of relocation is not logically planned. On the other hand, piecemeal on-site upgrading and improve-ment is slow and inefficient considering the rate of increase of global slum populations. Significant improvement can be made in the process of slum upgrades by using novel technologies and alternative delivery methods of prefabrication. This approach minimizes the disruption of residents' lives and livelihoods while keeping pace with the increasing population.

Traditional redevelopment schemes demolish old structures before replacing them with taller residential towers. This approach is often flawed because it fails to measure existing conditions. According to statistics published by the alliance of SPARC, Mahila Milan, and the National Slum Dwellers Federation, the substantial city sized settlement in Dharavi has developed over a long period of time with diverse communities, rather than being quickly formed as a temporary settlement [6]. The informal settlements also provide critical services to the greater city, and it is important for any redevelopment plan to minimize the impact on these dynamic activities. This is often ignored by conventional developer-oriented plans. Some case studies have shown that the existing housing was in good shape and the new public housing failed to improve the lives of the residents [7]. Eviction followed by mass housing construction typically requires multiple years. Any successful redevelopment strategy must measure the costs of migration and life disruption during this period. Therefore, clearing the slum site is not only inhumane but also technically difficult. In Dharavi, land ownership is very complicated, owned by the Municipal Corporation of Greater Mumbai (MCGM) and various private communities. Additionally, since the ground floor of typical units provide space for living and working, it is inevitable to expect serious resistance from the residents. Quality of construction and maintenance is also difficult to achieve through profit-driven and contractor-oriented construction without serious community interaction. We should note that the success of mass-development of high-rise apartment blocks in Singapore, South Korea, and some other global cities is based on countless community actions, reactions, mistakes, and modifications over 30–40 years.

In the site analysis and pre-design phase, one of the most important tasks architects do is to uncover strengths within current conditions that are worth preserving. We should continue and enrich the positive aspects of existing communities in the redevelopment plans. The goal of novel redevelopment plans is to maintain the dynamic urban activities while promoting larger economic development. In order to do that, we must avoid significant disruption resulting from the "upgrade." Rather than creating new socio-economic problems related to the costs of mass-migration due to demolition, we propose designing a prefabricated building system that could be easily constructed by low-skilled workers in a factory located close to the informal settlement. The factory would produce three dimensional modules or two dimensional components for the incremental and participatory urban upgrade process.

Considering the important financial factors in conventional slum upgrading has been subsidies and loans from governments, support from private banks, and international donations, the key contribution from the building industry is to help the state and agencies to support conditions for community-based slum upgrading [8]. And this local business for the prefabricated housing factory can support the incremental development. The factory's products would be optimized for a manufacturing process that requires minimal capital investment. Local residents would make up the bulk of the factory's workforce and management teams. The modules or components would be delivered to the existing informal settlement using the most cost efficient transportation methods. The process of site installation would be designed to minimize disruption to the informal settlement residents (Figs. 18.2, 18.3, 18.4, 18.5, 18.6, and 18.7).

Demolition would be minimized so as to provide basic services such as garbage and sewer lines. Demolition would also provide access for pile driving equipment

Fig. 18.2 Projected morphology of Dharavi. (Source: Jin Young Song)

Fig. 18.3 Minimum site clearance for basic public service. Prefabricated modules are stacked on top of micro piles. (Source: Jin Young Song)

Fig. 18.4 Local health center and public toilets are delivered as prefab-modules. (Source: Jin Young Song)

Fig. 18.5 Inner courtyard is planned for future amenity such as playground and green open space. (Source: Jin Young Song)

and a compact construction crane. The first residents impacted by the demolition would be the first group to occupy the new prefabricated dwelling units. The factory would prefabricate temporary housing and work spaces for the first group of slum dwellers. These units would be located near the factory or slum. The first group of residents would occupy these dwellings and use the work spaces while their homes were being demolished and replaced with new dwellings. They would move out

Fig. 18.6 The major road (16–20 m) is planned as future zoning plan. (Source: Jin Young Song)

Fig. 18.7 Schematic rendering of the street after the delivery of units on the left. (Source: Jin Young Song)

after the new dwellings were suitable for occupancy. The next impacted group would occupy the temporary structures and start the process over again. This approach would also allow the factory to refine their process before delivering units to the slum. Micro pile foundations, inserting structural piles by drilling machines, which are often used for underpinning historic buildings, would be used to provide an acupunctural structural solution. This approach allows the rich and complex community to continue functioning at the ground level while new dwelling units are added above. The new dwelling units would be combinations of either three dimensional structural modules or two dimensional components. The final choice between the two prefabricated methods would balance transportation logistics with site disruption. Ancillary structures like stairs, awnings, etc., would be prefabricated by the factory to the extent possible and shipped to the site for field installation.

Off-site prefabrication methods and inspections would assure structural integrity. Combining basic modules and components in different ways would create a varied, architecturally pleasing design. The upgrade process would be participatory, incremental, and respect local materials and customs whenever possible. The existing community would be engaged so that they can evaluate the first group of dwellings and offer improvements for the next group. The ground floor can be improved whenever the owners are ready to upgrade. This will provide another source of business for the local prefabrication company. Costs for this theoretical framework (business startup, construction materials, foundations, labor, etc.) are subject to the market strategy and will be offset by the demand resulting from the city's substantial size. This alternative project delivery has a potential to make a global impact on other slum sites by showing how construction firms, academia, and humanitarian agencies can use technology to improve the lives of the inhabitants.

The mode of slum upgrade should improve the wellness of the inhabitants by maintaining dynamic urban activities while promoting larger economic development. Alternative modes of housing delivery are the key concepts that can be used for any informal settlement case. In order to overcome both large scale slum clearance and slow piecemeal upgrade, the prefabrication and novel design configuration provide a "grand upgrade" without mass demolition of neighborhoods and communities.

References

1. Krieger J, Donna LH. Housing and health: time again for public health action. Am J Public Health. 2002;92(5):758–68.
2. Sharfstein J, Sandel M. Not safe at home: how America's housing crisis threatens the health of its children. Boston: Boston University Medical Center; 1998.
3. Marmot M. Social determinants of health inequalities. Lancet. 2005;365(9464):1099–104.
4. UN-Habitat. Slum almanac 2015–2016: tracking improvement in the lives of slum dwellers. 2015.
5. Patel S, Arputham J. Plans for Dharavi: negotiating a reconciliation between a state-driven market redevelopment and residents' aspirations. Environ Urban. 2008;20(1):243–53.

6. Patel S, Arputham J. Plans for Dharavi: negotiating a reconciliation between a state-driven market redevelopment and residents' aspirations. Environ Urban. 2008;20(1):245.
7. Patel S. Upgrade, rehouse or resettle? An assessment of the Indian government's Basic Services for the Urban Poor (BSUP) Programme. Environ Urban. 2013;25(1):177–88.
8. Burra S. Towards a pro-poor framework for slum upgrading in Mumbai, India. Environ Urban. 2005;17(1):67–88.

Chapter 19
Integrating Traditional and Modern Medicine: Perspectives from Ethnobotany, Medical Anthropology, Microbiology, and Pharmacy

Gail R. Willsky, Rainer W. Bussmann, Mayar L. Ganoza-Yupanqui, Gonzalo Malca-Garcia, Inés Castro, and Douglas Sharon

Project Goals

Traditional medicine (TM) as defined by the World Health Organization is "the sum total of the knowledge, skill, and practices based on the theories, beliefs, and experiences indigenous to different cultures, whether explicable or not, used in the maintenance of health as well as in the prevention, diagnosis, improvement or treatment of physical and mental illness" [1]. The objective of this study is to introduce more (TM) practices, central to the local culture, into the Peruvian public health system (PPHS) in Northern Peru. There are three interrelated parts: (1) to help public health service personnel understand the local TM practices of their patient population to supplement the therapy for these patients, (2) to inform the patient population that many of the herbal remedies of TM are the same as the medicines used in the allopathic medical tradition, (3) to promote conservation programs by educating the public and maintaining gardens of native medicinal plants. The work involves an interdisciplinary approach of ethnobotany, medical anthropology, microbiology, and pharmaceutical

G. R. Willsky (✉)
University at Buffalo, State University of New York (SUNY), Buffalo, NY, USA
e-mail: gwillsky@buffalo.edu

R. W. Bussmann · G. Malca-Garcia
Missouri Botanical Garden, St. Louis, MO, USA

M. L. Ganoza-Yupanqui · I. Castro
Universidad Nacional de Trujillo, Trujillo, Peru

D. Sharon
Missouri Botanical Garden, St. Louis, MO, USA

University at Buffalo, State University of New York (SUNY), Buffalo, NY, USA

© Springer Nature Switzerland AG 2020
K. H. Smith, P. K. Ram (eds.), *Transforming Global Health*,
https://doi.org/10.1007/978-3-030-32112-3_19

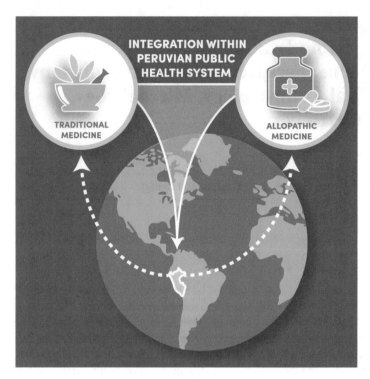

Fig. 19.1 Conceptual diagram depicting the integration of traditional and allopathic medicines in the Peruvian public health system. (Source: images: Gail R. Willsky, and Gonzalo Malca-Garcia; graphic design: Nicole C. Little)

sciences to improve patient satisfaction with public health services and to preserve local medicinal plants and the knowledge concerning their use (Fig. 19.1).

TM and the PPHS in Peru

Aspects of TM practiced by the indigenous healers (*curanderos*) are used extensively in Northern Peru. The evidence-based allopathic medical tradition aims to alleviate the pathological symptoms of disease with remedies (such as drugs or surgery) directly targeted to the diagnosed problem. This system is used in many medical schools in the developed world. Osteopathic medical training, based on similar principles, has a slightly different philosophy emphasizing the interrelationship of all systems in the body to promote good health. The PPHS provides care predominantly based on the allopathic system.

In response to World Health Organization (WHO) guidelines outlined in the Alma-Ata Declaration of 1978, Peru has become a major innovator in trying to promote health care equity relating to the integration of TM into primary health care. In 1979 and 1988, it hosted two WHO conferences on TM, leading to the formation of the National Institute of Traditional Medicine in 1991. In the late

1990s, Peru's Social Security Administration organized a public health program (EsSalud) that developed the National Program of Complementary Medicine (PRONAMEC) and opened three Centers of Complementary Medicine (CAMECs) in Peru's major cities (Lima, Arequipa, and Trujillo). Today there are 27.

In 2015, Peru's Ministry of Culture began an initiative to recognize north-coastal TM or *curanderismo* as "cultural patrimony of the nation". Shamanistic practices in the area involve folk healing embracing extensive use of medicinal plants (including the psychoactive San Pedro cactus) and grass-roots psychotherapy. If successful, after nearly 500 years of persecution, the shamanism of this area is officially protected under the General Law of Cultural Patrimony of the Nation, which resulted from Peru's ratification of the 2003 UNESCO Agreement for Safeguarding Intangible Culture (such as oral traditions). This law complemented the 2002 Law for the Protection of the Biological Knowledge of Indigenous Communities [2].

Robust TM and Public Health Facilities Exist Separately in Peru, with Integration Only Recently Attempted

Only beginning in the 1990 have there been efforts to incorporate aspects of TM, including the use of medicinal plants (Fig. 19.2) into PPHS in spite of the fact that large segments of the local population, especially in rural and peri-urban areas

Fig. 19.2 Medicinal plant use in the PPHS Clinic in Trujillo. Samples of medicinal plants growing in the clinic's garden. (Source: Gail Willsky)

Fig. 19.3 Medicinal plant use in the PPHS Clinic in Trujillo. Medicinal plants are dispensed in the clinic's pharmacy. (Source: Mayar Ganoza-Yupanqui)

appear to have more faith in traditional healers, *curanderos*, than in the practitioners of allopathic medicine. This is particularly true with regard to mental health and stress-related ailments that can be considered by modern science to be psychosomatic or psychosocial in nature. To further complicate the problem, no conservation programs exist for medicinal plants that are currently being lost due to climate change and development (Fig. 19.3).

In Trujillo's EsSalud-Complementary Medicine Center (CAMEC), one of the major treatment modalities (including Tai Chi, acupuncture, diet, aromatherapy, psychological evaluation, etc.) involves medicinal plant therapy (Fig. 19.3). The program's *Manual de Fitoterapia* surveys the scientific literature on the ethnobotany, phytochemistry, pharmacology, and sustainable agriculture of 70 Peruvian medicinal plants and their medical applications; however, it does not address the cultural content of medicinal plant use [3]. TM medicinal plant therapy includes psychosocial and cultural variables of respectable antiquity, i.e., dating back to at least 1500 B.C., that are not included in the CAMEC program. Medical anthropology and ethnobotany studies are being carried out to describe these indigenous practices, documenting their use by the general population in order to help the physicians in the PPHS treat their patients in a manner congruent with specific cultural contexts.

Characterization of TM Practices and Comparison with Techniques of Allopathic Medicine Will Help the Integration of TM into the PPHS

The authors of this chapter have utilized the US National Institutes of Health Minority Health & Health Disparities International Research Training grant (MHIRT) awarded to San Diego State University to characterize TM practices. This multi-disciplinary research has been conducted in Trujillo, Peru.

Ethnobotany

Ethnobotany is the study of plants from a specific region and their practical uses by the culture, i.e., essentially a systematic study of the relationships between plants and people. Initially working with practitioners of TM (*curanderos*), we published a database of approximately 500 medicinal plants that has been recently updated [4]. Of the plant remedies used 65% were mixtures [5]. The Peruvian medicinal plant set studied by MHIRT contains many species that are also recognized as medicinal plants in many other cultures. Our database, some facts from which are outlined below, was obtained by going into the field and collecting information from patients, *curanderos*, and herbalists. These data, shown below, indicate a heavy orientation to psychotherapeutic application.

Medicinal Plant Use in the North-Coastal Area of Peru

- 43% of all plants are used for *curandero* described "magical" & "nervous system" disorders
- 18% used for respiratory problems
- 16% for kidney & urinary tract ailments
- 10% for cardiac & circulatory disorders
- 9% for rheumatic & arthritic conditions
- 4% for female reproductive system problems

We found that environmental destruction (overharvesting of medicinal plants, deforestation, mining, changing climate) in the areas studied poses a grave threat to the use of native plants. These studies also provide a framework for trying to save plant species disappearing from native habitats. Support from the US government sources for the study of native plants in order to develop new pharmaceuticals must set aside funds to compensate the people who cultivate and use these plants for medicinal purposes. This process is hampered by the difficulty in tracing the parties who should be compensated [6]. For example, *curanderos* use remedies passed on through generations and are now known to many; the vendors selling the plants to the people may not be those who developed the cures. Many plants growing in the wild or by the roadside are harvested both by *curanderos,* who use and understand them, and vendors, who collect and sell plants for which there is a demand in the local markets. For documenting usage, it is much easier to locate the vendors rather than knowledgeable *curanderos.*

Our research has an educational and applied component. The experience gained with a medicinal plant garden we planted in 2010 at the archeological site museum for the nearby pre-Hispanic city of Chan Chan, as well as surveys from two community studies and the previous three clinics studied in Trujillo are being used to develop an inventory and database of the most frequently used herbal remedies applied in alleviating day-to-day ailments. A criterion for selection is that the efficacy of these remedies for treating specific ailments has been scientifically verified by Peru's National Program for Complementary Medicine (PRONAMEC) as published by the social security system in the *Manual de Fitoterapia* (3). Further verification is to be found in our plant book containing 510 north Peruvian plants and their uses as well as *Plantas Medicinales del Perú,* a survey of the taxonomy, ethnobotany, ecology, and phenology of 774 Peruvian medicinal plants assembled by botanists from the National University of Trujillo's Faculty of Biology [4, 7].

Given that "traceability" is a major problem in using plants from local markets, many of which grow in the sierra, another criterion for selection includes the recommendation that these plants can be cultivated in the soil and climate of the Trujillo Valley [8]. The same plant grown under different conditions (soil, sun, water) can have very different properties. Crucial to this assessment was the re-planting of the Chan Chan garden. The original plants—many of which had not regenerated—were obtained in local markets where vendors are unable to specify the origins of what they sell. Advancing the latter work and a study of market suppliers and sustainability, we also are documenting the efforts of a homeopathic herbalist and owner of one of the three clinics in Trujillo where previous studies were conducted [9]. On his farm in the Andean foothills east of Trujillo, he is gradually adapting plants from other ecozones to coastal conditions with the aim of becoming a major supplier for local markets.

There is also a sustainability goal for this undertaking. As noted above, Trujillo's social security-administered Center for Complementary Medicine (CAMEC) has a plant treatment modality via their certified natural pharmacy. Guidelines for the program are contained in CAMEC's *Manual de Fitoterapia* [3]. Currently, only 20 of these plants are obtainable on a large scale from natural plant product suppliers and no plant mixture remedies or cultural variables are included in the manual. Potentially, the gardens and possible offshoots could eventually become a supplier for CAMEC-Trujillo by demonstrating the viability of the medicinal plant market to local farmers. In addition, a seed bank element could provide a much-needed conservation function as demand increases over time.

Laboratory Science

Characterizing TM remedies using laboratory sciences important in pharmacology (microbiology, toxicology, and medicinal chemistry) that are involved in allopathic medicine helps PPHS personnel see how TM practices relate to what they have been taught. Employing these techniques to validate remedies of TM helps the patients of the PPHS see how connected TM practices are to those remedies used in allopathic medical practices. In 2005, we began laboratory research with the Clínica Anticona (CA) and the Universidad Privada Antenor Orrego (UPAO) followed in 2012 by starting work at the National University of Trujillo (UNT). The experience of MHIRT-Peru demonstrates the difficulties with laboratory studies in Northern Peru. Problems with reliable electricity and getting supplies in a timely fashion hinder laboratory work. Many laboratories are not climate-controlled, making it difficult to carry out studies requiring sterile techniques or other studies where contaminants can be airborne and deposited on samples. A lack of rules concerning the disposal of various chemicals is a problem for the environment, researchers, and the general population. Extensive paperwork requirements make it difficult to get samples and reagents into and out of the country.

Most Western research efforts are confined to the study of single plants. As reported above in our ethnobotany results, 65% of the herbal remedies in our studies were mixtures [4, 5]. The Western approach of searching for a "magic bullet" in looking at individual plant extracts appears to be the wrong research paradigm since synergistic effects may be found when plant mixtures are used.

Antibacterial activity of ethanol and water extracts obtained from dried, ground native plants was determined using standard serial dilution techniques in liquid cultures and with plate tests. *Curanderos* administer the plants mostly as a water decoction while concentrated ethanol extracts (sometimes used as treatments) contain more chemical components from the plants. Toxicity of the extracts was measured by adding them to growing brine shrimp larvae at room temperature. This assay can be done in areas where the strict sterility constraints and equipment for mammalian cell culture are difficult to obtain. Phytochemistry studies included determination of the basic chemical composition of plant extracts and some isolation of components using thin layer chromatography techniques. The data reported below for toxicology results of single plant extracts suggest that TM preparation of remedies should consider toxicity when preparing mixtures for treatments. Phytochemistry results suggest that looking only at extracts from single plants could miss some important antibacterial components [5, 10–12].

Sample Laboratory Results

- Microbiology: 81% of ethanol extracts from 141 plants used in TM to treat infectious disease showed antibacterial activity against *S. aureus*, while 36% showed activity against *E. coli* [13].

- Toxicology: 76% of ethanol extracts from 341 plants used in TM showed high toxicity against brine shrimp, while 24% of water extracts from the same plants were toxic to the brine shrimp [10].
- Phytochemistry: Data supporting the presence of a new compound with anti-bacterial activity was found in an ethanol extract of the mix of *Salix chilensis* Molina and *Prunus serotina* Ehrhart-subsp. *capuli* (Cav.) McVaugh that was not present in the separate ethanol extracts made from each plant [14].

Scientists throughout Peru are working to advance the study of the country's medicinal plants. The quality of the labs is improving and funding is being made available from the government and local corporations. Many of the plants used in Peru are also found in other cultures, and are actively being characterized in labs around the world. Although these studies will provide much-needed knowledge, it is difficult to relate this information to medicinal plants grown in Northern Peru.

Medical Anthropology

Medical Anthropology studies are needed to characterize how the patient relates to the TM healer or *curandero* and how the *curandero* relates to the patient, the earth and the cosmos. Studies with Trujillo's EsSalud-Complementary Medicine Center (CAMEC) included an evaluation of the Center's plant treatment program where we noted a need to include the cultural context of medicinal plant use. Supplementing ethnobotanical work with local healers and herbalists, medical anthropology surveys were conducted to determine knowledge and use of medicinal plants in city clinics and communities around Trujillo. These studies expanded the outreach efforts of our local partners at the National University of Trujillo and EsSalud-CAMEC. In this context, these two public entities began a collaborative effort to conduct phytochemical research on medicinal plants not yet available in the Center's Natural Pharmacy. Initially this research included cultural information from the field, a trend that we hope will continue.

Studies of three clinics (public, private, and herbal homeopathic) in the city of Trujillo demonstrated that, although modern pharmaceuticals were used by patients more frequently than medicinal plants in a private Western clinic, the use of medicinal plants was still relatively high [15, 16]. On the other hand, in an herbal homeopathic clinic, the situation was reversed, i.e., a slightly higher use of medicinal plants as compared to pharmaceutical products [17]. Studies at the public EsSalud-Complementary Medicine clinic which were applied in a more informed manner than in the other clinics also showed a higher use of medicinal plants in comparison to the use of pharmaceutical products [18].

To date 193 different types of plants have been listed by interviewees in the community. With this list we were able to determine the most commonly known plants. Even though we were not able to determine actual use, we have an idea where to start in future research specifically seeking to document usage. Working with a homeopathic herbalist and market supplier we were able to discover the origins of most of the preferred plants. This information should be very helpful to our research partners at EsSalud in their efforts to increase plant knowledge and use among their clientele through the Programa Nacional de Medicina Complementaria (PRONAMEC).

Regarding medical preference, comparison of the two communities studied revealed an almost equal split between preference for medicinal plants and medicine from a doctor or pharmacy [18]. This is similar to the findings from the three clinic studies discussed above. Many participants indicated that—even though they *prefer* one type of medicine—cost, severity of illness, and availability are key decision-making factors. However, belief in culturally bound illnesses ("fright," "shame," and "evil eye") was about 90% or higher for both communities and everyone explained that these are not illnesses that can be treated by a medical doctor. Some respondents even expressed the opinion that a doctor could accidentally harm a patient by treating the wrong ailment. We feel that these strongly held culturally bound beliefs demonstrate the unique coexistence of physical and psychosocial illnesses in Peru.

A worrisome statistic was the large percentage of participants stating that their parents knew more about medicinal plants than they did, i.e., 81% of participants in a newer community and 66% in an older one. This may indicate that we are witnessing a decline in the passing on or interest in the use of medicinal plants as communities continue to grow and urbanize, a trend also noted in the three clinics surveyed in prior studies [18]. This knowledge concerning the plants has not always been documented. Nonetheless, the overall knowledge of medicinal plants in the three clinics and two communities was impressive and speaks to a deeply rooted practice of plant medicine in this area as further indicated by the number and size of open-air medicinal plant markets which we have documented and continue to monitor. Packaged samples of these plants are also offered in modern supermarkets that first appeared in Trujillo in 2009 and even outside of Peru.

To date, our community studies are also showing that, in spite of increased use of modern medicine, the beliefs and practices of TM and its associated worldview are still very strong at grass-roots level, i.e., over 90% of community respondents held traditional beliefs regarding culturally bound illnesses which they felt could only be treated successfully by local *curanderos*.

In order to better characterize the mindset of the patient population of the PPHS more documentation is necessary. Recent work by our UB medical anthropology colleague, Ana Mariella Bacigalupo is documenting how *curanderos* in rural areas are counteracting poor health linked to environmental destruction by reinforcing local holistic health models based on the relationship between plants, sacred spaces, and people. They see people's health as inextricably linked to the environment. They reinforce notions that community spaces and medicinal plant gardens are inhabited by spirits, especially the spirits of their ancestors (los gentiles) and seek to reconnect locals to these sacred spaces which provide the context for traditional health practices. We feel that Peruvian public health programs need to include sensitivity to this cultural context along with the findings from the lab coupled with concerted conservation efforts.

Comparison of Specific Plants and Plant Mixtures Used in North Peruvian TM

Workers in the health field need to understand and respect the different cultural values of the patient population. An example of the value of such an open-minded approach is seen in the use of the *San Pedro* cactus. Easily prepared extracts of the *San Pedro* cactus (*Echinopsis pachanoi*), which is widely available in most open-air markets (Fig. 19.4), are used in healing ceremonies. In a MHIRT-Peru study describing 974 herbal mixtures used in TM in the area, the *San Pedro* cactus appears in five mixtures, only one of which is specifically used as a hallucinogen [5]. The evidence-based medical practices of the developed world have determined that mescaline is the active hallucinogenic compound in the *San Pedro* cactus [19]. Mescaline has been abused in Western cultures due to its hallucinogenic effects. In contrast, *San Pedro* extracts (easily made by boiling the cactus in water) are successfully used in TM in Northern Peru without general problems of abuse due to the respect for the sacred nature of the plant.

Buenas Tardes *(Mirabilis Jalapa L.)* and Flor de Arena *(Tequila paronychoides* [PHIL.] A. Richardson) were used in five two-plant mixtures, three of which were for diseases that could be caused by infectious agents: inflammation, kidney problems, and renal disease [5]. We tested the antibacterial activity of ethanol extracts made from the individual plants and from a mixture of both plants to measure the amount of extract needed to kill 50% of the bacterial population. Each extract made from the individual plants was able to kill half of the *S. aureus* at approximately 2 mg/ml. The extract made from a mixture of the two plants was twice as effective, killing half of the bacteria at a concentration of 0.8 mg/ml [11, 12].

Fig. 19.4 Open-air market showing San Pedro cactus (bottom right). (Source: Gail Willsky)

Hierba Buena (*Mentha spicata* L.) and Pimpinela (*Sanguisorba minor Scop*) were combined in five two-plant mixtures, four of which could be caused by an infectious agent: indigestion, colic, renal disease, and gastritis [5]. The activity of alcohol extracts of individual plants and the plant mixtures was tested using two strains of bacteria, *E. coli* and *S. aureus*. Using *E. coli*, the extracts made from the individual plants and the plant mixture all had slightly effective antibacterial activity, requiring between 5 and 6 mgs/ml of extract to kill half of the bacteria. Using *S. aureus*, none of the extracts made from the individual plants were considered to have antibacterial activity with over 36 mgs/ml of each extract needed to kill half of the bacteria. However, in the extract made from a mixture of these two plants only 0.6 mgs/ml was needed to kill half of *the S. aureus* bacteria which was over 60 times more effective than the extract made from either plant alone [12].

These results with the two-plant mixtures used to treat diseases potentially caused by infectious agents demonstrate that evaluating the composition and antibacterial activity of only one plant at a time may miss an important chemical component used in TM treatments. In our studies two common bacterial strains were used and all plant extracts were obtained using alcohol extracts. Our studies may not have involved the specific bacteria that might be causing the disease the TM mixture was intended to treat. Also, our scientifically required standardized preparation of extracts did not completely follow how the mixtures are used in the varied methods found in TM, i.e., our alcohol extracts contained more chemicals than the water-based teas used in the TM treatments.

Continuing Work in Ethnobotany, Laboratory Science, and Medical Anthropology Will Greatly Speed the Progress of Integrating TM into the PPHS

Obtaining more community information on these areas and educating both the patients and workers in the PPHS along with the people who provide the medicinal plants can be greatly beneficial. The workers in the PPHS need to understand the worldview of the patient population, requiring further documentation of indigenous knowledge and practice. The educational aspects of MHIRT-Peru involving gardens of native medicinal plants is also vital for the continued availability of these plants and the potential to get the native population involved in medicinal plant production and conservation.

MHIRT-Peru Program and Student Exposure to TM

An important part of the MHIRT-Peru program is exposure of our students to Peruvian TM via participant-observation in *curandero* "cleansing" rituals (*limpias*) involving the use of medicinal plants, including the psycho-active *San Pedro* cactus,

Fig. 19.5 Mesa used in a curing ceremony in Northern Peru. (Source: Doug Sharon)

conducted in conjunction with healing altars or *mesas*. A healing ceremony or cleansing ritual can take all night and often involves more than one patient. The people attending the ceremony gather around the mesa (Fig. 19.5) arranged on the ground in a symbolic format common to all north-coastal healers. During the ceremony the *curandero* invokes the spirits of sacred mountains, lagoons, pre-Hispanic temples (*huacas*), and Christian saints. The *curandero* passes the *San Pedro* concoction around to all who are present during the ceremony. Incense and the spraying of perfumes in the four cardinal directions and over the patient purify the environment and patient while stimulating the sense of smell. The *curandero* questions the patient, diagnoses the presenting problem, and rubs a staff symbolizing the targeted ailment over the patient's body. A final "cleansing" is performed by the *curandero* who orally sprays a perfume mixture over all present.

The *mesa* is an important tool of the *curandero* giving concrete expression to the basic cultural ideology underlying North Peruvian psychotherapy, an ethnographic fact verified by a large number of scholars working in the north. This metaphysical concept has been characterized the "complementarity of opposites." It is manifest in the spatial arrangement of the power objects on the *mesa* into three "fields" (*campos*): left (negative), right (positive), and middle (mediating). Up until midnight, the curer's rituals symbolically activate the healing energies of the *mesa* zones through the agency of the Middle Field. After midnight, this balanced energy is ritually focused on each patient seeking relief from "sorcery," "fright," "bad luck," or a variety of other culture-bound illnesses.

The following text is the testimony of a Latino archeology student regarding his interaction with one of the *curanderos* who works with us. After the "cleansing" ritual, the *curandero* said that he wanted to talk further about the barriers in the student's life which were impeding his academic progress. In previous conversations, the student had discussed the difficulties he was experiencing in his studies. He had come to the *curandero* in the hope of changing his luck.

Testimony of a Student Concerning Their Experience with a Curandero-Conducted Healing Ceremony
The maestro (a common term for addressing a curandero) listened to my personal story and the description of the impediments to my progress. I felt surprisingly comfortable in relating all this to him. The central problem was that I was at my father's side when he died in a car accident in 1998. Related to this and other matters, a sense of fear and anxiety had manifested within me. The maestro didn't necessarily resolve my problems, but I felt recharged with regard to my ability to deal with daily life and to renew my desire to transcend the tragedy and leave it in the past. Also, the maestro and his assistant worked very hard to achieve a resolution of my problems and make me remember my true state of being and the progress I had made since the death of my father. I began to look beyond the negative impression caused by this experience which had influenced me. Using my personal history, I felt that the maestro wasn't just helping me by means of the session, but that he was imparting to me a part of what he does as a curandero.

In the final analysis the session lasted about 4 h and, because of a cold I was experiencing, I did not drink the San Pedro. Despite this, the combination of fragrances, songs, and the rhythm of the rattle made me feel that I had reached a state of ecstasy and that my inhibitions had been removed. I thought that the session had lasted 2 h because I felt totally awake. Upon returning home, I couldn't sleep. In the year after this session, I re-initiated contact with my brothers, renewed my zeal for my studies, and I experienced an improved ability to deal with my anxieties. It isn't that the maestro performed a miracle for me, but he gave me the tools to manage my life in a more effective fashion.

Acknowledgments Author affiliations and research fields: University at Buffalo— GRW(Biochemistry), DS (Medical Anthropology); Missouri Botanical Garden— RWB(Ethnobotany), GM-G (Medicinal Chemistry), DS; National University of Truillo (UNT,Peru)–MG-Y and IC (Pharmacy), This work was funded by the MIRT/MHIRT grant program (# G0000613, Fund 54112B MHIRT and NIH/NIMHD # 5 T37 MDOO 1442-18) of the US National Institutes of Health awarded to San Diego State University as well as the Deutsches Forschungs Gemeinschaft (DFG) and the William L. Brown Center of the Missouri Botanical Garden. Unfortunately, we cannot name all the individuals who participated in this project. We gratefully acknowledge the support of Luis Fernandez, M.D. and the staff at the EsSalud-CAMEC PHS Trujillo site; Drs. Manuel Vera and Alberto Quezada at UNT; Dr. Noe Anticona of the Clinica Anticona; Dr. Fredy Perez at UPAO; Carolina Tellez our consulting botanist, Manuel Bejarano at Laboratorios BEAL; Ashley Glenn and other staff members of the Missouri Botanical Garden; Dr. Tom Love from Linfield College; and the numerous students who worked on the project both as MHIRT awardees and volunteers.

References

1. World Health Organization Traditional. Complementary, and integrative medicine.http://www. who.int/traditional-complementary-integrative-medicine/about/en/
2. Quevedo Z. Criterios para la Salvaguarda de la Medicina Tradicional Peruana. Pueblo Continente. 2012;1(23):100–4.
3. Villar M, Villavicencio O. Manual de Fitoterapia. Lima: OPS/OMS/EsSalud-Programa Nacional de Medicina Complementaria (PRONAMEC); 2001.
4. Bussmann R, Sharon D. Medicinal plants of the Andes and the Amazon: the magic and medicinal flora of Northern Peru. Trujillo: Graficart; 2015.
5. Bussmann R, Glenn A, Meyer K, et al. Herbal mixtures in traditional medicine in Northern Peru. J Ethnobot Ethnomed. 2010;6(10):1–11.
6. Hayden C. When nature goes public: the making and unmaking of bioprospecting in Mexico. Princeton and Oxford: Princeton University Press; 2001.
7. Mostacero J, Castillo F, Mejía F, Gamarra O, Charcape J, Ramírez R. Plantas Medicinales del Perú: Taxonomía, Ecogeografía, Fenología y Etnobotánica. Trujillo: Asamblea Nacional de Rectores: Instituto de Estudios Universitarios "José Antonio Encinas"; 2011.
8. Evans S, Tellez C, Vega C. Traceability of twenty medicinal plants in the markets of Northern Peru. Acta. Hort. 1030:143-50, ISHS. 2014.
9. Revene Z, Bussmann R, Sharon D. From Sierra to Coast: tracing the supply of medicinal plants in Northern Peru—a plant collector's tale. Ethnobot Res Appl. 2008;6:15–22.
10. Bussmann R, Malca G, Glenn A, et al. Toxicity of medicinal plants used in traditional medicine in Northern Peru. J Ethnopharmacol. 2011;137:121–14.
11. Cleland T, Galarza R, Barranis N, et al. Use of Artemia franciscana in evaluating the toxicity of plants used to treat infectious disease in Northern Peru: Research Day for Biochemistry Dept at University at Buffalo; 2014.
12. Lau C, Willsky G, Besch A, et al. Effects of alcohol extracts made from mixtures of plants and individual plants used by the curanderos of Northern Peru to treat infectious disease: ABRCMS; 2012.
13. Bussmann R, Malca-Garcia G, Glenn A, et al. Minimum inhibitory concentrations of medicinal plants used in Northern Peru as antibacterial remedies. J Ethnopharmacol. 2010;132:101–8.
14. Peréz F, Rodriguiz F, León G, et al. Phytochemical and antibacterial study of medicinal plant mixtures. In search of new components. Pueblo Continente. 2011;23(2):339–43.
15. Bussmann R, Sharon D, Garcia M. From Chamomile to Aspirin? Medicinal plant use among clients of Laboratorios Beal in Trujillo, Peru. Ethnobot Res Appl. 2009;7:399–407.
16. Bussmann R, Sharon D, Lopez A. Blending traditional and Western medicine, medicinal plant use among patients at Clinica Anticona in El Porvenir, Peru. Ethnobotany Res Appl. 2007;5:185–99.
17. Fajardo S, Sours A, Love T. Patient surveys at EsSalud's complementary medicine clinic in Trujillo, Peru. Indian J Tradit Know. (Under Revision) 2020.
18. Alvarez M. Herbal healing: A comparison of medicinal plant and pharmaceutical medication preference in two communities in La Libertad, Peru. M.A. thesis, Latin American Studies and Epidemiology, San Diego State University; 2017.
19. Poisson J. The presence of Mescaline in a Peruvian Cactus. Annales Pharmaceutiques Francaises. 1960;18:764–5.

Index

Printed in the United States
by Baker & Taylor Publisher Services